Wastewater Treatment

This book comprehensively explores advanced methodologies and practical applications in the treatment and valorization of wastewater. This book synthesizes cutting-edge research and practical insights into various techniques used to address the environmental challenges posed by liquid waste streams. From fundamental principles to applied solutions, each chapter delves into specific methodologies such as adsorption, electrochemical methods, biological treatments using microalgae, advanced oxidation processes, and integrated treatment systems.

Key Features:

- Bridges the gap between theoretical advancements and real-world applications in wastewater treatment.
- Discusses the importance and challenges of treating liquid wastes, including emerging contaminants, setting the stage for detailed discussions on innovative techniques.
- Explores in-depth studies on adsorption mechanisms, the effectiveness of electrochemical treatments, and the utilization of biological processes for waste remediation.
- Features case studies from different geographical regions that illustrate practical implementations and highlight the diverse approaches to wastewater management worldwide.

The intended audience includes researchers, engineers, and practitioners in environmental and chemical engineering seeking to enhance their understanding of modern techniques and their application in liquid waste treatment. By consolidating knowledge from global experts, this book serves as a valuable resource for academics and professionals striving to address the growing concerns of environmental sustainability through innovative waste treatment solutions.

Reda Elkacmi has a Ph.D. in Chemical Engineering from Hassan II University, Morocco. He is affiliated with the Faculty of Sciences and Techniques, Sultan Moulay Slimane University, Morocco. His research interests include chemical and process engineering, industrial process modeling and simulation, and the development of innovative and sustainable technologies for wastewater treatment and resource recovery.

Sudip Chakraborty has a PhD in Chemical Engineering from the University of Calabria, Italy. He is Abilitazione Scientifica Nazionale – full professor in sector – ING-IND 24 at the Laboratory of Transport Phenomena and Biotechnology, University of Calabria. He was also a visiting researcher at the Massachusetts Institute of Technology. His major fields of interest are membrane separation, plasmonic nanoparticles, composite materials, energy, and process intensification.

Catia Algieri has been a researcher at the Institute on Membrane Technology, National Research Council of Italy (ITM-CNR) since 2001. She has collaborated on several international and national projects focused on membrane preparation and characterization for application in water treatment, catalytic reactions, and gas separations. In her research, she prepares and characterizes zeolite membranes for water treatment, catalytic reactions, and gas separations, as well as mixed matrix membranes for drug-controlled release and gas separations.

Wastewater Treatment

Emerging Contaminants and Case Studies

Edited by

Reda Elkacmi, Sudip Chakraborty, and Catia Algieri

CRC Press
Taylor & Francis Group
Boca Raton London New York

CRC Press is an imprint of the
Taylor & Francis Group, an **informa** business

Designed cover image: © 2026 Shutterstock

First edition published 2026
by CRC Press
2385 NW Executive Center Drive, Suite 320, Boca Raton FL 33431

and by CRC Press
4 Park Square, Milton Park, Abingdon, Oxon, OX14 4RN

CRC Press is an imprint of Taylor & Francis Group, LLC

ISBN: 978-1-041-03696-8 (hbk)
ISBN: 978-1-041-03697-5 (pbk)
ISBN: 978-1-003-62490-5 (ebk)

DOI: 10.1201/9781003624905

Typeset in Times
by codeMantra

Contents

Preface

Wastewater Treatment: Emerging Contaminants and Case Studies presents a comprehensive and up-to-date exploration of recent advancements in innovative wastewater treatment technologies. This book addresses pressing environmental challenges related to water pollution and is designed for professionals in water and wastewater management, including civil, industrial, environmental, and chemical engineers. It also serves as a valuable reference for undergraduate and graduate students in related engineering disciplines.

As emerging contaminants such as pharmaceuticals, microplastics, heavy metals, and persistent organic pollutants become more prevalent, wastewater treatment grows increasingly complex and urgent. In response, global treatment technologies have advanced significantly. For example, the global market for wastewater treatment equipment was valued at over USD 64 billion in 2018 and is projected to grow at a rate exceeding 5% annually, driven by water scarcity, sustainability concerns, and stricter environmental regulations. These trends underscore the need for adaptable, high-performance treatment solutions.

This book features contributions from leading researchers worldwide, covering a diverse array of cutting-edge technologies. Topics include membrane filtration, advanced adsorption methods, micro- and nanoplastics, biochar applications, electrochemical and photo-Fenton processes, ultrasonic techniques, hybrid adsorptive–photocatalytic systems, and microalgae-based treatment strategies. Additionally, it explores economic feasibility, environmental impacts, and sustainability assessments critical for large-scale implementation.

A unique strength of this volume lies in its emphasis on detailed case studies across various industries and regions. These real-world examples from petrochemical, textile, pharmaceutical, food processing, and municipal sectors illustrate how specific wastewater challenges are addressed globally. Each chapter integrates theoretical concepts with experimental findings, optimization methods, and regulatory discussions, offering a comprehensive view from laboratory research to field applications.

Beyond pollution control, the technologies discussed aim to recover resources and generate value-added products such as clean water, bioenergy, and fertilizers, contributing to both environmental protection and circular economy goals. By bridging cutting-edge science with practical implementation, this book serves as an essential resource for researchers, practitioners, policymakers, and industry stakeholders dedicated to sustainable water management for future generations.

Editors
Reda Elkacmi
Sudip Chakraborty
Catia Algieri

Contributor List

Hafid Abderrafia
Laboratory of Molecular Chemistry – Materials and Catalysis
Faculty of Science and Technology
Sultan Moulay Slimane University
Beni Mellal, Morocco

Brahim Achiou
Laboratory of Materials, Membranes and Environment
Faculty of Sciences and Technologies of Mohammedia
Hassan II University of Casablanca
Mohammedia, Morocco

Mounir Akhatar
Laboratory of Environmental, Ecological and Agro-Industrial Engineering
Faculty of Science and Technology
Sultan Moulay Slimane University
Beni Mellal, Morocco

Saad ALjlil
Water Management & Treatment Technologies Institute
King Abdulaziz City for Science and Technology
Riyadh, Saudi Arabia

Fatima Ezzahra Atmani
Team of Analytical & Computational Chemistry Nanotechnology and Environment
Faculty of Science and Technologies
Sultan Moulay Slimane University
Beni Mellal, Morocco

Belkacem Benadda
DEEP Laboratory
National Institute of Applied Sciences
Lyon, France

Redouane Beniazza
Institute of Science, Technology & Innovation (IST&I)
Mohammed VI Polytechnic University
Ben Guerir, Morocco

Hamza Bellouk
Processes, Materials and Environment Laboratory
Faculty of Sciences and Technology of Fez
Sidi Mohamed Ben Abdellah University
Fez, Morocco

Zineb Bencheqroun
Processes, Materials and Environment Laboratory
Faculty of Sciences and Technology of Fez
Sidi Mohamed Ben Abdellah University
Fez, Morocco

Mourad Benzina
Laboratory "Water Energy and Environment"
National Engineering School of Sfax
University of Sfax
Sfax, Tunisia

Mohamed Berkani
Team of Analytical & Computational Chemistry Nanotechnology and Environment
Faculty of Science and Technologies
Sultan Moulay Slimane University
Beni Mellal, Morocco

Samira Bnikkou
National Thematic Institute for Scientific Research-Water
Ibn Zohr University
Agadir, Morocco

Otmane Boudouch
Laboratory of Sciences and Technologies Team
Higher Education and Training School
Chouaib Doukkali University
El Jadida, Morocco

Majda Breida
Laboratory of Materials, Membranes and Environment
Faculty of Sciences and Technologies of Mohammedia
Hassan II University of Casablanca
Mohammedia, Morocco

Sudip Chakraborty
Department of DIMES
University of Calabria
Cosenza, Italy

Zineb Chaouki
National Thematic Institute for Scientific Research-Water
Ibn Zohr University
Agadir, Morocco

Siham Dabali
Laboratory of Environmental, Ecological and Agro-Industrial Engineering
Faculty of Science and Technology
Sultan Moulay Slimane University
Beni Mellal, Morocco

Ikram Daou
Laboratory of Chemistry/Biology Applied to the Environment
Faculty of Sciences
Moulay Ismaïl University
Meknes, Morocco

Yassine Darmane
Polydisciplinary Faculty of Ouarzazate
Ibn Zohr University
Ouarzazate, Morocco

Mounir El Achaby
MSN Department
Mohammed VI Polytechnic University
Ben Guerir, Morocco

Radouane El Amri
Environmental Ecological and Agro-Industrial Engineering Laboratory
Faculty of Sciences and Technology
University Sultan Moulay Slimane
Beni Mellal, Morocco

Abdelali El Gaidoumi
Laboratory of Biotechnology, Bioresources and Bioinformatics
Higher School of Technology of Khénifra
Sultan Moulay Slimane University
Beni Mellal, Morocco

Nour-Eddine El Mansouri
Laboratory of Molecular Chemistry, Materials and Catalysis
Faculty of Science and Technology of Beni Mellal
University Sultan Moulay Slimane
Beni Mellal, Morocco

Imane El Mrabet
Team of CAG2M
Polydisciplinary Faculty of Ouarzazate
Ibn Zohr University
Ouarzazate, Morocco

Oumayma El Wafi
Team of Analytical & Computational Chemistry Nanotechnology and Environment
Faculty of Science and Technologies
Sultan Moulay Slimane University
Beni Mellal, Morocco

Reda Elkacmi
Environmental, Ecological and Agro-Industrial Engineering Laboratory
Faculty of Sciences and Technology
University Sultan Moulay Slimane
Beni Mellal, Morocco

Mohamed Elmernissi
Laboratory of Environmental, Ecological and Agro-Industrial Engineering
Faculty of Science and Technology
Sultan Moulay Slimane University
Beni Mellal, Morocco

Khaoula Esslimani
Laboratory of Tribology and System Dynamics
National Engineering School of Saint-Étienne
Saint-Étienne, France

Corinne Ferronato
Université de Lyon
Villeurbanne, France

Ahmed Gamouh
Team of Analytical & Computational Chemistry Nanotechnology and Environment
Faculty of Science and Technologies
Sultan Moulay Slimane University
Beni Mellal, Morocco

Mohssine Ghazoui
Laboratory of Environmental, Ecological and Agro-Industrial Engineering
Faculty of Science and Technology
Sultan Moulay Slimane University
Beni Mellal, Morocco

Anas Hafdi
Team of Analytical & Computational Chemistry Nanotechnology and Environment
Faculty of Science and Technologies
Sultan Moulay Slimane University
Beni Mellal, Morocco

Maryeme Hamdouch
Laboratory of Sciences and Technologies Team
Higher Education and Training School
Chouaib Doukkali University
El Jadida, Morocco

Aziz Hasib
Environmental, Ecological and Agro-Industrial Engineering Laboratory
Faculty of Sciences and Technology
University Sultan Moulay Slimane
Beni Mellal, Morocco

Abdelghani Hsini
Laboratory of Advanced Materials and Process Engineering
Faculty of Sciences
Ibn Tofaïl University
Kenitra, Morocco

Manal Idgharnane
Laboratory of Materials, Membranes and Environment
Faculty of Sciences and Technologies of Mohammedia
Hassan II University of Casablanca
Mohammedia, Morocco

Noureddine Kamil
Laboratory of Materials Engineering for the Environment and Valorization
Faculty of Sciences Aïn Chock – Maarif
Hassan II University of Casablanca
Casablanca, Morocco

Meriem Kasbaji
MSN Department
Mohammed VI Polytechnic University
Ben Guerir, Morocco

Mohamed Amine Kasbaji
Team of Analytical & Computational Chemistry Nanotechnology and Environment
Faculty of Science and Technologies
Sultan Moulay Slimane University
Beni Mellal, Morocco

Khaoula Khaless
Mohammed VI Polytechnic University
Ben Guerir, Morocco

Mohamed M'barki
Team of Analytical & Computational Chemistry Nanotechnology and Environment
Faculty of Science and Technologies
Sultan Moulay Slimane University
Beni Mellal, Morocco

Nouhaila Makrouni
Laboratory of Materials, Membranes and Environment
Faculty of Sciences and Technologies of Mohammedia
Hassan II University of Casablanca
Mohammedia, Morocco

Mohammad Saood Manzar
Department of Environmental Engineering
College of Engineering
Imam Abdulrahman Bin Faisal University
Dammam, Saudi Arabia

Mouad Mouhsin
Team of Analytical & Computational Chemistry Nanotechnology and Environment
Faculty of Science and Technologies
Sultan Moulay Slimane University
Beni Mellal, Morocco

Mohamed Ouammou
Laboratory of Materials, Membranes and Environment
Faculty of Sciences and Technologies of Mohammedia
Hassan II University of Casablanca
Mohammedia, Morocco

Mustapha Oubenali
Team of Analytical & Computational Chemistry Nanotechnology and Environment
Faculty of Science and Technologies
Sultan Moulay Slimane University
Beni Mellal, Morocco

Salah Rafqah
Laboratory of Analytical and Molecular Chemistry
Faculty Poly Disciplinary of Safi
Safi, Morocco

Oussama Salhi
Laboratory of Materials, Membranes and Environment
Faculty of Sciences and Technologies of Mohammedia
Hassan II University of Casablanca
Mohammedia, Morocco

Abdelillah Shaim
Laboratory of Advanced Materials and Process Engineering
Faculty of Sciences
Ibn Tofaïl University
Kenitra, Morocco

Aboubacar Sidigh Syllaa
Laboratory of Molecular Chemistry, Materials and Catalysis
Faculty of Sciences and Technics
Sultan Moulay Slimane University
Beni Mellal, Morocco

Karim Tanji
Laboratory of Advanced Materials and Process Engineering
Faculty of Sciences
Ibn Tofaïl University
Kenitra, Morocco

Hector Valdes
Clean Technology Laboratory
Facultad de Ingeniería
Universidad Católica de la Santísima Concepción
Concepción, Chile

Mezlini Wiem
University of Tunis El Manar
Department of Geology
Faculty of Sciences of Tunis
Tunis, Tunisia

Saad Alami Younssi
Laboratory of Materials, Membranes and Environment
Faculty of Sciences and Technologies of Mohammedia
Hassan II University of Casablanca
Mohammedia, Morocco

Rajaa Zahnoune
Laboratory of Sciences and Technologies Team
Higher Education and Training School
Chouaib Doukkali University
El Jadida, Morocco

Hicham Zaitan
Processes, Materials and Environment Laboratory
Faculty of Sciences and Technology of Fez
Sidi Mohamed Ben Abdellah University
Fez, Morocco

About the Editors

Prof. Reda Elkacmi is a Chemical Engineer and academic specializing in process engineering and sustainable environmental technologies. He earned his Ph.D. in Chemical Engineering from Hassan II University, Morocco, and serves as Professor at the Faculty of Sciences and Techniques, Sultan Moulay Slimane University, Morocco. His research focuses on industrial process modeling and simulation, adsorption science, and the design of innovative technologies for wastewater treatment, pollution control, and resource recovery. Professor Elkacmi's work bridges theoretical insights with practical applications, contributing to the advancement of cleaner production and circular economy strategies. He has authored numerous scientific papers and book chapters in high-impact journals and serves as a reviewer and editorial board member for several international publications. Actively engaged in global research collaborations, he promotes interdisciplinary approaches to environmental sustainability and the development of eco-efficient industrial processes.

Dr. Sudip Chakraborty is a leading Professor in Chemical Engineering at the University of Calabria, Italy, with over a decade of experience advancing materials for energy harvesting, membrane technology, water treatment, and bioprocesses, including process intensification and **sustainable development goals** (SDGs). He is also an associate senior researcher at CNR-NANOTEC (Italy), affiliated guest faculty at the Technische Hochschule Mittelhessen, University of Applied Sciences, Germany, and visiting researcher at MIT, UCLA. He contributes globally through editorial roles in high-impact journals. Recognized four times among the world's top 2% scientists by Stanford and Elsevier, he is also a Senior Member and Technical Activity Chair of IEEE-TEMS. He has spearheaded numerous international, EU-funded projects and published extensively in top-quality journals. He has over 210 scientific journal publications with an h-index of 54.

Catia Algieri is a Senior Researcher at the Institute on Membrane Technology (ITM-CNR), Italy. Her primary research focuses on membrane science and separation processes, including wastewater treatment, water desalination, and gas separations. She has authored over 80 publications, including peer-reviewed journal articles, book chapters, and encyclopedia entries, and holds two granted patents. She collaborated on numerous international and national research projects dedicated to membrane development and applications in water treatment, catalytic reactions, and gas separation technologies. She is an adviser to foreign and Italian students; an external adviser for international projects; a member of evaluation boards and of scientific and organizing committees of international conferences; a reviewer of scientific journals, research projects, and PhD theses; and an associate editor of different scientific international journals.

1 Bio-Based Activated Carbons for Wastewater Remediation

A Comparative Case Study on Dyes and Heavy Metals

Mohssine Ghazoui, Rajaa Zahnoune, Radouane El Amri, Aboubacar Sidigh Sylla, Siham Dabali, Mohamed Elmernissi, and Reda Elkacmi

1.1 INTRODUCTION

Water pollution by synthetic dyes and heavy metals is a major environmental and health issue worldwide. These two classes of contaminants are ubiquitous in industrial waste, particularly from the textile, mining, metallurgy, pharmaceutical, and agri-food sectors. Their high toxicity, chemical stability, and non-biodegradable nature give them a high pollution potential, capable of sustainably altering aquatic ecosystems and threatening human health (Briffa et al. 2020; Hama Aziz et al. 2023; Farghaly et al. 2024). The development of effective, sustainable, and economically viable processes for treating these effluents is therefore becoming a strategic priority, particularly in developing countries where access to cutting-edge technologies remains limited.

Several conventional techniques have been developed for eliminating inorganic and organic contaminants from industrial wastewater (Wei et al. 2023). These include chemical precipitation, ion exchange, membrane filtration, advanced oxidation processes, electrocoagulation, flotation, and biological processes. Although some of these approaches are effective, they often have significant limitations, particularly in terms of investment and operating costs, by-product generation, and operational complexity. For example, membranes require costly maintenance and are susceptible to clogging, while biological processes are not very effective in dealing with toxic or poorly biodegradable substances (Dou et al. 2021; Pabby et al. 2022).

In this context, adsorption has emerged as one of the most promising techniques due to its ease of implementation, low cost, rapidity of action, and ability to eliminate a wide range of pollutants, even at low concentrations (Chaari et al. 2021). However,

DOI: 10.1201/9781003624905-1

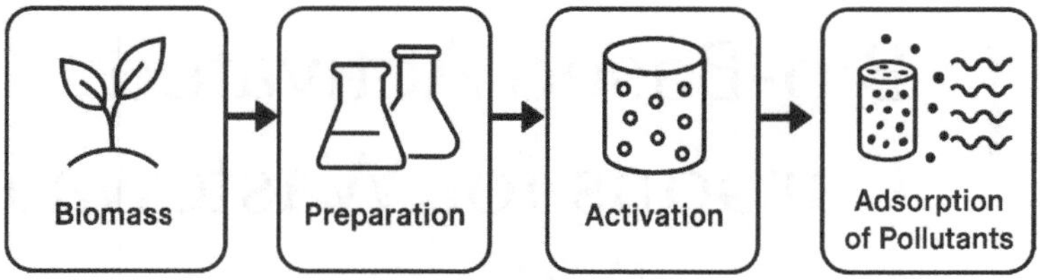

FIGURE 1.1 Overview of the biomass valorization process of plants with significant cellulose and hemicellulose content for the development of activated adsorbents for wastewater treatment.

the choice of adsorbent remains a key factor in guaranteeing the performance of the process. Using local biomass to manufacture activated carbons is an environmentally friendly, cost-effective, and circular alternative. Lignocellulosic biomass from medicinal plants and agricultural or forestry residues can be recovered using pyrolysis or chemical activation processes to produce porous carbonaceous materials with a high specific surface area and reactive functional groups favourable to adsorption (Prahas et al. 2008; Tripathi et al. 2022). The overall biomass recovery process is summarized in the conceptual diagram shown in Figure 1.1.

This chapter is primarily based on two experimental studies recently published by our group, focusing on bio-based activated carbons derived from *Saponaria officinalis* and *Rhus pentaphylla*, respectively, applied to the removal of organic dyes and cadmium ions from aqueous solutions (Ghazoui et al. 2024, 2025).

In the development of adsorption processes, the optimization of experimental parameters plays a central role in maximizing efficiency while reducing costs. To this end, RSM has emerged as a powerful statistical tool for exploring the cross effects of several variables. It facilitates the modelling of responses (adsorption capacity, production yield, overall efficiency) as a function of factors such as pH, contact time, initial concentration, temperature, and material preparation conditions (Sun et al. 2019; Ahmadi et al. 2024). By limiting the number of tests, this approach makes it possible to identify optimum operating conditions, helping to rationalize processes and transfer them to larger-scale applications.

As part of our experimental work, we have developed two biosourced activated carbons from plants: *Saponaria officinalis* and *Rhus pentaphylla*. These plant species were chosen for their availability, but also for their ethnobotanical interest and their richness in lignocellulosic components. Historically, *Saponaria officinalis* is known for its high content of natural saponins, while *Rhus pentaphylla* is traditionally used as a vegetable tannin in the treatment of skin or for medicinal purposes. These two characteristics bear witness not only to their functional potential but also to the ease with which they can be integrated into an approach that makes the most of local resources.

From a chemical point of view, both plants have a significant cellulose, hemicellulose, and lignin content – key constituents for the production of high-porosity activated carbons. The choice of these biosourced precursors is part of a circular economy strategy aimed at transforming raw materials that are little used, or even considered waste, into functional materials for water treatment (Sayğili and Güzel

2016; Jawad et al. 2016). This approach not only reduces adsorbent production costs but also strengthens the local roots of the proposed solution by exploiting renewable, accessible, and inexpensive plant resources.

1.2 PREPARATION AND OPTIMIZATION OF BIO-BASED AC

1.2.1 Synthesis of Activated Carbon derived from Saponaria officinalis Root

Roots of *Saponaria officinalis*, collected from the Beni Mellal–Khénifra region (Morocco), were thoroughly washed with deionized water to remove surface impurities and then oven-dried at 103°C for 24h. The dried biomass was ground and sieved to obtain a fine and homogeneous powder. Chemical activation was conducted using three different activating agents: phosphoric acid (H_3PO_4), sulphuric acid (H_2SO_4), and potassium hydroxide (KOH). The biomass was impregnated with each agent at a fixed weight ratio of 1:1 (w/w) and maintained under ambient conditions for 24h. Following impregnation, neutralization was carried out using either 1M HCl (for basic treatment) or saturated Na_2HCO_3 solution (for acidic agents), and the samples were washed repeatedly with deionized water until a neutral pH was achieved.

Subsequently, the neutralized materials were oven-dried at 103°C overnight and thermally activated in a muffle furnace at either 550°C or 650°C for 90min. The obtained carbonaceous materials were then ground, sieved, and stored in desiccators. Final products were coded according to the activating agent used: SOR-H_3PO_4, SOR-H_2SO_4, and SOR-KOH.

1.2.2 Synthesis of AC-RP

Fruits of *Rhus pentaphylla* were harvested from the Beni Mellal region (Morocco), washed with deionized water, and oven-dried at 100°C for 24h. The dried biomass was then pulverized using a mechanical grinder and sieved to achieve a particle size below 2mm, ensuring consistency during activation.

Chemical activation was performed using concentrated sulphuric acid (H_2SO_4) at varying impregnation ratios (1:1, 2:1, and 3:1 w/w), with contact times of 8, 16, and 24h. The impregnation process was conducted under constant agitation at room temperature. After acid treatment, the biomass was filtered and rinsed thoroughly with distilled water until pH neutrality was reached and then dried overnight at 100°C.

Carbonization was carried out in a tubular furnace under nitrogen flow at temperatures ranging from 550°C to 650°C, with a residence time of 90min. The resulting carbon was ground, sieved, and stored in sealed containers under the code Rp-H_2SO_4-AC.

1.2.3 Optimization of Preparation Conditions

To enhance the structural and adsorptive properties of the bio-based activated carbons, the synthesis conditions were optimized using statistical experimental design. Two complementary approaches were employed for each precursor: a Box–Behnken design for activated carbon derived from Saponaria officinalis root (AC-SOR) and a full factorial design for AC-RP. The RSM was applied in both cases to identify the optimal combination of factors that maximized the iodine number, used as a proxy for microporosity and adsorption potential.

1.2.4 Optimization of Preparation Conditions for AC-SOR

A complete factorial design was applied to optimize the synthesis parameters of the AC-SOR. Three variables, chemical activating agent, impregnation time, and carbonization temperature, were investigated due to their crucial influence on porosity development and iodine adsorption capacity. The iodine number was selected as the response variable, reflecting the degree of microporosity and adsorptive potential. Statistical analysis confirmed the strong significance of the model, with a determination coefficient (R^2) of 0.992, indicating excellent agreement between predicted and experimental values. The best performance was obtained using sulphuric acid as the activating agent, a 24-h impregnation period, and a carbonization temperature of 650°C, achieving an iodine number of 1,250.9 mg g^{-1}. These conditions favoured the generation of well-developed porosity and high surface functionality. The interaction between the chemical agent and carbonization temperature was found to be the most influential, as visualized in Figure 1.2. The use of sulphuric acid under elevated temperatures significantly enhanced the iodine value, demonstrating the synergistic

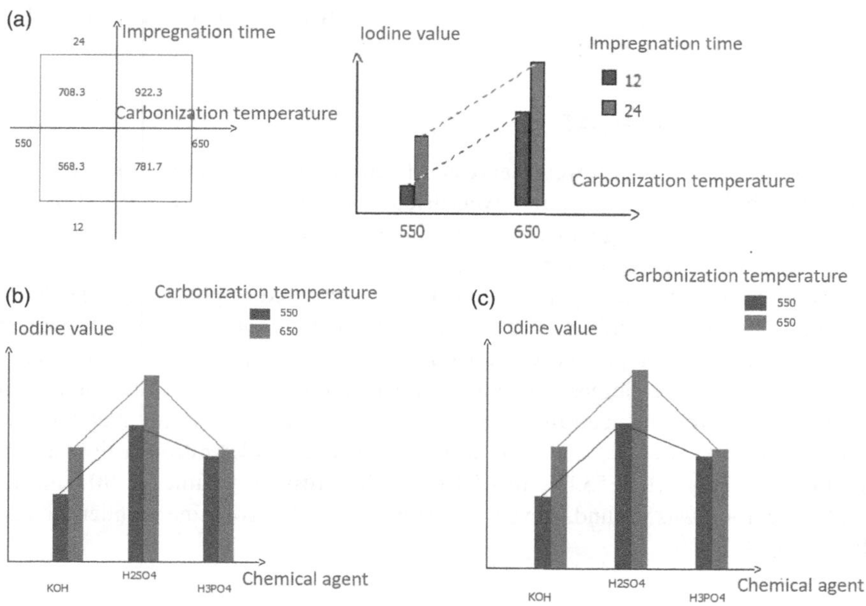

FIGURE 1.2 Interaction plots illustrating the combined effects of key preparation variables on the iodine number of AC-SOR: (a) Impregnation time versus carbonization temperature, (b) activating agent versus carbonization temperature, and (c) activating agent versus impregnation time.

effect of strong acid activation and thermal treatment. In contrast, the interaction between impregnation time and temperature had a less pronounced impact but still contributed positively to performance.

Based on these findings, the selected preparation conditions for optimized AC-SOR were validated and used for further material characterization and adsorption evaluations in the subsequent sections.

1.2.5 Optimization of AC-RP Preparation

The preparation conditions of the AC-RP were optimized using RSM based on a Box–Behnken design. Three influential parameters were selected: impregnation time (A: 8–24 h), activation temperature (B: 550–650°C), and the H_2SO_4-to-biomass ratio (C: 1:1–3:1 w/w). The iodine number (Y_1), a key indicator of micropore development, was chosen as the response variable.

Model fitting resulted in a second-order polynomial with excellent predictive performance ($R^2=0.9984$, Adj. $R^2=0.9955$, Pred. $R^2=0.9745$), and a low coefficient of variation (1.64%), confirming the model's reliability. Analysis of variance results highlighted the significant contribution of linear terms (A, B, C), their interaction (AB), and the quadratic effects A^2 and B^2. The AC and BC interactions, as well as C^2, were found to be statistically insignificant.

As shown in Figure 1.3, the three-dimensional response surface plot illustrates the combined effect of impregnation time and activation temperature on the iodine number. The red zone corresponds to optimal conditions where porosity was maximized. Notably, the iodine number increased substantially when both temperature and contact time were high, demonstrating their synergistic role in pore development.

The optimal synthesis conditions for AC-RP were determined as 16 h impregnation time, a carbonization temperature of 650°C, and a 3:1 H_2SO_4-to-biomass ratio, achieving an iodine number exceeding 1,040 mg g^{-1}. These conditions are ideal for producing highly porous bio-based carbons for environmental applications.

1.3 PHYSICOCHEMICAL CHARACTERIZATION OF THE BIO-BASED ACTIVATED CARBONS

The bio-based AC-SOR and AC-RP were optimized and thoroughly characterized to enhance their physicochemical properties and adsorption performance. Their structural and surface features were investigated using various analytical techniques (thermogravimetric analysis, fourier transform infrared spectroscopy (FTIR), SEM, Brunauer–Emmett–Teller [BET], XRD, and pHpzc) to establish correlations with their adsorption behaviour.

Thermogravimetric analysis revealed three main decomposition stages in both raw biomasses: initial water evaporation, degradation of lignocellulosic constituents (cellulose, hemicellulose, and lignin), and final carbonization. The absence of significant weight loss above 550°C confirmed the relevance of this temperature as the optimal threshold for activated carbon production.

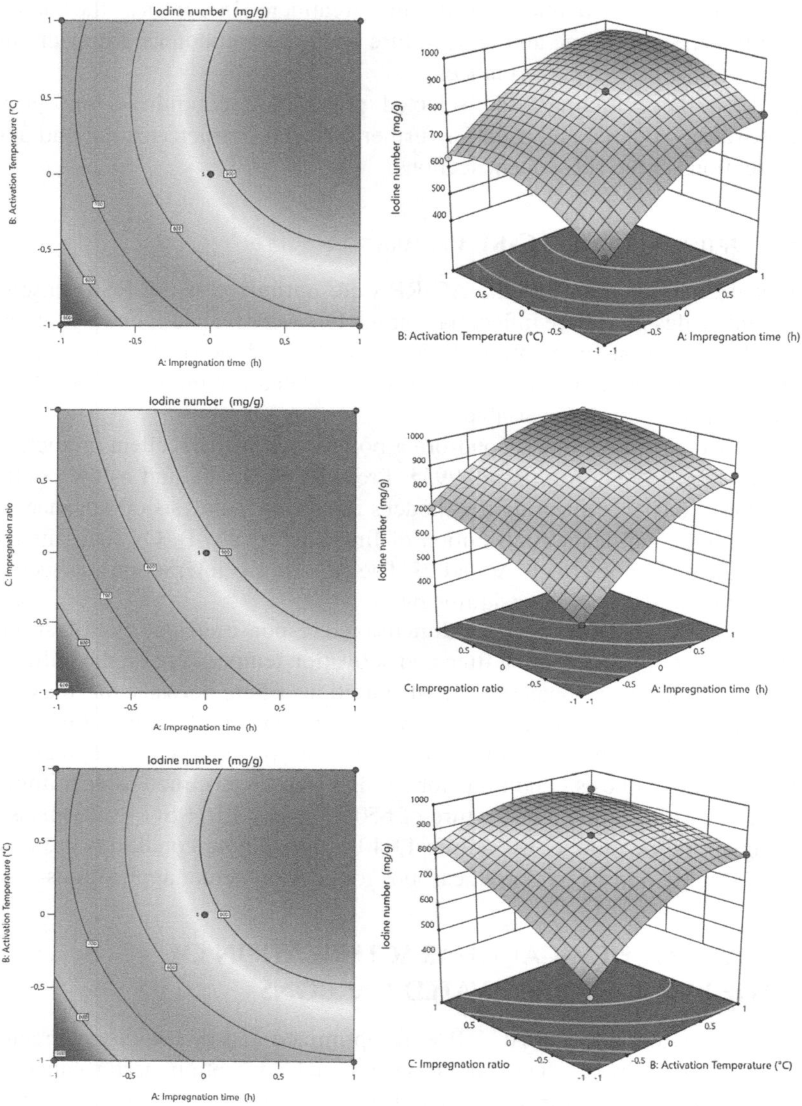

FIGURE 1.3 Response surface and contour plots for the iodine number (mg g^{-1}).

SEM showed a remarkable transformation of surface morphology following activation. Both AC-SOR and AC-RP displayed highly porous structures with well-distributed pores of various shapes and sizes. This enhanced porosity increases the number of accessible active sites, thereby promoting adsorption processes.

Nitrogen adsorption–desorption isotherms confirmed the predominant microporous nature of both carbons. BET-specific surface areas were 516.25 m^2 g^{-1} for optimized AC-SOR and 545.01 m^2 g^{-1} for optimized AC-RP. The corresponding total pore volumes were 0.219 cm^3 g^{-1} and 0.2494 cm^3 g^{-1}, respectively, indicating excellent pore accessibility.

FTIR spectroscopy revealed significant alterations in functional groups post-activation. Characteristic bands related to lignocellulosic structures (e.g., C–H, C=O) were diminished or disappeared, while new oxygenated groups (e.g., C–O, C–O–C) emerged, enhancing surface polarity and favouring interactions with polar or cationic contaminants.

XRD patterns exhibited broad, low-intensity peaks around $2\theta \approx 25°$ and 43°, indicating an amorphous structure typical of activated carbons. This structural disorder increases the density of unorganized active sites, which is advantageous for adsorption applications. Finally, the pHpzc was determined to be 6.6 for optimized AC-SOR and 7.35 for AC-RP. Below these pH values, the carbon surfaces are positively charged, promoting the adsorption of anions (e.g., dyes), whereas at pH levels above the pHpzc, the surfaces become negatively charged, enhancing the uptake of cationic species (e.g., Cd (II)).

Together, these physicochemical insights underscore the complementary structural, textural, and chemical features of the two bio-based activated carbons, which explain their demonstrated efficacy in the adsorption tests discussed in subsequent sections.

1.4 ADSORPTION STUDY

1.4.1 Adsorption of Dyes onto Optimized AC-SOR

1.4.1.1 Effect of Operational Parameters

The adsorption performance of the optimized AC-SOR was evaluated by investigating the influence of key operational parameters, particularly the initial dye concentration and contact time. These two variables play a crucial role in determining adsorption efficiency. A higher initial concentration creates a greater driving force for mass transfer, accelerating the diffusion of dye molecules into the porous structure of the adsorbent. Meanwhile, contact time governs the equilibrium state and overall treatment efficiency. Experimental results illustrated in Figure 1.3 reveal that for both MB and MR, the adsorption proceeds rapidly during the initial phase, followed by a slower approach toward equilibrium. For MB (Figure 1.4a), equilibrium

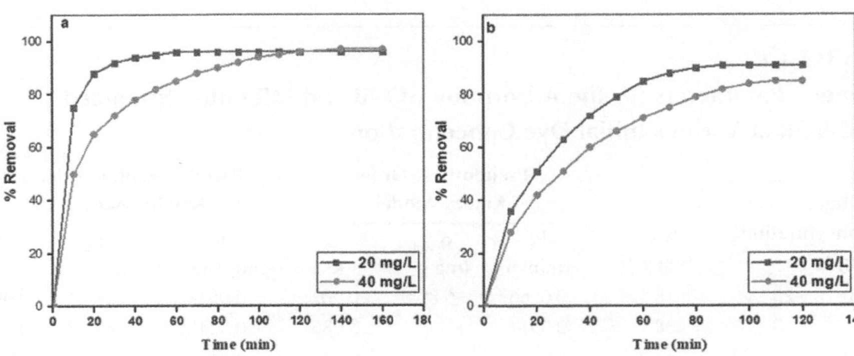

FIGURE 1.4 Adsorption behaviour of optimized AC-SOR toward MB and MR: Effect of initial concentration and contact time.

is achieved at approximately 50 min for an initial concentration of 20 mg L^{-1} and around 110 min for 40 mg L^{-1}. Similarly, MR adsorption (Figure 1.4b) reaches equilibrium at 100 and 110 min for 20 mg L^{-1} and 40 mg L^{-1}, respectively, indicating a slightly slower kinetic behaviour. This variation can be attributed to the structural and physicochemical differences between the two dyes. MB, being smaller and possessing a more stable cationic structure, interacts more rapidly and efficiently with the functional groups on the AC-SOR surface. In contrast, the bulkier molecular size of MR may hinder its diffusion into narrower micropores, limiting the rate and extent of adsorption. Overall, the combined effect of contact time and dye concentration underscores the need for careful optimization in real-world applications, where varying pollutant loads and process durations can significantly affect treatment outcomes.

1.4.1.2 Kinetic Modelling of Dye Adsorption

A thorough understanding of adsorption kinetics is essential for the rational design and scale-up of wastewater treatment systems (Mitrogiannis et al. 2015). In this study, the experimental data were fitted using the pseudo-first-order and pseudo-second-order kinetic models to describe the adsorption dynamics of MB and MR onto activated carbon derived from optimized AC-SOR. The pseudo-first-order model, which assumes that the rate of occupancy of adsorption sites is proportional to the number of unoccupied sites, exhibited poor agreement with the experimental data for both dyes, particularly MB. This was evidenced by the discrepancy between calculated and experimental equilibrium capacities ($q_{e\ calq}$ vs. $q_{e\ exp}$) and relatively low correlation coefficients ($R^2 < 0.99$).

In contrast, the pseudo-second-order model, which is based on the assumption that chemisorption is the rate-limiting step involving valence forces through electron sharing or exchange, provided an excellent fit for both dyes. The correlation coefficients exceeded 0.99 across all concentrations, and the calculated q_e values closely matched the experimental data, indicating that this model reliably describes the kinetic behaviour of MB and MR adsorption onto AC-SOR. Table 1.1 summarizes the kinetic parameters obtained from both models, including rate constants (k_1 and k_2) and corresponding q_e values. These results suggest that the adsorption mechanism is predominantly governed by chemisorption, particularly under the tested conditions.

TABLE 1.1
Kinetic Parameters for the Adsorption of MB and MR onto Optimized AC-SOR at Various Initial Dye Concentrations

Initial Concentration (mg L^{-1})		$q_{e,exp}$ (mg g^{-1})	Pseudo-First-Order Kinetic Model			Pseudo-Second-Order Kinetic Model		
			k_1 (min^{-1})	$q_{e,cal}$ (mg g^{-1})	R^2	k_2 [g($mg^{-1}min^{-1}$)]	$q_{e,cal}$ (mg g^{-1})	R^2
MB	20	18.5	0.0866	6.80	0.902	0.0528	18.91	0.999
	40	38.7	0.0314	19.94	0.986	0.0243	41.10	0.999
MR	20	14.2	0.0324	13.50	0.935	0.0523	19.10	0.990
	40	19.4	0.0265	17.76	0.988	0.0417	23.95	0.992

1.4.1.3 Adsorption Isotherm Modelling

Adsorption isotherms provide critical insight into the interactions between adsorbate molecules and the surface of the adsorbent (Falconer et al. 2021) and are essential for optimizing adsorbent utilization in real-world applications. In this study, the equilibrium data for the adsorption of MB and MR onto the optimized AC-SOR were fitted to four common isotherm models: Langmuir, Freundlich, Temkin, and Dubinin–Radushkevich (D–R).

The parameters derived from the Langmuir and Freundlich models, as shown in Table 1.2, offer valuable information on maximum adsorption capacities and surface heterogeneity. The Langmuir model, which assumes monolayer adsorption onto a homogeneous surface, provided the best fit to the experimental data, with high correlation coefficients ($R^2=0.998$ for MB and 0.992 for MR), and predicted maximum adsorption capacities of 74.96 mg g^{-1} and 46.27 mg g^{-1} for MB and MR, respectively. Moreover, the separation factor R_L values (between 0 and 1) indicated favourable adsorption.

The Freundlich model also described the data reasonably well ($R=0.949$ for MB; 0.975 for MR), suggesting the presence of heterogeneous binding sites. However, the better performance of the Langmuir fit highlights a predominantly monolayer adsorption mechanism.

The Temkin model, which considers interactions between adsorbate molecules and a linear decrease in adsorption heat with surface coverage, revealed a higher affinity of MB for optimized AC-SOR ($A=4.31$ L mg^{-1}) compared to MR ($A=0.703$ L mg^{-1}). The calculated adsorption energies (1.38 kJ mol^{-1} for MB and 1.01 kJ mol^{-1} for MR) suggest that physisorption is the dominant mechanism.

The D–R model, which provides insight into the adsorption energy and potential mechanism (physisorption vs. chemisorption), yielded maximum capacities of

TABLE 1.2
Isotherm Model Parameters for the Adsorption of MB and MR onto Optimized AC-SOR

		Dye Type	
Isotherm	**Isotherm Constants**	**MB**	**MR**
Langmuir	q_{max} (mg g^{-1})	74.96	46.27
	b (L mg^{-1})	0.252	0.074
	R^2	0.998	0.992
Freundlich	K_F	18.25	7.39
	1/n	0.404	0.409
	R^2	0.949	0.974
Temkin	K_T (L mg^{-1})	4.31	0.703
	B (J mol^{-1})	177.1	241.7
	R^2	0.985	0.994
D–R	q_{max} (mg g^{-1})	60.5	36.6
	B (mol^2 J^{-2})	-6.12×10^{-7}	-5.37×10^{-6}
	R^2	0.895	0.974

60.5 mg g^{-1} for MB and 36.6 mg g^{-1} for MR. The mean adsorption energy (E) was estimated at 28.6 kJ mol^{-1} for MB, indicating possible chemisorption, and 9.65 kJ mol^{-1} for MR, consistent with physisorption. These findings suggest that the stronger structural and electronic properties of MB contribute to a more robust interaction with the surface functional groups of optimized AC-SOR.

Taken together, the results demonstrate that MB has a higher affinity, greater capacity, and more intense interaction with AC-SOR than MR across all models. While Langmuir provides the best global fit, Temkin and D–R models contribute additional information on the energetic and mechanistic aspects of the process.

These findings position optimized AC-SOR as a highly efficient adsorbent for dye removal, particularly for MB, and suggest its strong potential for application in industrial wastewater treatment. Future research should investigate regeneration efficiency and adsorption performance in real wastewater conditions to confirm the long-term feasibility of its use. Table 1.2 presents the isotherm model parameters for MB and MR adsorption onto AC-SOR, and compares the maximum capacities with other bio-based activated carbons reported in the literature.

1.4.2 Adsorption of Cd(II) onto Optimized AC-RP

1.4.2.1 Effect of Operational Parameters on Cd(II) Adsorption

The adsorption performance of AC-RP was investigated under varying operational conditions, including solution pH, adsorbent dosage, contact time, and initial Cd(II) concentration, all of which play a critical role in determining removal efficiency. Among these, pH emerged as a particularly influential parameter due to its effect on both the ionization state of Cd(II) and the surface charge of the adsorbent. The removal efficiency of Cd(II) increased significantly with pH, peaking at approximately 64% at pH 7. Under highly acidic conditions, the competition between H^+ ions and Cd(II) for active adsorption sites limits uptake, while at alkaline pH levels (above 8), the formation of cadmium hydroxide precipitates reduces the effective availability of Cd^{2+} ions in solution. Thus, pH 7 was chosen as the optimal value for further experiments.

In parallel, the influence of adsorbent dosage was explored by varying the amount of AC-RP from 0.1 to 1.5 g L^{-1}, under fixed pH and initial concentration. The results indicate a progressive enhancement in removal efficiency with increasing adsorbent dose, attributed to the greater number of available active sites and expanded surface area for adsorption. Furthermore, higher dosages likely promote better dispersion of the adsorbent, facilitating more effective interaction between Cd(II) ions and the porous matrix.

The effect of contact time and initial Cd(II) concentration was also studied to assess adsorption dynamics. As depicted in Figure 1.5, the process followed a typical adsorption profile with a rapid initial uptake during the first 30 min, followed by a slower approach to equilibrium. Equilibrium was reached at approximately 55 min for 50 mg L^{-1}, 80 min for 100 mg L^{-1}, and 110 min for 200 mg L^{-1} initial concentrations. This trend reflects the influence of concentration gradients and the gradual saturation of active sites at higher solute concentrations, which increases the resistance to mass transfer.

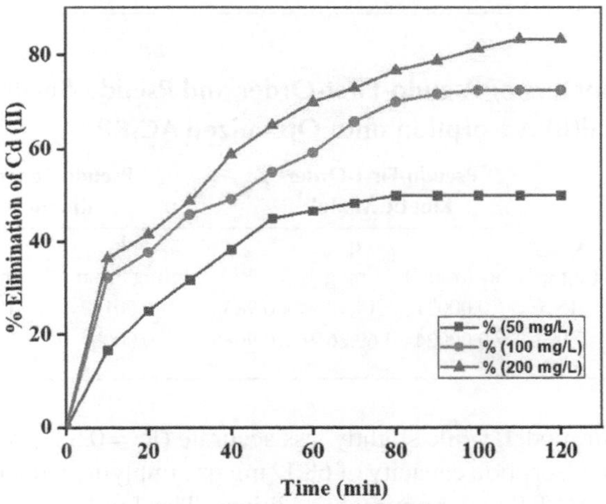

FIGURE 1.5 Effect of contact time and initial Cd(II) concentration on removal efficiency using optimized AC-RP (conditions: pH = 7, adsorbent dose = 1 g L^{-1}, temperature = 25°C).

Collectively, these results demonstrate that the adsorption efficiency of AC-RP for Cd(II) is significantly governed by operational parameters. The optimal conditions identified, pH 7, a moderate adsorbent dose, and sufficient contact time, are critical for maximizing removal performance and should be considered in any scale-up or practical application of this bio-based adsorbent.

1.4.2.2 Kinetic Modelling of Cd(II) Adsorption

The kinetic behaviour of Cd(II) adsorption onto AC-RP was evaluated using the pseudo-first-order and pseudo-second-order models. The pseudo-first-order model provided a better fit to the experimental data, particularly at lower initial concentrations, as evidenced by higher correlation coefficients and closer agreement between calculated and experimental adsorption capacities. For example, at 50 mg L^{-1}, the calculated q_e was 14.93 mg g^{-1} compared to the experimental 15.5 mg g^{-1}, with $R^2 = 0.983$. At 200 mg L^{-1}, a similar trend was observed ($q_{e,cal} = 69.27$ mg g^{-1}; $R^2 = 0.968$).

In contrast, the pseudo-second-order model showed higher $q_{e,cal}$ values but lower correlation coefficients, suggesting that it was less suitable under the tested conditions. These results suggest that the Cd(II) adsorption process onto AC-RP is mainly governed by a pseudo-first-order mechanism, indicating a physisorption-dominated process. The kinetic parameters for both models are summarized in Table 1.3.

1.4.2.3 Adsorption Isotherm Modelling of Cd(II)

The equilibrium data for Cd(II) adsorption onto AC-RP were fitted to four classical isotherm models: Langmuir, Freundlich, Temkin, and D–R, to investigate the adsorption mechanism and surface characteristics. Among these, the Freundlich model provided the best correlation ($R^2 = 0.9876$), indicating a heterogeneous adsorption process involving multilayer formation and a distribution of active sites with varying binding energies.

TABLE 1.3
Kinetic Parameters of Pseudo-First-Order and Pseudo-Second-Order Models for Cd(II) Adsorption onto Optimized AC-RP

Cadmium (II) Initial Concentration (mg L^{-1})	$q_{e,exp}$ (mg g^{-1})	Pseudo-First-Order Kinetic Model			Pseudo-Second-Order Kinetic Model		
		k_1 (min^{-1})	$q_{e,cal}$ (mg g^{-1})	R^2	k_2 [g(mg^{-1} min^{-1})]	$q_{e,cal}$ (mg g^{-1})	R^2
50	15.5	0.00021	14.9269	0.9832	0.00122	19.7628	0.9427
200	80	0.00024	69.2659	0.9685	0.00004	89.0471	0.9192

The Langmuir model, while slightly less accurate (R^2=0.889), suggested a maximum monolayer adsorption capacity of 68.12 mg g^{-1}, implying a relatively high retention potential of AC-RP under optimal conditions. The Temkin model also showed a satisfactory fit (R^2=0.920), reflecting moderate interactions between the adsorbent and Cd(II), with adsorption heat decreasing linearly with coverage.

The isotherm parameters obtained for Cd(II) adsorption onto AC-RP are summarized in Table 1.4. In contrast, the D–R model yielded a much lower correlation (R^2=0.79) and an adsorption energy below 1 kJ mol^{-1} (E=0.085kJ mol^{-1}), supporting a physisorption-dominated mechanism. Overall, the isotherm analysis reveals that Cd(II) adsorption onto AC-RP occurs through a predominantly physical, heterogeneous process with favourable uptake capacity.

These findings highlight the promising adsorption performance of both AC-SOR and AC-RP under optimized conditions. To better contextualize these results, a comparative analysis with other bio-based activated carbons is presented in the following section.

TABLE 1.4
Isotherm Model Parameters for Cd(II) Adsorption onto Optimized AC-RP

Isotherm	Isotherm Constants	Values
Langmuir	q_{max} (mg g^{-1})	68.119
	b (L mg^{-1})	0.011
	R^2	0.8892
Freundlich	K_F	2.3599
	1/n	0.5944
	R^2	0.9876
Temkin	K_T (L mg^{-1})	0.114
	B (J mol^{-1})	33.98
	R^2	0.9202
D–R	q_{max} (mg g^{-1})	34.99
	B (mol^2 J^{-2})	6.84×10^{-5}
	R^2	0.79588

1.5 COMPARATIVE CROSS-ANALYSIS OF BIO-BASED ACTIVATED CARBONS

The performance of activated carbons (AC) derived from plant biomass is governed by several interdependent factors, including the chemical nature of the precursor, the conditions of chemical and thermal activation, and the physicochemical parameters applied during the adsorption process. Based on the experimental findings detailed in the previous section, it is essential to place the materials developed in this study (AC-SOR and AC-RP) within a broader comparative context by incorporating recent literature data on other bio-based activated carbons.

Table 1.5 below presents a structured comparative summary of eight bio-based activated carbons, including their synthesis conditions (activating agent, impregnation ratio, activation temperature, and time), textural properties (BET surface area, iodine number), and adsorption performance toward various pollutants. The comparison includes both organic dyes (e.g., MB, MR) and inorganic contaminants (e.g., cadmium, lead), along with optimal adsorption conditions (pH, equilibrium time), best-fit isotherm and kinetic models, and corresponding bibliographic references.

This compilation allows not only the benchmarking of the intrinsic efficiencies of the studied adsorbents but also reveals general trends connecting structural features to adsorption performance. It also highlights how process parameters such as activation temperature or chemical treatment influence the final adsorption behaviour. This section serves as a foundation for the cross-analysis developed in Section 1. 4 where the selectivity, stability, regenerability, environmental sustainability, and feasibility of large-scale deployment are discussed.

A critical examination of the comparative data reveals several key insights. Certain activated carbons, such as bamboo fibre–AC or prickly pear seed cake–AC, exhibit exceptional BET surface areas (up to 2,892 $m^2\ g^{-1}$) and iodine numbers exceeding 2,500 $mg\ g^{-1}$. However, these properties are often achieved through intensive activation conditions such as high temperatures (>800°C), long treatment durations, and the use of highly corrosive agents, which may significantly increase the energy footprint and production cost, thus limiting scalability in resource-constrained settings.

By contrast, the ACs developed in this study (*Saponaria officinalis* and *Rhus pentaphylla*) were synthesized under milder and more sustainable conditions (650°C, $\leq$90 min, H_2SO_4), yet demonstrated competitive adsorption capacities: 74.96 $mg\ g^{-1}$ for MB and 46.27 $mg\ g^{-1}$ for Cd(II), with removal efficiencies exceeding 95% and 85%, respectively. These values are comparable or superior to other adsorbents prepared under more demanding protocols, underscoring the value of optimized, low-cost, bio-based precursors.

Notably, the AC-SOR shows excellent affinity toward organic dyes, attributable to its high iodine value (1,250.9 $mg\ g^{-1}$) and the abundance of polar surface groups that favour chemisorption via π–π interactions and hydrogen bonding. In contrast, AC-RP exhibits higher selectivity for divalent metal ions, consistent with Freundlich-type adsorption behaviour and the presence of specific oxygenated surface functionalities.

One of the major distinguishing features of AC-RP is its competitive operating cost, estimated at $\approx$0.998 USD per litre of treated water, including raw material collection, chemical activation, pyrolysis, and application in batch adsorption.

TABLE 1.5
Comparative Summary of Preparation Conditions, Textural Properties, and Adsorption Performance of Bio-based AC

Parameter	*Saponaria officinalis*-AC	*Rhus pentaphylla*-AC	Peanut Hull-AC	Prickly Pear Seed Cake–AC	Areca Catechu Nut-AC	Caraway Seed-AC	Bamboo Fibre-AC	Cotton Wood-AC
Activating Agent	H_2SO_4	H_2SO_4	H_3PO_4	H_3PO_4	H_3PO_4	Na_2CO_3	KOH	NaOH
Impregnation Ratio (w/w)	1:1	1:2	1:1	1:2	1:1	1:2	1:5	1:1
Activation Temp. (°C)	650	650	600	500	400	700	850	100
Activation Time (min)	90	90	12	60	180	120	45	1,440
BET Surface Area (m^2 g^{-1})	684	545	952.6	551	1705	926	2,892	395.48
Iodine Number (mg g^{-1})	1,250.9	1,042.5	1,002	2,527.3	888	1,107	3,754.32	-
Adsorbate	MB, MR	Cd(II)	Brilliant Blue R	Lead, Cd (II)	MB	MB, MR	Crystal violet	MB, CR
q_{max} (mg g^{-1})	74.96 46.27	68.11	149.5	170 180	369	296 208	7	40.875 45.97
Optimal pH	pH of the solution	7	pH of the solution	6	10	7 6	1,353.09	10 2
Adsorption Equilibrium (min)	30 90	90	-	240 220	180	100	40	30 150
Isotherm Model	Langmuir	Freundlich	Langmuir	-	Langmuir	Langmuir	Langmuir	Temkin
Isotherm Kinetic	Pseudo-second-order	Pseudo-first-order	Pseudo-second-order	-	-	Pseudo-second-order	Pseudo-second-order	Bangham
References	This study	This study	Zhong et al. (2012)	Dhahri et al. (2022)	Joshi et al. (2022)	Paluch et al. (2023)	Sun et al. (2023)	Nguyen Thi Hong et al. (2024)

For comparison, Jawad et al. (2016) reported a treatment cost of approximately 1.8 USD L^{-1} for H_3PO_4-activated carbon derived from biomass waste. Other advanced materials, such as graphene oxide-based composites or metal nanoparticles, can exceed 3 USD L^{-1}, due to expensive synthesis procedures and limited reusability (Sun et al. 2019; Ahmadi et al. 2024). These observations reinforce the economic viability of the present materials, especially AC-RP, for decentralized water treatment systems.

Furthermore, from an environmental perspective, the lower energy consumption, absence of hazardous byproducts, and valorization of underused local biomass contribute to a sustainable and circular approach. Both materials can be integrated into modular filtration systems, functionalized membranes, or fixed-bed columns, offering context-adapted solutions, particularly in rural or resource-limited regions (Ghazoui et al. 2025).

In summary, this comparative assessment confirms that *Saponaria officinalis* and *Rhus pentaphylla* are not only effective and accessible precursors but also align with the goals of green chemistry, circular economy, and environmental sustainability. These findings justify their in-depth valorization in the concluding section.

1.5.1 Limitations and Perspectives

While the results presented in this study confirm the promising adsorption performance of the bio-based activated carbons derived from *Saponaria officinalis* and *Rhus pentaphylla*, certain limitations must be acknowledged to guide future research and practical implementation.

First, all adsorption experiments were conducted in controlled laboratory conditions using synthetic solutions of dyes and metal ions. Although such conditions enable mechanistic elucidation and reproducibility, they do not fully replicate the complexity of real industrial effluents, which often contain a mixture of organic and inorganic contaminants, competing ions, surfactants, and variable pH or ionic strength. As noted in the literature, the performance of adsorbents in real wastewater is often significantly lower due to matrix effects and competitive adsorption phenomena (Crini and Lichtfouse 2019). Therefore, the selectivity and efficiency of the developed materials should be further validated under real-field conditions.

Second, although preliminary cost estimations and activation protocols were optimized for feasibility, the regeneration and reuse potential of the adsorbents was not experimentally addressed in this work. Sustainable application requires not only high removal efficiency but also mechanical stability and adsorption retention over multiple cycles. Regenerability is a critical factor influencing the economic viability of adsorbents in real operations (Foo and Hameed 2010). Future investigations should include desorption studies, regeneration agent screening, and cycle-to-cycle performance evaluation.

Third, the characterization of surface chemistry and interaction mechanisms, while comprehensive, could be further refined by advanced techniques such as X-ray photoelectron spectroscopy, Raman spectroscopy, or zeta potential analysis. These tools would allow deeper insights into the bonding nature and electron transfer processes involved during adsorption, especially in cases where chemisorption dominates (Zhang et al. 2021).

Finally, from an engineering perspective, the current study focuses on batch-scale evaluation. For real-world deployment, it is essential to assess column performance, breakthrough curves, flow rate influence, and long-term stability in fixed-bed or membrane-integrated systems. Such configurations are more representative of field conditions and essential for scale-up design (Kamran et al. 2019).

In summary, this study provides a strong foundation for the valorization of underutilized biomass into efficient adsorbents. However, to bridge the gap between lab-scale validation and real-world application, a multidisciplinary approach involving materials science, environmental engineering, and techno-economic modelling will be essential.

1.6 CONCLUSION

This chapter presented a comparative study between two bio-based activated carbons derived from indigenous Moroccan biomass: AC-SOR and AC-RP. These adsorbents were synthesized under statistically optimized conditions and applied for the removal of synthetic dyes and heavy metals from aqueous media. The use of RSM enabled the efficient optimization of synthesis parameters while minimizing the number of experimental runs.

The AC-SOR, prepared by sulphuric acid activation at 650°C for 24h, exhibited a high iodine number (1,250.9 mg g^{-1}) and excellent adsorption capacities for MB and MR dyes. Removal efficiencies exceeded 95%, and adsorption data fitted well with the Langmuir isotherm and pseudo-second-order kinetics, suggesting a monolayer chemisorption process facilitated by a homogeneous and well-developed porous structure.

Conversely, AC-RP, also activated with sulphuric acid, was optimized for Cd(II) removal. The material exhibited a BET surface area of 545 m^2 g^{-1} and an iodine number ranging from 483 to 992 mg g^{-1}. It achieved over 85% Cd(II) removal within 90 min. Kinetic and equilibrium modelling revealed a predominant pseudo-first-order behaviour and a better fit to the Freundlich isotherm, indicating a heterogeneous adsorption mechanism involving both physical and chemical interactions. Thermodynamic parameters confirmed the endothermic nature of the process. Moreover, the estimated treatment cost (0.9984 USD L^{-1}) underlines its viability for scalable applications.

A cross-analysis highlighted the complementarity of the two materials. AC-SOR demonstrated a greater affinity for organic pollutants, owing to the abundance of polar functional groups and a high pore volume. In contrast, AC-RP showed enhanced selectivity for metal ions, coupled with better chemical resilience and lower synthesis costs. These features position both materials as suitable candidates for targeted wastewater remediation depending on the effluent type.

Overall, these findings underscore the potential of bio-derived activated carbons in the treatment of complex industrial wastewater. Their low-cost production, local availability, and high efficiency make them excellent alternatives for decentralized or hybrid treatment systems. Practical implementations may include their integration into fixed-bed columns, membrane supports, or cartridge filters, particularly in resource-limited regions.

Future research should explore the performance of these materials in real wastewater matrices, their regeneration and reuse potential, and their synergy with other treatment technologies such as photocatalysis or membrane separation. Such integrated strategies could pave the way for advanced and sustainable pollutant remediation frameworks.

In summary, this study reaffirms the viability of valorizing local biomass residues like *Saponaria officinalis* and *Rhus pentaphylla* to produce high-performance activated carbons, offering a sustainable and scalable solution for environmental decontamination aligned with circular economy and green chemistry principles.

REFERENCES

Ahmadi M, Bahmanzadegan F, Qasemnazhand M, et al (2024) Experimental, RSM modelling, and DFT simulation of CO_2 adsorption on Modified activated carbon with LiOH. *Sci Rep* 14:1–19. https://doi.org/10.1038/s41598-024-64503-9

Briffa J, Sinagra E, Blundell R (2020) Heavy metal pollution in the environment and their toxicological effects on humans. *Heliyon* 6:e04691. https://doi.org/10.1016/j.heliyon.2020.e04691

Chaari I, Touil A, Medhioub M (2021) Adsorption-desorption of phenolic compounds from olive mills wastewater using Tunisian natural clay. *Chin J Chem Eng* 40:287–292. https://doi.org/10.1016/j.cjche.2020.12.020

Crini G, Lichtfouse E (2019) Advantages and disadvantages of techniques used for wastewater treatment. *Environ Chem Lett* 17:145–155. https://doi.org/10.1007/s10311-018-0785-9

Dhahri R, Yılmaz M, Mechi L, et al (2022) Optimization of the preparation of activated carbon from prickly pear seed cake for the removal of lead and cadmium ions from aqueous solution. *Sustainability (Switzerland)* 14:1–17. https://doi.org/10.3390/su14063245

Dou H, Xu M, Wang B, et al (2021) Microporous framework membranes for precise molecule/ion separations. *Chem Soc Rev* 50:986–1029. https://doi.org/10.1039/d0cs00552e

Falconer RJ, Schuur B, Mittermaier AK (2021) Applications of isothermal titration calorimetry in pure and applied research from 2016 to 2020. *J Mol Recognit* 34:e2901

Farghaly S, Alaswad M, Fiad N, et al (2024) The impact of fast fashion on sustainability and eco-friendly environment in fashion design world. *J Text Color Polym Sci* https://doi.org/10.21608/jtcps.2024.258355.1256

Foo KY, Hameed BH (2010) Insights into the modeling of adsorption isotherm systems. *Chem Eng J* 156:2–10. https://doi.org/10.1016/j.cej.2009.09.013

Ghazoui M, Boudouch O, Miyah Y, et al (2025) Assessment of Rhus Pentaphylla-sulfuric acid activated carbon performance for cadmium ions adsorption: mechanism, response surface methodology optimisation, and cost estimation. *Indian Chem Eng* 1–22. https://doi.org/10.1080/00194506.2025.2462953

Ghazoui M, Elkacmi R, Sylla AS, et al (2024) Efficient removal of methylene blue and methyl red dyes using a novel adsorbent derived from Saponaria officinalis Root via H_3PO_4, H_2SO_4, and KOH-activation: Optimization, kinetics, and isotherm study. *Desalin Water Treat* 318. https://doi.org/10.1016/j.dwt.2024.100378

Hama Aziz KH, Mustafa FS, Omer KM, et al (2023) Heavy metal pollution in the aquatic environment: Efficient and low-cost removal approaches to eliminate their toxicity: A review. *RSC Adv* 13:17595–17610. https://doi.org/10.1039/d3ra00723e

Jawad AH, Rashid RA, Ishak MAM, Wilson LD (2016) Adsorption of methylene blue onto activated carbon developed from biomass waste by H_2SO_4 activation: Kinetic, equilibrium and thermodynamic studies. *Desalin Water Treat* 57:25194–25206. https://doi.org/10.1080/19443994.2016.1144534

Joshi S, Shrestha RG, Pradhananga RR, et al (2022) High surface area nanoporous activated carbons materials. *J Carbon Res* 8:2. https://doi.org/10.3390/c8010002

Kamran M, Malik Z, Parveen A, et al (2019) Biochar alleviates Cd phytotoxicity by minimizing bioavailability and oxidative stress in pak choi (Brassica chinensis L.) cultivated in Cd-polluted soil. *J Environ Manage* 250:109500. https://doi.org/10.1016/j.jenvman.2019.109500

Mitrogiannis D, Markou G, Çelekli A, Bozkurt H (2015) Biosorption of methylene blue onto Arthrospira platensis biomass: Kinetic, equilibrium and thermodynamic studies. *J Environ Chem Eng* 3:670–680. https://doi.org/10.1016/j.jece.2015.02.008

Nguyen Thi Hong H, Le Quoc H, Ngo TCQ, et al (2024) Synthesis of activated carbon from cotton plant by-products in Vietnam to remove methylene blue and congo red. *J Chem Res* 48. https://doi.org/10.1177/17475198241255288

Pabby AK, Swain B, Sonar NL, et al (2022) Radioactive waste processing using membranes: State of the art technology, challenges and perspectives. *Sep Purif Rev* 51:143–173. https://doi.org/10.1080/15422119.2021.1878221

Paluch D, Bazan-Wozniak A, Nosal-Wiercińska A, Pietrzak R (2023) Removal of methylene blue and methyl red from aqueous solutions using activated carbons obtained by chemical activation of caraway seed. *Molecules* 28. https://doi.org/10.3390/molecules28176306

Prahas D, Kartika Y, Indraswati N, Ismadji S (2008) Activated carbon from jackfruit peel waste by H_3PO_4 chemical activation: Pore structure and surface chemistry characterization. *Chem Eng J* 140:32–42. https://doi.org/10.1016/j.cej.2007.08.032

Sayğili H, Güzel F (2016) High surface area mesoporous activated carbon from tomato processing solid waste by zinc chloride activation: Process optimization, characterization and dyes adsorption. *J Clean Prod* 113:995–1004. https://doi.org/10.1016/j.jclepro.2015.12.055

Sun Q, Zhu L, Aguila B, et al (2019) Optimizing radionuclide sequestration in anion nanotraps with record pertechnetate sorption. *Nat Commun* 10:1–9. https://doi.org/10.1038/s41467-019-09630-y

Sun S, Zhu Y, Gu Z, et al (2023) Adsorption of crystal violet on activated bamboo fiber powder from water: Preparation, characterization, kinetics and isotherms. *RSC Adv* 13:6108–6123. https://doi.org/10.1039/d2ra08323j

Tripathi AD, Mishra PK, Khosarvi-Darani K, et al (2022) Hydrothermal treatment of lignocellulose waste for the production of polyhydroxyalkanoates copolymer with potential application in food packaging. *Trends Food Sci Technol* 123:233–250. https://doi.org/10.1016/j.tifs.2022.03.018

Wei F, Jin S, Yao C, et al (2023) Revealing the combined effect of active sites and intra-particle diffusion on adsorption mechanism of methylene blue on activated red-pulp pomelo peel biochar. *Molecules* 28. https://doi.org/10.3390/molecules28114426

Zhang Y, Zhao M, Cheng Q, et al (2021) Insights into the modeling of adsorption isotherm systems. *Chemosphere* 279:130927. https://doi.org/10.1016/j.chemosphere.2021.130927

Zhong ZY, Yang Q, Li XM, et al (2012) Preparation of peanut hull-based activated carbon by microwave-induced phosphoric acid activation and its application in Remazol Brilliant Blue R adsorption. *Ind Crops Prod* 37:178–185. https://doi.org/10.1016/j.indcrop.2011.12.015

2 Membrane Technologies in Wastewater Treatment

Fundamentals, Industrial Applications, and Innovations

Manal Idgharnane, Nouhaila Makrouni, Oussama Salhi, Majda Breida, Redouane Beniazza, Mohamed Ouammou, Saad Alami Younssi, and Brahim Achiou

2.1 INTRODUCTION

Despite covering most of the Earth's surface, freshwater remains scarce, with less than 1% readily available for human use and not easily replenished. The statistics are striking: 2.1 billion people lack access to safe drinking water in their homes, and almost 4 billion face severe water shortages for at least one month each year. This situation has resulted in intense competition and an unequal allocation of the limited freshwater resources and their use across different sectors. In addition, industrialization and rapid urbanization generate wastewater composed of about 99.9% water and a small fraction of solids, dramatically increasing both the purification complexity of the water due to its diverging chemical composition and the volume of the generated wastewater [1]. Further, tons of solid and liquid waste are produced each day throughout the world, much of them being disposed of into water sources.

Among them, industrial wastewaters from pharmaceutical production, electroplating, and petroleum refineries are among the most difficult to treat due to their complex mixtures of hazardous contaminants, including HMs, persistent organic pollutants, and emerging substances (such as drugs) that can be responsible for the development of antibiotic resistance genes (ARGs) [2,3]. In fact, electroplating wastewater is of particular concern for its high concentrations of toxic, bioaccumulative metals such as chromium, cadmium, and lead [3]. Petroleum refinery wastewater, produced in vast quantities, contains a challenging blend of hydrocarbons, carcinogenic compounds, and HMs [4]. Mining operations, particularly those generating acid mine drainage, also release highly acidic, metal-rich wastewater. The increasing presence of pharmaceutical residues, notably antibiotics, in wastewater contributes to the spread of antibiotic resistance, a significant public health risk. Furthermore, this

DOI: 10.1201/9781003624905-2

wastewater generated in large quantities impacts the chemical oxygen demand (COD), biochemical oxygen demand (BOD), suspended and dissolved solids, and other toxicity parameters of the water bodies where they are discharged [4–6]. Regulatory agencies like the U.S. Environmental Protection Agency have established stringent maximum contaminant levels to mitigate these risks [7]. Despite severe regulations, conventional treatment methods often fail to adequately remove these contaminants, highlighting the need for more effective solutions. Traditional wastewater treatment methods, while effective at removing some contaminants, often struggle with high energy use, sludge generation, and the treatment of emerging pollutants. This has driven interest in advanced technologies that can address these challenges and support environmental sustainability [1]. The design of treatment technologies must consider several factors, including chemical composition and flow patterns of wastewater, which are crucial, as well as temperature and flow conditions, which can also vary with the seasons.

Membrane filtration technologies have emerged as a keystone in the treatment of complex wastewaters, offering distinct advantages over conventional methods due to their high removal efficiency, scalability and relatively low energy requirements. Membranes can be organic or inorganic depending on the base material. For instance, polypropylene, polysulfone, polyether sulfone, polyvinylidene fluoride (PVDF), and cellulose acetate each have different characteristics and are used for filtering various types of pollutants [8]. Pressure-driven membrane processes, namely microfiltration (MF), ultrafiltration (UF), nanofiltration (NF), and reverse osmosis (RO), are widely implemented to address the diverse contaminant profiles of wastewaters. For example, RO and NF have demonstrated high removal rates for toxic HMs (e.g. chromium, cadmium, and lead) in electroplating wastewater, with studies reporting 99% rejection of Cr(VI) and Cd(II), mitigating their bioaccumulation and associated health risks [9]. UF and RO membranes have been effectively used to reduce COD, BOD, dissolved solids (TDS), and specific pollutants such as organic compounds to below regulatory limits [1]. Although membranes are widely applied for HM removal and desalination, several challenges remain, including limited selectivity, membrane fouling, materials-based limitations, layer detachment, high capital costs, and reproducibility issues, all of which highlight areas for further improvement. Addressing current knowledge gaps is essential for the development of innovative and efficient wastewater treatment plants.

This chapter explores membrane technologies, with a focus on their growing significance in advanced wastewater treatment. It examines the scientific principles underpinning pressure-driven membrane processes and highlights recent innovations that have expanded their effectiveness across diverse and complex wastewater streams. Emphasis is placed on the ability of membrane systems to efficiently target a broad range of contaminants, including HMs, organic compounds, pathogens, and emerging pollutants offering reliable solutions where conventional methods often fall short. The chapter also considers the strategic role of membrane technologies in promoting circular water use, improving effluent quality, and supporting long-term sustainability objectives in industrial and municipal contexts. Finally, it addresses the key challenges faced in membrane applications, such as fouling and energy demands, and provides recommendations and future perspectives to guide further research and practical implementation.

2.2 PRESSURE DRIVING-BASED MEMBRANE PROCESSES

Pressure-driven membrane processes are widely employed in water and wastewater treatment and include MF, UF, NF, and RO [10]. These processes differ mainly in membrane structure, molecular weight cut-off, operating pressure, and separation mechanisms, as summarized in Table 2.1.

MF membranes have relatively large pores, typically between 0.1 and 5 μm, which efficiently remove suspended solids, bacteria, and larger colloidal particles primarily based on the physical sieving mechanism [11]. Due to their low operating pressures, usually between 0.5 and 3 bar, MF membranes offer high permeability, though their rejection is limited to larger particles. This makes MF an ideal first step in filtration, preparing the water for finer treatment [12,13].

UF membranes follow, with membranes possessing smaller pores ranging from 2 to 100 nm. Operating at moderate pressures from 2 to 6 bar, UF membranes capture smaller macromolecules such as proteins, viruses, and oil emulsions that would pass through MF. This increased selectivity protects subsequent filtration stages and improves overall treatment effectiveness.

NF membranes take selectivity further, featuring pores approximately 0.001–0.01 μm in size and operating under higher pressures between 4 and 15 bar. Beyond simple size exclusion, NF membranes leverage membrane surface charge and the

TABLE 2.1
Characteristics and Performance Parameters of Pressure-Driven Membrane Processes

Membrane Process	Membrane Structure	Molecular Weight Cut-off (kDa)	Pore Diameter (μm)	Operating Pressure (bar)	Permeability ($L\ m^{-2}h^{-1}\ bar^{-1}$)	Rejected Substances
MF	Symmetric or asymmetric porous	100–500	~0.1–5	0.5–2	400–600	Bacteria, emulsified oils, suspended solids, macromolecular organic compounds
UF	Asymmetric microporous	20–100	~0.002–0.1	2–6	120–200	Proteins, colloids, microplastics, polysaccharides
NF	Semi-dense thin-film composite	3–15	~0.001–0.01	4–12	12–25	Dyes, divalent ions, sugars, pharmaceuticals
RO	Dense thin-film composite	0.2–1.5	~0.0001–0.001	30–70	4–10	Salt (mono/divalent), HMs, trace contaminants

Donnan exclusion effect to repel ions with similar charges. This enhances rejection of multivalent ions like sulfate and calcium while allowing some monovalent ions to pass, enabling applications like water softening, color removal, and partial desalination of brackish water.

At the finest scale, RO membranes have the smallest effective pores, around 0.0001–0.001 μm, and require the highest pressures, ranging from 15 to 70 bar to overcome the osmotic pressure of saline feedwater. RO separates water molecules by the solution diffusion mechanism, effectively rejecting nearly all dissolved salt, including monovalent ions. Although charge effects play a lesser role here, they still impact fouling and ion transport. RO is the leading technology for seawater desalination and advanced wastewater reuse, despite challenges such as membrane scaling and brine disposal [14].

2.3 WASTEWATER TREATMENT USING PRESSURE DRIVING-BASED MEMBRANE PROCESSES

2.3.1 Electroplating Wastewater

Electroplating wastewater is generally loaded with heavy metals that significantly threaten environmental integrity in aquatic, terrestrial, and atmospheric systems. Industrial wastewater discharges into freshwater are estimated at approximately $697 \times 10\ 6\,m^3\,year^{-1}$. The total volume of untreated industrial wastewater treatment has reached 547 million $m^3\,year^{-1}$ with 4.5 million $m^3\,year^{-1}$ standardized to equivalent discharge and 145.5 million $m^3\,year^{-1}$ classified as noncompliant discharge [15]. The accumulation of HMs within ecosystems presents serious risks to human health, including respiratory disorders, neurological diseases, and carcinogenic impacts.

To mitigate these dangers, regulatory agencies like the U.S. Environmental Protection Agency set maximum contaminant levels for HMs in water [3]. A comprehensive assessment of their origins, chemical composition and interactions, and environmental impacts is necessary for the development of effective remediation strategies.

2.3.1.1 Characteristics of HMs

HMs are naturally occurring elements with high densities ($7\,g\ cm^{-3}$) and atomic weights. Their physical composition determines how they behave in various environmental situations, impacting their solubility, motility, and interactions with other substances. This, in turn, facilitates the formation of complexes that have an impact on their toxicity and longevity. HMs can be divided into three categories based on their chemical properties: (i) transition metals (such as chromium (Cr), copper (Cu), zinc (Zn)), (ii) post-transition metals (e.g., lead (Pb), cadmium (Cd)), and (iii) metalloids (such as arsenic (As) with oxidation states of III or V) [3].

2.3.1.2 Advancing Membrane Technologies for Removing HMs

Optimizing filtration procedures while improving durability and selectivity is a key component of developing membrane technology for HM removal (Figure 2.1) [16].

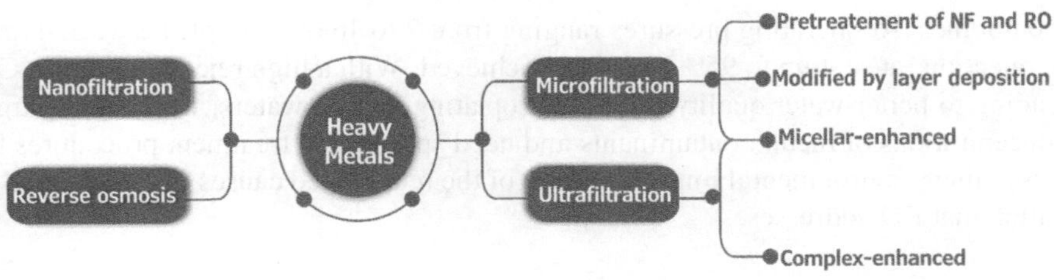

FIGURE 2.1 Different strategies for removing HMs in water.

MF is vital for wastewater treatment and outperforms sand filtration and other pretreatment methods by removing suspended and organic matter when used for plating wastewater. Integrating MF with RO, NF, and UF as a pretreatment strategy has proven effective in removing pollutants and enabling the reutilization of plating wastewater along with minimizing fouling and enhancing the lifespan and consistent performance of NF and RO [16].

Recent advancements have further expanded the capabilities of MF technology for targeted HM removal by incorporating surfactants. For example, nickel (Ni^{2+}) and cadmium (Cd^{2+}) have been successfully separated from aqueous solutions with higher pH using an MF membrane modified with 8-hydroxyquinoline, resulting in efficient removal of both metals. The metal ion rejection was improved using the modified MF membrane [16]. These findings demonstrate how MF technology can be tailored to manage a variety of contaminants in water treatment procedures. Tailored MF could be effective at removing HMs under specific conditions.

The efficiency of micellar-enhanced UF in drastically lowering metal pollutants has been shown in numerous investigations. For example, micellar-enhanced UF removes Cd^{2+}, Co^{2+}, Ni^{2+}, Cu^{2+}, and Zn^{2+} from wastewater with an efficiency of above 96%, at an operating feed concentration of metal ions (50 mg L^{-1}) at the pH range of 1–12 [17]. Beside micellar-enhanced UF, 99% of Cd^{2+} has been removed from aqueous solutions with the use of poly(ammonium acrylate) in complex-enhanced UF with a transmembrane pressure of 2 bar [18]. These findings demonstrate how effective filtration methods may be in lowering HM pollution and enhancing water quality for use in environmental and industrial settings.

NF with polyamide nanofilm membranes effectively removes HMs from water by achieving high rejection rates for different HMs, significantly reducing contamination levels. Research shows that Cu^{2+} (93.3%), Ni^{2+} (93.0%), Zn^{2+} (91.8%), and Pb^{2+} (88.3%) are all efficiently removed from aqueous solutions containing 200 mg L^{-1} of each HM using an NF membrane with a water permeability of 19.3 L $m^{-2}h^{-1}$ bar^{-1} [19]. The selective permeability of polyamide membranes enables for the rejection of these metal ions while enabling the flow of smaller molecules, making NF a promising technology for water purification and wastewater treatment.

RO is a highly effective method for separating HM pollutants from wastewater. By filtering out contaminants using a semipermeable membrane, this method successfully lowers the amounts of metals in treated water. Effective Cd removal from model water and battery industrial wastewater with an initial concentration of 7 ppm has been demonstrated in experimental studies using thin-film composite commercial membranes. At operating pressures ranging from 2 to 16 bar and pH between 4 and 6, rejection rates of up to 95% have been achieved. With a high rejection rate for Cr, leading to better water quality [20]. Electroplating rinsing waters, which have large concentrations of metal contaminants and need specialized treatment procedures to lessen their environmental impact, are one of the recognized causes of HM contamination that RO addresses.

2.3.2 Petroleum Refinery Wastewater

Petroleum refineries produce wastewater volumes which range from 0.4 to 1.6 times the amount of crude oil they manage. It is estimated that in 2025 the daily global petroleum consumption was 103.9 million barrels daily, producing an estimated 6.6 billion liters of wastewater [21]. The refining process converts crude oil into gasoline and diesel and jet fuel and feedstock for petrochemical products while producing wastewater volumes that exceed 1.6 tons per ton of crude oil processed, thus creating billions of cubic meters annually [21]. This wastewater contains toxic HMs and organic pollutants that can harm ecosystems and human health if untreated. These substances may accumulate in aquatic life, damage coral reefs, contaminate drinking water, and affect human neurological, reproductive, and respiratory systems. Advanced technologies like membrane filtration, photocatalysis, and bioremediation are essential for removing these pollutants and ensuring sustainable water management [4].

2.3.2.1 Characteristics of Petroleum Refinery Wastewater

The wastewater generated from petroleum refineries can vary significantly based on the properties of the crude oil used and the refinery processes employed. It typically includes suspended solids (approximately 1.414 g L^{-1}), such as sand, rust particles, and polymer fragments. The colloidal fraction consists of emulsified hydrocarbons (C_6–C_{20}), ferric hydroxide ($Fe(OH)_3$), zinc hydroxide ($Zn(OH)_2$), and surfactant micelles, which are stabilized by electrostatic and steric forces, making conventional treatment methods challenging. Biological components present in the wastewater include bacteria like Pseudomonas spp. and Enterobacteriaceae, as well as enzymes, extracellular polymeric substances, and albumins, all of which contribute to membrane fouling [22]. Organic pollutants in the wastewater consist of BTEX compounds (benzene, toluene, ethylbenzene, xylene), phenols (phenol, o-, m, and p-cresols), and polycyclic aromatic hydrocarbons such as naphthalene and anthracene. The wastewater has a high COD of up to 3,200 mg L^{-1}, indicating a significant organic load. In terms of inorganic constituents, the wastewater contains HMs like lead, cadmium, zinc, copper, iron, nickel, and selenium, primarily from corrosion and catalyst breakdown. Major cations (Na^+, Ca^{2+}, Mg^{2+}, and NH_4^+) and anions (Cl^-, SO_4^{2-}, HCO_3^-, and PO_4^{3-}) contribute to high ionic strength and an electrical conductivity of approximately 4.2 mS. The wastewater exhibits an alkaline pH of approximately 9.64 and a TDS concentration of around 1.268 g L^{-1}, indicative of its substantial physicochemical complexity [23].

2.3.2.2 Membrane Technologies for Petroleum Refinery Wastewater Treatment

MF membranes are essential in treating petroleum refinery wastewater, effectively removing nearly all suspended solids and a significant portion of oil and grease due to their pore sizes ranging from 0.1 to 10 μm. These membranes operate at low pressure (around 1.6 bar) and provide high permeate flux. MF membranes are typically made from polymers like PVDF (polyfluoride from vinyl), polysulfone, and polyethersulfone (PES), and ceramic options are also used. Recent studies highlight the advantages of silicon carbide ceramic membranes, which offer high oil rejection rates (89%–93%), exceptional water permeability, and super hydrophilicity, chemical resistance, and antifouling properties compared to polymeric membranes [24]. Often used as a pretreatment step, MF protects downstream processes such as UF, NF, or RO by minimizing fouling and removing larger contaminants. MF can be combined with methods like activated carbon adsorption to further enhance water quality.

UF membranes are widely used in petroleum refinery wastewater treatment due to their ability to efficiently remove colloidal hydrocarbons, emulsified oils, suspended solids, and chemical oxygen demand. UF membranes are mainly used as a pretreatment step before RO and are commonly made from polysulfone, PVDF, and polyacrylonitrile via non-solvent induced phase separation and can achieve less than 5 ppm of oil and grease concentrations in the permeate. While membrane fouling remains a challenge, advances such as the use of amphiphilic comb copolymer additives and optimized cleaning protocols have improved fouling resistance and membrane longevity. Overall, UF offers a cost-effective and robust solution for meeting stringent discharge standards in refinery wastewater treatment.

NF membranes can remove small organic particles and multivalent ions by steric and electrostatic mechanisms. Common NF membrane materials include polyamide thin-film composites and ceramics such as alumina or silicon carbide. For instance, ceramic NF membranes with a pore size of 0.9 nm have been tested and achieved up to 95%–98% phenols and BTEX (benzene, toluene, ethylbenzene, xylene), 90%–99% HM (Cr^{3+}, Ni^{2+}) rejection, and 100% removal of total suspended solid (TSS) with separation efficiency strongly influenced by factors like recovery rate and transmembrane pressure. Polyamide-based NF membranes, especially thin-film composite membranes, have also demonstrated effective removal of organics and scale-forming ions, making them suitable for water reuse and reducing environmental impact. These membranes operate at a pressure range of 10–34.5 bar, the pH ranging from 4 to 7, with a flow of approximately $144\,L\,m^{-2}h^{-1}$ [4].

RO offers the highest level of selectivity separation among membrane technologies. It uses membranes with a pore size of less than 0.001 μm, which removes almost exclusively solutes in solution by diffusion. Its efficiency allows for the removal of 98%–99% of ions (Na^+, Cl^-, Ca^{2+}, and SO_4^{2-}), as well as organic species (e.g., phenols, benzene), silica, ammonium (NH_4^+), and fluoride (F^-). RO removal of COD/BOD is very high (98%–99%), allowing the reuse of water in boilers and cooling towers. Typically, RO membranes are polyamide composites on a polysulfone support, produced by interfacial polymerization; they can generally be used under operating pressures between 20 and 48 bars (and up to 70 bars in some cases), under pH between 4 and 7, and require pretreatment methods (in particular UF, AOP) guaranteeing good control of fouling [4]. For a clearer comparison of the pollutant removal efficiencies of various membrane technologies, Figure 2.2 summarizes the overall efficiencies across the major categories of relevant contaminants found in petroleum refinery wastewater.

2.3.3 Pharmaceutical Wastewater

The extensive and continuous use of antibiotics in human medicine, veterinary care, and agriculture has led to their widespread presence in pharmaceutical wastewater. Since antibiotics are only partially metabolized by humans and animals, significant quantities of active compounds and their metabolites are discharged into wastewater treatment plants. For example, influents at wastewater treatment plants have been found to contain antibiotics such as oxytetracycline, enoxacin, levofloxacin, norfloxacin, and ciprofloxacin at measurable concentrations. These residual antibiotics exert selective pressure that accelerates the proliferation of ARGs and antibiotic-resistant bacteria, which have become emerging contaminants of global concern. Conventional treatment processes are often insufficient for removing these compounds and genetic materials, resulting in their release into aquatic ecosystems and posing serious risks to human health and environmental safety [2].

In parallel, pharmaceutical wastewater can be heavily contaminated with pathogenic microorganisms responsible for numerous waterborne diseases, including

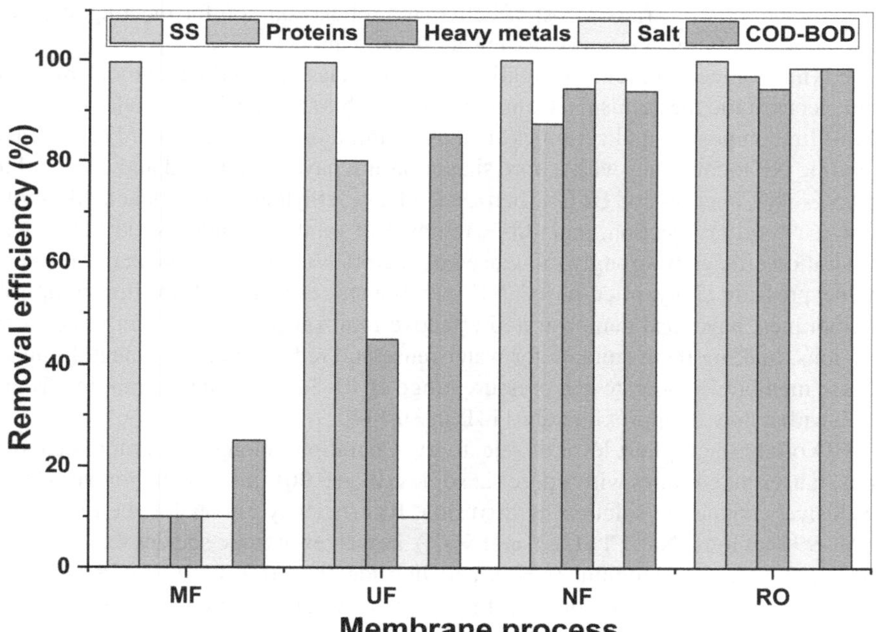

FIGURE 2.2 Effectiveness of pressure-driven membrane processes in the removal of primary contaminants from petroleum refinery wastewaters.

cholera, cryptosporidiosis, diarrhea, gastroenteritis, giardiasis, hepatitis, meningitis, and poliomyelitis. Such pathogens contribute significantly to global morbidity and mortality, especially in regions with limited access to clean water, where treated or untreated wastewater is reused for hygiene and even consumption purposes [25].

Given the limitations of traditional treatment methods, pressure-driven membrane processes offer promising solutions for the effective removal of both antibiotics and pathogens [2,25]. These membranes leverage physical size exclusion, electrostatic repulsion, and molecular sieving mechanisms to reject micropollutants, ARGs, and microbial cells with high efficiency. NF and RO, in particular, can reject molecules down to the nanometer scale, making them ideal for eliminating low-molecular-weight antibiotics and dissolved contaminants, as illustrated in Figure 2.3. Their application is crucial for reducing the environmental release of hazardous pharmaceutical compounds and mitigating the spread of antibiotic resistance, thus safeguarding public health.

2.3.3.1 MF

MF effectively removes larger pathogens such as bacteria and protozoa from pharmaceutical wastewater through size exclusion, often achieving removal efficiencies above 99% [25]. However, most antibiotics, being dissolved organic molecules with sizes much smaller than MF membrane pores, are not effectively retained by MF membranes and thus largely pass through [2]. This limits MF's ability to directly remove antibiotic compounds from wastewater. However, MF can capture antibiotic-associated particulates or bacteria-bound antibiotics, providing partial removal of antibiotic contaminants. Due to these limitations, MF is primarily used as a pretreatment step to reduce suspended solids and microbial loads, improving the performance and lifespan of subsequent pressure-driven membranes such as UF, NF, and RO, which offer higher removal rates for dissolved antibiotics and viruses. Innovations in MF membrane surface modifications and hybrid systems aim to enhance antibiotic and virus rejection, but MF alone is insufficient for comprehensive antibiotic elimination in pharmaceutical wastewater treatment.

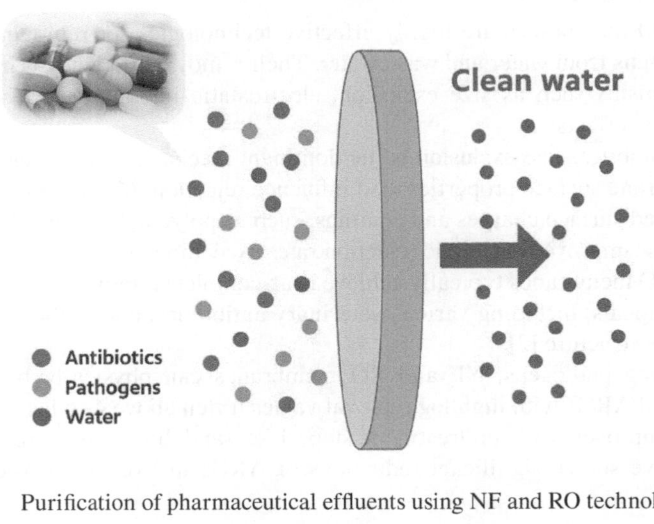

FIGURE 2.3 Purification of pharmaceutical effluents using NF and RO technologies.

2.3.3.2 UF

UF membranes show limited efficiency in removing antibiotics when used alone, primarily due to the mismatch between the molecular weight cut-off of UF membranes (typically 10–100 kDa) and the relatively low molecular weight of most antibiotic compounds, which are usually below 1 kDa. As a result, removal through size exclusion is minimal. However, some improvement can be achieved through adsorption mechanisms, especially in the presence of natural organic matter or by modifying membrane surfaces. For example, membranes enhanced with hydrophilic nanomaterials like SiO_2 or functional polymers such as chitosan polyelectrolytes have demonstrated improved antibiotic rejection, including compounds like amoxicillin, ciprofloxacin, and tetracycline. These improvements are largely due to secondary mechanisms such as hydrophobic interactions, hydrogen bonding, and cake layer formation, rather than pure size exclusion. However, due to their limited standalone performance, UF membranes are often integrated into hybrid systems or used as a pretreatment step to reduce particulate and microbial loads before advanced treatment processes more suited for micropollutant removal [2].

In terms of pathogen removal, UF membranes are highly effective against bacteria and moderately effective against viruses. Removal mechanisms include size exclusion, where pathogens larger than membrane pores are physically retained, and electrostatic interactions that prevent negatively charged viruses from approaching similarly charged membrane surfaces. Membrane modifications, such as grafting zwitterionic hydrogels, have significantly enhanced virus rejection, achieving up to 6-log reductions. Additionally, fouling by natural organic matter or suspended solids can form secondary barriers that unintentionally improve virus rejection by narrowing membrane pores or creating cake layers [2]. However, UF membranes alone may not consistently remove all viruses due to their small size and potential membrane imperfections, thereby highlighting the importance of pretreatment strategies or membrane enhancements for effective pathogen control.

2.3.3.3 NF and RO

NF and RO membranes are highly effective technologies for removing antibiotics and pathogens from water and wastewater. Their removal efficiency is mainly driven by mechanisms such as size exclusion, electrostatic interactions, adsorption, and hydrophobic effects.

For antibiotics, size exclusion is the dominant mechanism, but molecular charge and membrane surface properties also influence rejection. Modified NF membranes with tailored surface charges and coatings, such as polyethylene glycol or dopamine layers, show improved antibiotic rejection rates exceeding 90%, even at high concentrations. RO membranes typically achieve near-complete removal of a broad range of pharmaceuticals, including various veterinary antibiotics, due to their dense, nearly non-porous structure [2].

Regarding pathogens, NF and RO membranes can physically block bacteria, viruses, and ARGs with high log removal values (often above 4–6 logs). Their effectiveness improves with pretreatment steps like sand filtration to reduce fouling. Studies have shown significant reductions of ARGs in livestock wastewater using

NF and RO, whereas conventional biological treatments fail to remove these genes effectively [25].

The main operational challenge for both NF and RO is membrane fouling caused by microorganism growth and organic matter accumulation, which reduces performance over time. Advances in membrane material design aim to enhance fouling resistance and chlorine tolerance.

2.3.4 Textile Wastewater

According to the World Bank, the textile industry is the third largest consumer of water in the world, contributing 20% of industrial wastewater [26]. The wastewater comes from different fabric treatment processes, such as sizing, washing, bleaching, dyeing, printing, and finishing. These processes involve the use of synthetic dyes, surfactants, salt, and additives, resulting in wastewater that typically has an alkaline pH ranging from 9 to 11. This wastewater has a high COD and contains toxic substances such as azo dyes, polyvinyl alcohol, phenolic compounds, HMs (such as Cr^{3+}, Zn^{2+}, Pb^{2+}), and complexing agents like ethylenediaminetetraacetic acid, which are often resistant to degradation. The wide variety of fibers (including cotton, polyester, and wool) and types of dyes (such as reactive, azo, vat, and disperse) used in these processes can lead to significant variations in the composition of pollutants. The high levels of dyes and salt needed for fixing fibers create challenges for traditional biological treatments, which struggle to break down stable dyes and tough organic matter. This is why membrane separation technologies are becoming increasingly popular for meeting discharge regulations and allowing for water reuse, ultimately promoting sustainable practices in the circular economy [27,28].

2.3.4.1 Characteristics of Textile Wastewater

Textile wastewater is a complex wastewater containing a variety of physical, biological, organic, and inorganic pollutants. The mixture consists of solids in suspension, such as fibers and mineral particles, as well as colloids like dye aggregates and surfactant micelles. These elements are responsible for turbidity, coloration, and a higher potential for fouling [27,29,30]. Microorganisms such as Pseudomonas, Bacillus, and Enterococcus can result in elevated levels of COD and BOD, along with the generation of biofouling extracellular polymeric substances. Enzymatic residues smaller than 100 nm can also exacerbate membrane fouling [5]. The pH of textile wastewater typically ranges from 4 to 12, with alkaline conditions often seen post-dyeing. COD levels can reach up to 20,000 mg L^{-1}, BOD levels range from 200 to 3,000 mg L^{-1}, TDS levels exceed 6,000 mg L^{-1}, and conductivity falls between 2,000 and 12,000 μS cm^{-1} [31]. Organic pollutants found in textile wastewater include various synthetic dyes (such as Reactive Blue 19, Acid Orange 7, azo, vat, and disperse dyes), polyvinyl alcohol, solvents, phenolics, surfactants, and chelants like ethylenediaminetetraacetic acid and diethylenetriaminepentaacetic acid. The inorganic pollutants present in textile wastewater consist of high salt levels (e.g. NaCl, Na_2SO_4 exceeding 50 g L^{-1}), major ions (such as Na^+, Ca^{2+}, Mg^{2+}, Fe^{2+}, Cl^-, SO_4^{2-}, CO_3^{2-}, and PO_4^{3-}), HMs (including Cr^{3+}, Zn^{2+}, Cu^{2+}, and Pb^{2+}), and silica originating from dye additives [4,29].

TABLE 2.2
Analyzing Pollutant Profiles and Wastewater Characteristics at Different Stages of Textile Wet-Processing Operations

Processing Phase	Key Parameters	Principal Contaminants
Desizing	COD up to 20,000 mg L^{-1}	Starch and surfactants
Bleaching	Alkaline pH, presence of peroxides	Hydrogen peroxide and oxidizing agents
Dyeing	pH range of 4–12 and under high salinity conditions	Utilization of azo/reactive dyes alongside NaCl and Na_2SO_4
Washing/Finishing	High turbidity and suspended solids	Oils, surfactants, and residual dyes

To provide an overview of the diversity in pollutant levels and wastewater characteristics from different wet-processing phases, a summary of key textile wastewater-generating operations is presented in Table 2.2.

2.3.4.2 Membrane Technologies for Textile Wastewater Treatment

MF is widely used in textile wastewater treatment using a mechanical serving mechanism. This method is capable of achieving approximately 100% removal of TSS and around 99% removal of bacterial cells, as well as effectively eliminating oils and greases from softeners and lubricants with an efficiency of 85%–90%. However, MF is not suitable for removing dissolved dyes and salt [32]. Its main purpose is to serve as a pretreatment method to prevent fouling of finer membranes.

UF is designed to target macromolecules and colloids through size exclusion and limited charge repulsion using membranes with pores ranging from 1 to 100 nm in size. UF is efficient in removing pigment dye aggregates (90%–98%), proteins, and enzymes (95%–99%), and can reduce COD by 72%–98%, depending on the strength of the influent. Viral particles and fine colloids are retained at a rate greater than 99%. UF typically operates at a pressure of 2–5 bar and a pH of 6–9, serving as an intermediate pretreatment step before NF or RO processes.

NF combines sieving and electrostatic repulsion mechanisms to reject low-molecular-weight dyes (with a molecular weight of 350–800 Da) and multivalent ions. NF is effective in removing azo/reactive dyes (>98%), divalent ions such as Ca^{2+}, Mg^{2+}, SO_4^{2-}, and PO_4^{3-} (95%–99%), and HMs like Cr^{2+}, Zn^{2+}, and Fe^{2+} (90%–99%). NF performs well under alkaline pH conditions and is commonly used for color removal, dye recovery, and partial desalination processes.

RO employs dense membranes, and it relies on the solution diffusion mechanism to remove nearly all dissolved species. RO can achieve removal rates greater than 98%–99% for ions such as Na^+, Cl^-, SO_4^{2-}, F^-, SiO_2, NH_4^+, NO_3^-, ethylenediaminetetraacetic acid, and glycol, as well as ensuring near-total rejection of COD and color. RO typically operates at a pressure of 30–70 bar and a pH of 5–7, making it an ideal choice for final polishing and zero-liquid discharge applications, provided that effective pretreatment is implemented [27].

2.3.5 Tannery Wastewater

Approximately 90% of water utilized in tanning operations is discharged as wastewater [6]. Tanneries undertake various chemically intensive processes to convert raw animal hides into leather, including soaking, liming, dehairing, deliming, pickling, tanning, dyeing, and finishing. These procedures result in wastewater containing high levels of pollutants such as organics, nitrates, HMs, and sulfides. The wastewater is typically characterized by its intense coloration, elevated levels of COD, BOD, and the presence of toxic and malodorous substances. The composition of tannery wastewater can vary based on factors such as animal species (bovine, ovine, caprine), type of tanning agents employed (chrome vs. vegetable), and processing techniques. Wastewater from chromium-based tanning processes presents distinct hazards due to the potential oxidation of trivalent chromium (Cr^{3+}) into the carcinogenic hexavalent chromium (Cr^{6+}). This risk extends beyond wastewater to include treated sludge utilized in agricultural practices. Additionally, components like decomposed proteins, fats, surfactants, sulfates, and chlorides contribute to membrane fouling and complicate treatment procedures.

2.3.5.1 Characteristics of Tannery Wastewater

Tannery wastewater is characterized by significant pH variability (ranging from 3 to 11), high levels of COD (1,500–9,000 mg L^{-1}), BOD (500–4,000 mg L^{-1}), and TDS (exceeding 10,000 mg L^{-1}) [6]. The composition of the wastewater includes a variety of organic compounds, such as tannins, surfactants, fatty acids (e.g., oleic and stearic), volatile organic compounds, and degraded proteins like keratin and albumin. The relative contribution of the contaminants from varying processing stages in the tannery process is depicted in Figure 2.4. The figure synthesizes the significant effect of some operations within the process, including unhairing and chrome tanning, on wastewater pollution. Furthermore, inorganic contaminants like salt and HMs, notably trivalent chromium (Cr^{3+}) at concentrations up to 250 mg L^{-1}, are also present [33]. The presence of suspended solids (>1 μm) like leather residues, $Ca(OH)_2$, and hair fragments, as well as colloidal particles (10–1,000 nm) such as emulsified fats and gelatinous proteins, contributes to turbidity and physical fouling of membranes. Additionally, biological pollutants, including bacterial species like Bacillus, Clostridium, and Enterococcus (0.5–10 μm), viruses (20–200 nm), and enzymatically hydrolyzed proteins, increase the BOD/COD ratio and promote the formation of biofilms through extracellular polymeric substances. These constituents act in different ways of pore blockage, adsorption, scaling, and biological activity to contribute to membrane fouling.

2.3.5.2 Membrane Technologies for Tannery Wastewater Treatment

The process of MF utilizes membranes made of PES, PVDF, or ceramic materials. MF primarily removes larger suspended solids and colloidal particles as a pretreatment step for tannery wastewater through size exclusion. A PES membrane with 0.4 μm pore size demonstrated removal rates of 92% for TSS, 92% for BOD, 98% for COD, 83% for Cr^{3+}, and 100% for total coliforms in tannery wastewater [34].

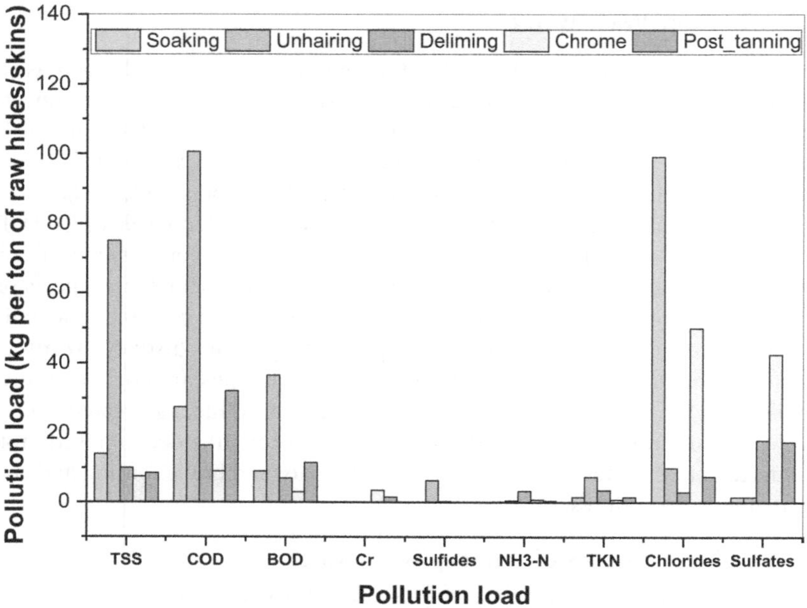

FIGURE 2.4 Estimation of the contribution of pollutants by treatment stages in tannery wastewater.

UF membranes made from PES, polysulfone, or cellulose triacetate combine size exclusion with electrostatic and hydrophobic interactions to remove proteins, fatliquors, and viruses from tannery wastewater. Removal efficiencies reported include 95%–99% for structural proteins, 94.5% for fats, >99% for viruses, 72%–98% for COD and BOD, and 30%–40% for Cr^{3+} from tannery wastewater.

NF uses thin-film composite membranes (polyamide or cellulose acetate) that selectively retain metal ions, sulfates, phosphates, phenolic compounds, and tannins from tannery wastewater through restricted diffusion and Donnan exclusion mechanisms. Performance data shows 90%–99% removal of various pollutants from tannery wastewater using NF technology.

RO employs thin-film composite polyamide membranes that remove dissolved solutes, HMs, salt, and organic compounds specific to tannery wastewater through the solution diffusion mechanism. RO consistently achieves >98%–99% removal of contaminants like HMs, salt, and organic compounds from tannery wastewater. Integrated treatment systems combining NF and RO produced final permeates with significantly reduced levels of contaminants in tannery wastewater [6].

3.3.6 Food Industry Wastewater

The food sector generates wastewater from a variety of sources, each with its own set of wastewater characteristics. Fruit processing, seafood, slaughterhouses, olive mills, and dairy processing are important industries. Effective treatment techniques are required to reduce the environmental impact and guarantee regulatory compliance because of the complexity of wastewater [35].

2.3.6.1 Characteristics of Food Industry Wastewater

Wastewater generated from the food industry comes from a variety of sources, each with unique traits. TDS, which can amount to 253.6 mg L^{-1}, and organic load are both high in dairy wastewater. Wastewater from olive mills contains a high COD between 55,730 and 156,000 mg L^{-1} and is high in phenols (2,439–8,300 mg L^{-1}), both of which are signs of serious organic contamination. The majority of the high total nitrogen 427 mg L^{-1} found in slaughterhouse waste comes from animal byproducts. Mineral salts and carbohydrates are characteristics of fruit processing wastewater, which moderately raise BOD/COD levels. Lastly, the high fat and salt concentrations (2%–5% w/v) of wastewater from the fish industry affect conductivity and BOD/COD. The best treatment plans for any industry are determined by these factors [35].

2.3.6.2 Membrane Technologies for Food Industry Wastewater Treatment

The goal of membrane technology (Figure 2.5) for wastewater treatment in the food sector is to provide a sustainable solution while lowering organic matter and suspended solids.

Wastewater from the food industry, especially from dairy processing, can be effectively treated by combining MF with UF because dairy wastewater contains proteins, lipids, and waste processing chemicals. For example, this hybrid membrane technique with ceramic membranes was operated at a pressure of 0.15 MPa (MF) and 0.3 MPa (UF) and has demonstrated encouraging results. The sequential filtering approach greatly improves water quality with initial characteristics of BOD = 890 mg L^{-1}, COD = 3,536 mg L^{-1}, turbidity = 623 NTU, TSS = 1,860 mg L^{-1}, and pH of 7.3. After filtration, the process achieves a COD reduction of 89%, color removal of 93%, and turbidity reduction of 98%. The separation process's performance indicators show excellent pollutant removal [36]. These removal efficiencies demonstrate how well MF-UF integration works for treating dairy wastewater, offering a cost-effective and environmentally friendly way to lower the organic load and enhance wastewater quality prior to release or reuse.

Polyamide NF membranes are a potential method for treating food industry wastewater, particularly restaurant wastewater, which is recognized for its high organic content and turbidity. This wastewater usually has a pH range of 4.49–6.15, turbidity (402.67–1,208 NTU), TSS (1,860 mg L^{-1}), BOD (816.17–1,097.25 mg L^{-1}), and COD (10,356.67–16,443.33 mg L^{-1}). After filtration, at an operating pressure of 2 bar, the results show a removal of COD (99.4%), BOD (86.8%), and turbidity (99.9%), demonstrating the effectiveness of the NF process in treating restaurant wastewater [37].

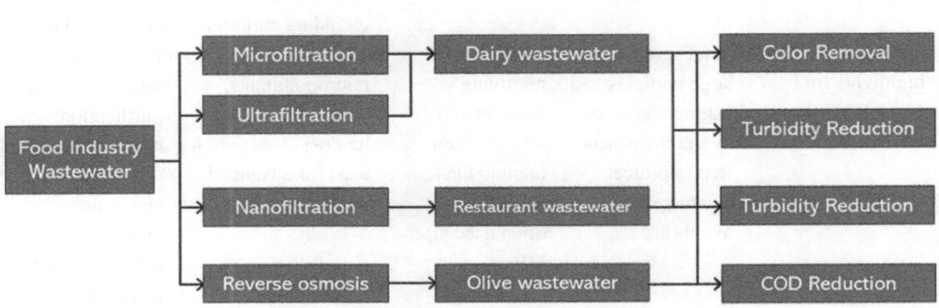

FIGURE 2.5 Different methods for treating different types of food industry wastewater.

RO removes a variety of impurities, such as organic pollutants, salt, inorganic contaminants, and chemical substances. RO has been successfully used, for instance, on olive wastewater, which has COD values of 120.5–226.6 mg L^{-1} and suspended matter of 14–16 mg L^{-1}. With an incredible 99.8% COD removal rate with a filtration pressure of 20 bar and a temperature of 22°C, this treatment successfully lowers organic pollutants and improves water clarity [38].

2.3.7 Municipal Wastewater

Municipal wastewater contains a diverse mix of contaminants that pose risks to both human health and the environment. It includes pathogenic microorganisms, micropollutants such as pharmaceuticals and pesticides, suspended solids, biodegradable organic matter, HMs, nutrients, and inorganic compounds. These substances can lead to oxygen depletion in water bodies, toxicity, bioaccumulation, eutrophication, and other environmental problems. To effectively treat this wastewater and mitigate its impacts, pressure-driven membrane technologies like NF and RO are increasingly employed [39]. These advanced membranes offer varying mechanisms and efficiencies for removing specific contaminants, as summarized in Table 2.3. The table

TABLE 2.3
Application of Pressure-Driven Membranes for Municipal Wastewater Treatment

Membrane Type	MF	UF	NF	RO
Key separation mechanism	Sieve	Sieve	Sieve, solution/diffusion, exclusion, electric repulsion	Solution/diffusion, exclusion
Typical transmembrane pressure	<5 bar	<10 bar	<20 bar	<100 bar
Main contaminants removed in municipal wastewater	Suspended solids, turbidity, protozoan cysts, some bacteria, and viruses	Macromolecules, proteins, colloids, bacteria, viruses	Low-molecular-weight compounds, polyvalent ions, some viruses, organic micropollutants including antibiotics	Low-molecular-weight compounds, ions, salt, hardness, most pathogens including viruses and bacteria
Application highlights for municipal wastewater	Used for removing large particles and pathogens; often as a pretreatment step to protect downstream membranes	Effective for removing bacteria and viruses; reduces organic matter and turbidity; often used after MF	Removes micropollutants, some pathogens, hardness ions; often used for advanced wastewater treatment	Provides the highest level of purification; removes nearly all dissolved contaminants including antibiotics and pathogens

highlights the key characteristics, targeted pollutants, and operational parameters of MF, UF, NF, and RO membranes, illustrating how each membrane process addresses specific treatment goals.

In addition to their selective separation capabilities, the practical performance of these membranes is influenced by several operational factors. For instance, membranes with smaller pores require higher operating pressures. A major challenge in membrane filtration is fouling, which can be caused by particulate matter, organic compounds, inorganic scaling, or biofilm formation. These types of fouling can be either reversible or irreversible and often occur simultaneously, leading to increased filtration resistance and reduced membrane efficiency. Therefore, optimizing fouling control strategies is critical for ensuring sustainable and cost-effective operation. One key concept is the threshold flux, the flow rate below which fouling remains minimal, which helps balance operational efficiency with maintenance needs. Membranes used for wastewater reuse are typically made from polymeric or inorganic materials, with the choice depending on factors such as pore size, mechanical strength, fouling resistance, durability, and cost [39].

2.3.8 Mining Wastewater

Acid mine drainage is caused by the oxidative dissolution of sulfide minerals, resulting in high quantities of dissolved metals and metalloids. This process creates low pH, exceptionally acidic water, which accelerates contamination and puts water bodies in the area at risk. The impacts on the environment are significant because toxic metals add to long-term pollution, and fine particle turbidity reduces water quality. By turning HMs more soluble, acidification increases contamination even more, damaging aquatic life and disrupting the natural equilibrium [40].

2.3.8.1 Composition of Mining Wastewater

Complex interactions between geological materials and chemical conditions are reflected in the presence of metals, dissolved minerals, and sulfates. These chemicals' movement and ecological impact are influenced by the dynamics of dissolution and precipitation. For example, in acid mine drainage, pyrite (FeS_2) dissolves, releasing ferrous iron (Fe^{2+}) and sulfate (SO_4^{2-}) [41].

2.3.8.2 Innovative Approaches to Acid Mine Drainage Treatment using Membrane Technology

Advancement of pressure-driven membrane processes for mining discharge treatment is focused on improving filtration efficiency while assuring long-term sustainability (Table 2.4).

Ceramic MF membranes effectively remove contaminants from wastewater, using various conditioning processes to improve performance. One approach uses ceramic MF membranes (α-Al_2O_3; pore size of 2.0 μm) that have been chemically conditioned with 0.1 w/w polymeric flocculant and 5% lime milk; the process was carried out in tangential flow mode as a downstream treatment, attaining a steady-state flux of 590 LMH and a rejection efficiency of >99.9% [42]. This method is ideal for sludge dewatering since it removes iron hydroxide sediments. The recovery of copper sulfate

TABLE 2.4
Summary of Membrane Technologies for Mining Wastewater Treatment

Membrane Technology	Mining Discharge	Rejection (%)
Ceramic MF membrane with flocculation	Iron hydroxide sediments	99.9
Ceramic MF membranes with metal bioprecipitation	Copper sulfate nanoparticles	92
Silica-modified alumina UF membranes	TSS	99
	Iron and manganese	99
Ceramic NF membranes with an amorphous TiO_2 layer	Arsenic	20–33
	Sulfate	62.8
Polymeric NF membranes (polypiperazine amide)	Sulfate	93
	Selenium	45
	Arsenic	81.5
	Aluminum	80–90
MF and RO	Arsenic, copper, iron, nickel	100

nanoparticles from acid mine drainage is another example of how ceramic MF membranes in conjunction with metal bioprecipitation show a high separation efficiency of 92%, aiding in pollutant recovery and lowering HM contamination.

Silica-modified alumina UF membranes (100 nm pore size), in combination with ceramic and inorganic polymer UF membranes, effectively remove suspended particles and dissolved metals from mining wastewater. The TSS in open stone cutting mine wastewater is 99% removed using silica-modified alumina membranes at a pressure of 1.1 bar and a pH of 6.8, which also lowers iron and manganese concentrations to less than 0.03 and 0.07 mg L^{-1}, respectively, from initial concentrations of 17.5 mg L^{-1} for iron and 0.7 mg L^{-1} for manganese. Using a new γ-Al_2O_3 ceramic membrane with a pore size of 0.01 μm and the same filtering settings, the procedure maintained excellent efficiency, achieving >99.8% rejection of total iron and >91% removal of total manganese, substantially improving wastewater quality.

NF membranes, whether ceramic or polymeric, are effective treatment options for contaminated water [43]. Ceramic NF membranes with an active layer of amorphous TiO_2 at a pressure range of 6–13 bar and a pH of 1–12 demonstrated rejection rates of 20% and 33% for arsenic, up to 62.8% for sulfate removal, and 55%–67% for divalent cations. In the meantime, polymeric NF membranes made of polypiperazine amide achieve 93% sulfate removal and 96% rejection of multivalent cations, with the exception of aluminum, which is retained at 80%–90% efficiency. Also, polymeric NF membranes improve the removal of selenium by 45% and arsenic by 81.5%. Operating at a pressure between 6–10 bar and a pH range of 1–12, exhibiting excellent pollutant mitigation performance in the treatment of drainage from abandoned coal mines in southwest Pennsylvania.

Advanced membrane filtration methods, including RO combined with MF, are crucial for effectively treating mining wastewater with HM pollutants. By eliminating dissolved contaminants, including 5 mg L^{-1} of arsenic (As), copper (Cu), iron (Fe), and nickel (Ni), these procedures guarantee total purification [40]. After MF and RO filtration under operating conditions of pH 7.59 and temperature 19.7°C, the water's metal ion concentrations were completely removed (Table 2.4), indicating the effectiveness of RO in removing HMs and improving water quality.

2.4 CHALLENGES, RECOMMENDATIONS, AND PERSPECTIVES

Membrane filtration technologies face several common challenges across the studied wastewater streams. Membrane fouling remains the primary obstacle, manifesting as organic and biofouling in NF and tight UF membranes triggered by proteins, polysaccharides, surfactants, and azo dyes, which adsorb onto pores given their large molecular structure and stability [5]. These fouling mechanisms significantly reduce permeate flux, system efficiency, and membrane lifespan while increasing operational costs [4,16]. For MF and UF membranes, fouling is primarily caused by the accumulation of suspended solids, colloidal emulsion, and biological matter, forming a cake layer that reduces permeate flux and requires frequent cleaning [35,44]. Further, the presence of emulsified fats, oils, and sulfated compounds in wastewater intensifies organic fouling, while residual lime and silica promote inorganic scaling. Silica control is especially complex due to its polymerization potential, which, if unaddressed, leads to irreversible fouling. Further, NF and RO membranes are vulnerable to inorganic scaling, mainly from sparingly soluble salts such as $CaSO_4$, $MgSiO_3$, and SiO_2 [4,33].

It is worth noting that tannins and hydrolyzed proteins, from food and other wastewater streams, contribute to the formation of dense fouling layers that fill membrane pores and shift surface properties from hydrophilic to hydrophobic, increasing fouling severity over time. The extent of fouling is influenced by the design of the membrane module and the characteristics of the membrane material. Although several studies have focused on the membrane itself, overlooking its interaction with feed spacers and other components, it would be valuable to investigate whether the presence of these various components affects the salt deposition mechanism.

In addition, the application of both MF and UF as pretreatment for wastewater contaminated with viruses (pharmaceutical sectors) results in the accumulation and entrapment of live viruses on the surface and pores of these membranes, which increases the chance of viruses' elusion (when pressure is applied) and poses health risks. Filtration of Cr^{3+} in electroplating and tannery wastewaters induces fouling caused by Cr precipitation into heavy insoluble hydroxides that deposit on the membrane surface, reducing permeability and necessitating frequent cleaning cycles. In a similar way, the presence of sulfides (S^{2-}) in wastewater can lead to sulfide (H_2S) gas formation, which corrodes membrane housings and shortens lifespan. Low-pressure membranes find it difficult to separate dissolved HMs without further chemical treatments, which is another drawback. High energy costs are also an issue since operations like RO and NF necessitate large pressure gradients, which raise prices. Moreover, high conductivity and osmotic pressure in tannery wastewater restrict recovery rates in RO systems, increasing energy consumption. Even though membrane filtration has proven successful in separating pharmaceutical compounds, issues like fouling, management of concentrated retentate, and the study of the transformation products, their degradation pathways, and possible toxicities remain major concerns.

Pressure-driven membranes alone are unreliable for viral removal without suitable pre- or post-treatment, while advanced membrane systems like membrane bioreactors, membrane distillation with photocatalysis, and pressure-driven membranes can enhance filtration by integrating adsorption and photocatalysis, though challenges such as leaching and biofouling persist. To overcome these issues, researchers are applying: (i) surface modifications with nanoparticles (e.g., TiO_2, ZnO, graphene oxide, and others) to improve hydrophilicity and antimicrobial properties, (ii) adopting hybrid system configuration (like UF-RO or membrane bioreactor-NF) for optimized removal, and (iii) leveraging artificial intelligence-assisted membrane control systems for predictive maintenance and real-time fouling detection. Circular economic strategies, such as membrane recycling and repurposing, along with advanced pretreatment methods like coagulation-flocculation, further enhance membrane performance. Future advancements are expected from innovative fabrication techniques like 3D printing, nature-inspired engineering, and osmotic membrane bioreactors with renewable energy integration, reducing environmental impact and operational costs. For instance, desalination plants with integrated renewable energy have reduced environmental impacts and reduced costs to approximately USD 0.6 m^{-3} [16]. By combining these technological advances with sustainable materials and circular economy principles, membrane filtration systems can overcome current limitations to improve efficiency, lifespan, and economic sustainability of membrane filtration in wastewater treatment.

2.5 CONCLUSION

This chapter has demonstrated that the analysis of membrane processes in wastewater treatment reveals the versatility and effectiveness of these technologies in removing pollutants across various industrial wastewater streams. Although the use of membrane technology for the treatment of urban and industrial wastewater is relatively recent, its high effectiveness compared to other classic methods has promoted its worldwide application and spread. From electroplating to petroleum refining, as well as pharmaceutical and textile industries, membrane processes have proven capable of efficiently eliminating a wide array of contaminants, including HMs, organic pollutants, and pathogenic microorganisms. The combination of different membrane techniques allows for targeted treatment tailored to the specific chemical and physical properties of the wastewater.

However, these membrane technologies face significant challenges, particularly concerning membrane fouling, energy efficiency, and the need for effective pretreatment strategies. Innovative approaches, such as surface modification of membranes with nanomaterials, the development of hybrid membrane systems, and the use of artificial intelligence to optimize operations, offer promising prospects for enhancing the efficiency and sustainability of these technologies. The necessity to minimize the environmental impact of industrial wastewater and to manage water resources sustainably will only increase in the future. Therefore, ongoing research and development in the field of membrane processes are crucial to address the challenges of wastewater treatment and to improve water quality. The integration of sustainable practices and innovative technologies will play a key role in maximizing the efficiency and effectiveness of industrial wastewater treatment systems while minimizing ecological impacts.

REFERENCES

[1] J. Shamshad, R. Ur Rehman, Innovative approaches to sustainable wastewater treatment: A comprehensive exploration of conventional and emerging technologies, *Environ. Sci. Adv.* 4 (2025) 189–222. https://doi.org/10.1039/D4VA00136B.

[2] N. Nasrollahi, V. Vatanpour, A. Khataee, Removal of antibiotics from wastewaters by membrane technology: Limitations, successes, and future improvements, *Sci. Total Environ.* 838 (2022) 156010. https://doi.org/10.1016/j.scitotenv.2022.156010.

[3] S. Rajoria, M. Vashishtha, V.K. Sangal, Treatment of electroplating industry wastewater: A review on the various techniques, *Environ. Sci. Pollut. Res.* 29 (2022) 72196–72246. https://doi.org/10.1007/s11356-022-18643-y.

[4] G.R. Dos Anjos Silva, V.R. Moreira, M.C.S. Amaral, Applications and advancements in membrane technologies for sustainable petroleum refinery wastewater treatment, *J. Environ. Chem. Eng.* 13 (2025) 115199. https://doi.org/10.1016/j.jece.2024.115199.

[5] X. Wang, J. Jiang, W. Gao, Reviewing textile wastewater produced by industries: Characteristics, environmental impacts, and treatment strategies, *Water Sci. Technol.* 85 (2022) 2076–2096. https://doi.org/10.2166/wst.2022.088.

[6] F. Yang, X.-B. Wang, Y. Shan, C. Wu, R. Zhou, N. Hengl, F. Pignon, Y. Jin, Research recap of membrane technology for tannery wastewater treatment: A review, *Collagen Leather* 5 (2023) 24. https://doi.org/10.1186/s42825-023-00132-8.

[7] S. Foorginezhad, M.M. Zerafat, A.F. Ismail, P.S. Goh, Emerging membrane technologies for sustainable water treatment: A review on recent advances, *Environ. Sci. Adv.* 4 (2025) 530–570. https://doi.org/10.1039/D4VA00378K.

[8] S.P. Bera, M. Godhaniya, C. Kothari, Emerging and advanced membrane technology for wastewater treatment: A review, *J. Basic Microbiol.* 62 (2022) 245–259. https://doi.org/10.1002/jobm.202100259.

[9] A.E.D. Mahmoud, E. Mostafa, Nanofiltration membranes for the removal of heavy metals from aqueous solutions Preparations and applications, *Membranes* 13 (2023) 789. https://doi.org/10.3390/membranes13090789.

[10] K. Khaless, B. Achiou, R. Boulif, R. Benhida, Recycling of spent reverse osmosis membranes for second use in the clarification of wet-process phosphoric acid, *Minerals* 11 (2021) 637. https://doi.org/10.3390/min11060637.

[11] B. Achiou, S. Adlane, A. Harrati, A. Belgada, A. Aaddane, Z. Chafiq Elidrissi, M. Ouammou, C. Sadik, S. Alami Younssi, Low-cost ceramic peridotite membrane developed with geomaterial-based sintering aids for textile wastewater treatment, *J. Environ. Chem. Eng.* 13 (2025) 116375. https://doi.org/10.1016/j.jece.2025.116375.

[12] O. Samhari, S.A. Younssi, M. Rabiller-Baudry, P. Loulergue, M. Bouhria, B. Achiou, M. Ouammou, Fabrication of flat ceramic microfiltration membrane from natural kaolinite for seawater pretreatment for desalination and wastewater clarification, *Desalination Water Treat.* 194 (2020) 59–68. https://doi.org/10.5004/dwt.2020.25859.

[13] D. El Machtani Idrissi, Z.C. Elidrissi, B. Achiou, M. Ouammou, S. Alami Younssi, Fabrication of low-cost kaolinite/perlite membrane for microfiltration of dairy and textile wastewaters, *J. Environ. Chem. Eng.* 11 (2023) 109281. https://doi.org/10.1016/j.jece.2023.109281.

[14] M.K. Selatile, S.S. Ray, V. Ojijo, R. Sadiku, Recent developments in polymeric electrospun nanofibrous membranes for seawater desalination, *RSC Adv.* 8 (2018) 37915–37938. https://doi.org/10.1039/C8RA07489E.

[15] R.S. Ahmed, M.E. Abuarab, M.M. Ibrahim, M. Baioumy, A. Mokhtar, Assessment of environmental and toxicity impacts and potential health hazards of heavy metals pollution of agricultural drainage adjacent to industrial zones in Egypt, *Chemosphere* 318 (2023) 137872. https://doi.org/10.1016/j.chemosphere.2023.137872.

[16] K. Castro, R. Abejón, Removal of heavy metals from wastewaters and other aqueous streams by pressure-driven membrane technologies: An outlook on reverse osmosis, nanofiltration, ultrafiltration and microfiltration potential from a bibliometric analysis, *Membranes* 14 (2024) 180. https://doi.org/10.3390/membranes14080180.

[17] J. Huang, F. Yuan, G. Zeng, X. Li, Y. Gu, L. Shi, W. Liu, Y. Shi, Influence of pH on heavy metal speciation and removal from wastewater using micellar-enhanced ultrafiltration, *Chemosphere* 173 (2017) 199–206. https://doi.org/10.1016/j.chemosphere.2016.12.137.

[18] D. Jellouli Ennigrou, L. Gzara, M. Ramzi Ben Romdhane, M. Dhahbi, Cadmium removal from aqueous solutions by polyelectrolyte enhanced ultrafiltration, *Desalination* 246 (2009) 363–369. https://doi.org/10.1016/j.desal.2008.04.053.

[19] X. Guo, B. Zhao, L. Wang, Z. Zhang, J. Li, Asymmetrically charged polyamide nanofilm of intrinsic microporosity containing quaternary ammonium groups for heavy metal removal, *J. Membr. Sci.* 698 (2024) 122620. https://doi.org/10.1016/j.memsci.2024.122620.

[20] J. Kheriji, D. Tabassi, B. Hamrouni, Removal of Cd(II) ions from aqueous solution and industrial effluent using reverse osmosis and nanofiltration membranes, *Water Sci. Technol.* 72 (2015) 1206. https://doi.org/10.2166/wst.2015.326.

[21] R. Khan, Z. Haram, W. Ahmad, S. Sohni, J. Xu, M. Ilyas, Removal of hydrocarbon pollutants from refinery wastewater using N-hexadecylchitosan as an efficient adsorptive platform, *Sci. Rep.* 14 (2024) 17236. https://doi.org/10.1038/s41598-024-66429-8.

[22] B.B. Negi, Umesh, C. Das, Novel insights into mycoremediation mechanism of petroleum refinery wastewater using Phanerochaete chrysosporium: An experimental and computational approach, *J. Water Process Eng.* 70 (2025) 107119. https://doi.org/10.1016/j.jwpe.2025.107119.

[23] M.S. Lawan, R. Kumar, J. Rashid, M.A.E.-F. Barakat, Recent advancements in the treatment of petroleum refinery wastewater, *Water* 15 (2023) 3676. https://doi.org/10.3390/w15203676.

[24] D. Das, N. Kayal, G.A. Marsola, L.A. Damasceno, M.D.D.M. Innocentini, Permeability behavior of silicon carbide-based membrane and performance study for oily wastewater treatment, *Int. J. Appl. Ceram. Technol.* 17 (2020) 893–906. https://doi.org/10.1111/ijac.13463.

[25] A.M. Nasir, M.R. Adam, S.N.E.A. Mohamad Kamal, J. Jaafar, M.H.D. Othman, A.F. Ismail, F. Aziz, N. Yusof, M.R. Bilad, R. Mohamud, M. A. Rahman, W.N. Wan Salleh, A review of the potential of conventional and advanced membrane technology in the removal of pathogens from wastewater, *Sep. Purif. Technol.* 286 (2022) 120454. https://doi.org/10.1016/j.seppur.2022.120454.

[26] M.T. Islam, Md.A. Al Mamun, A.F.M.F. Halim, R. Peila, D.O. Sanchez Ramirez, Current trends in textile wastewater treatment—Bibliometric review, *Environ. Sci. Pollut. Res.* 31 (2024) 19166–19184. https://doi.org/10.1007/s11356-024-32454-3.

[27] Z. Ma, H. Chang, Y. Liang, Y. Meng, L. Ren, H. Liang, Research progress and trends on state-of-the-art membrane technologies in textile wastewater treatment, *Sep. Purif. Technol.* 333 (2024) 125853. https://doi.org/10.1016/j.seppur.2023.125853.

[28] Z. Chafiq Elidrissi, A. Essate, B. Achiou, S. Rode, B. Belaissaoui, S. Alami Younssi, M. Ouammou, High-performance MoS_2-based composite membrane for efficient dye/salt separation in textile industry, *Sep. Purif. Technol.* 376 (2025) 133901. https://doi.org/10.1016/j.seppur.2025.133901.

[29] D.A. Yaseen, M. Scholz, Textile dye wastewater characteristics and constituents of synthetic effluents: A critical review, *Int. J. Environ. Sci. Technol.* 16 (2019) 1193–1226. https://doi.org/10.1007/s13762-018-2130-z.

[30] B. Achiou, A. Bouazizi, Y. Kouzi, M. Brieda, J.I. Calvo, M. Ouammou, S. Alami Younssi, Low-cost ceramic clay-based ultrafiltration membranes for dyes removal, in: S. Bandyopadhyay (Ed.), *Appl. Low Cost Ceram. Membr. Wastewater Treat*, Springer Nature Singapore, Singapore, 2025: pp. 81–110. https://doi.org/10.1007/978-981-97-5842-5_3.

[31] K.B. Dhameliya, C. Ambasana, Assessment of wastewater contaminants caused by textile industries, *J. Pure Appl. Microbiol.* 17 (2023) 1477–1485. https://doi.org/10.22207/JPAM.17.3.09.

[32] A. Harrati, Y. Arkame, S. Adlane, A. Essate, B. Achiou, A. El Bouari, A. Aaddane, S. Alami Younssi, C. Sadik, Preparation of low-cost peridotite ceramic microfiltration membrane for treating industrial wastewater, *Int. J. Appl. Ceram. Technol.* 21 (2024) 4366–4379. https://doi.org/10.1111/ijac.14852.

[33] A. Bhardwaj, S. Kumar, D. Singh, Tannery effluent treatment and its environmental impact: A review of current practices and emerging technologies, *Water Qual. Res. J.* 58 (2023) 128–152. https://doi.org/10.2166/wqrj.2023.002.

[34] Sadek Chahredine, N. Medjemem, S. Boulanacer, F. Bouzerara, Efficient filtration of tannery wastewater using ceramic membranes, *J. Water Chem. Technol.* 47 (2025) 53–59. https://doi.org/10.3103/S1063455X25010047.

[35] Md.N. Pervez, M.R. Mishu, G.K. Stylios, S.W. Hasan, Y. Zhao, Y. Cai, T. Zarra, V. Belgiorno, V. Naddeo, Sustainable treatment of food industry wastewater using membrane technology: A short review, *Water* 13 (2021) 3450. https://doi.org/10.3390/w13233450.

[36] M. Zielińska, M. Galik, Use of ceramic membranes in a membrane filtration supported by coagulation for the treatment of dairy wastewater, *Water. Air. Soil Pollut.* 228 (2017) 173. https://doi.org/10.1007/s11270-017-3365-x.

[37] S. Zulaikha, W.J. Lau, A.F. Ismail, J. Jaafar, Treatment of restaurant wastewater using ultrafiltration and nanofiltration membranes, *J. Water Process Eng.* 2 (2014) 58–62. https://doi.org/10.1016/j.jwpe.2014.05.001.

[38] J.M. Ochando-Pulido, S. Rodriguez-Vives, G. Hodaifa, A. Martinez-Ferez, Impacts of operating conditions on reverse osmosis performance of pretreated olive mill wastewater, *Water Res.* 46 (2012) 4621–4632. https://doi.org/10.1016/j.watres.2012.06.026.

[39] J. Yang, M. Monnot, L. Ercolei, P. Moulin, Membrane-based processes used in municipal wastewater treatment for water reuse: State-of-the-art and performance analysis, *Membranes* 10 (2020) 131. https://doi.org/10.3390/membranes10060131.

[40] S.L. Dashtban Kenari, S. Mortazavi, S. Mosadeghsedghi, C. Atallah, K. Volchek, Advancing ceramic membrane technology for sustainable treatment of mining discharge: Challenges and future directions, *Membranes* 15 (2025) 112. https://doi.org/10.3390/membranes15040112.

[41] J. Retka, G. Rzepa, T. Bajda, L. Drewniak, The use of mining waste materials for the treatment of acid and alkaline mine wastewater, *Minerals* 10 (2020) 1061. https://doi.org/10.3390/min10121061.

[42] K. Meschke, V. Herdegen, T. Aubel, E. Janneck, J.-U. Repke, Treatment of opencast lignite mining induced acid mine drainage (AMD) using a rotating microfiltration system, *J. Environ. Chem. Eng.* 3 (2015) 2848–2856. https://doi.org/10.1016/j.jece.2015.10.013.

[43] J. López, M. Reig, X. Vecino, O. Gibert, J.L. Cortina, Comparison of acid-resistant ceramic and polymeric nanofiltration membranes for acid mine waters treatment, *Chem. Eng. J.* 382 (2020) 122786. https://doi.org/10.1016/j.cej.2019.122786.

[44] S. Cairone, S.W. Hasan, K.-H. Choo, C.-W. Li, A.A. Zorpas, M. Ksibi, T. Zarra, V. Belgiorno, V. Naddeo, Enhancing membrane fouling control in wastewater treatment processes through artificial intelligence modeling: Research progress and future perspectives, *Euro-Mediterr. J. Environ. Integr.* 9 (2024) 1887–1905. https://doi.org/10.1007/s41207-024-00659-0.

3 Membrane-Based Separation Processes in Wastewater Treatment

Principles, Materials, and Sustainability Perspectives

Mohamed Amine Kasbaji, Mouad Mouhsin, Fatima Ezzahra Atmani, Oumayma El Wafi, Anas Hafdi, Meriem Kasbaji, Mohamed M'Barki, Mounir El Achaby, and Mustapha Oubenali

3.1 INTRODUCTION

The increase and growth of anthropogenic, industrial, and agricultural activities are directly linked to population growth, which causes the expansion of pollution of natural resources, particularly water, as wastewater discharges reach alarming levels daily [1,2]. However, freshwater resources do not recycle quickly enough to meet the population growth and the increasing demand for water [3]. Therefore, many people around the world, especially in developing countries, do not have access to drinking water. The consequences of this shortage are remarkable worldwide, including in the Middle East, Asia, Africa, and Latin America [3]. According to statistical studies, nearly 4 billion people suffer from this water scarcity for at least one month annually. Moreover, 2.1 billion people do not have access to safe drinking water at home. Furthermore, studies indicate that there will be a 40% decrease in accessible freshwater worldwide by 2030, posing a major risk to socioeconomic development, which will be exacerbated by the growth in demand, highlighting the urgent need to develop sustainable approaches and solutions for water management [4,5].

Wastewater can generally contain various types of pollutants, such as inorganic substances like heavy metals and infectious agents like bacteria and viruses, as well as organic compounds such as dyes and chlorobenzenes [6–8], which leads to negative effects on human health and the environment, such as the disruption of the nervous system due to their accumulation in biological circuits, and the disruption of aquatic fauna by damaging DNA cells, resulting in genetic modifications. Moreover, these pollutants are difficult to degrade due to their chemical stability [9–11]. There are also other emerging contaminants such as perfluoroalkyl compounds,

DOI: 10.1201/9781003624905-3

pharmaceuticals, and cosmetic products [12,13]. The presence of these substances can cause a high demand for oxygen, both chemical oxygen demand (COD) and biological oxygen demand, as well as the presence of trace metals. These elements can have a negative impact on water quality and on various ecosystems [14]. These contaminants can also reduce dissolved oxygen levels in water, an essential element for the breathing of aquatic life. In addition, they restrict the entry of light, thus reducing photosynthesis by aquatic plants. This has an impact on the entire food chain and threatens the overall ecological balance [15,16].

However, current environmentalist approaches have raised awareness of the need to reduce the impact of pollution, prompting both scientists and industry to design reliable, economically attractive, and environmentally friendly systems for the simultaneous treatment of water contaminants and the protection of the health of those considered vulnerable. Over the years, a wide variety of wastewater treatment technologies, including conventional filtration, coagulation-flocculation treatment [17], and biological treatment [18], have become increasingly popular. All wastewater treatment systems have benefited from continuous improvements, both to comply with increasingly stringent landfill standards and to promote water reuse [19].

Among the most rapidly advancing approaches are membrane technologies such as NF [20], UF [21], RO [22], ED [23], and membrane distillation [24], which have grown considerably in the last 20 years, partly due to the array of advantages these technologies have to offer in water treatment applications. Advantages of membrane technologies, reduced equipment size, lower energy requirements, and low to moderate installation costs make them an attractive and efficient method of wastewater management [25,26]. Membrane technologies can make considerable contributions to economic and environmental visions, as they allow treatment processes with little to no chemical input, environmentally friendly designs, and implementation by a wider audience. Because of these advantages, membrane technologies are becoming recognized as a viable and sustainable alternative to conventional treatment processes [27,28].

Although membrane technology is not a recent innovation, it continues to evolve in order to address the increasing complexity and variability of wastewater. Current research focuses on enhancing membrane performance, reducing system footprint, lowering energy consumption, improving permeate quality, and simplifying the technical skills required for operation. At the same time, considerable efforts are being made to optimize membrane modules and components to mitigate fouling, which remains one of the main limitations of these processes. Moreover, the integration of membrane systems with other treatment technologies, such as coagulation or adsorption, in hybrid configurations is a growing area of development and is already being implemented in many treatment facilities [29,30].

This chapter provides an understanding of membrane technologies used for wastewater treatment. It first introduces the major types of membrane processes, followed by the main operating principles of each and the materials used to construct the membranes. It highlights their applications, practical aspects, advantages, and drawbacks for each type of treatment and integrates discussions on recent developments and future outlook. Its conclusions emphasize the increasing importance of membrane technologies, with practical successes, to achieve a sustainable and efficient water future.

3.2 TYPES OF MEMBRANE PROCESSES

There are several membrane processes, including UF, MF, NF, RO, ED, and membrane distillation. These technologies differ in terms of pore size, driving force (pressure, electrical potential, temperature), and the type of filtrate obtained, as shown in Table 3.1.

3.2.1 MF

MF is one of the oldest commercially available membrane separation techniques [31]. MF can separate micrometer-sized particles such as suspended solids, large

TABLE 3.1
Comparison of Membrane Types Used in Water and Wastewater Treatment Processes [65,66]

Process	Pore Size (μm)	Driving Force	Filtrate	Membrane Materials	Applications
MF	0.1–5	Applied Pressure (0.1–2 bar)	Suspended solids, asbestos, cells (e.g., red blood cells, bacteria)	PP, PVDF, PVC, PTFE	Pretreatment for desalination, turbidity removal
UF	0.01–0.1	Applied Pressure (1–10 bar)	Microorganisms, proteins, viruses, dissolved macromolecules	PSU, PES, PVC, PTFE, CA	Wastewater purification, food and biotech industries, recovery of valuable compounds
NF	0.001–0.01	Applied Pressure (4–30 bar)	Organic compounds, viruses, divalent salts	PSU, PES (with support layers), CA	Water softening, removal of organic matter, wastewater reuse
RO	0.0001–0.001	Applied Pressure (30–80 bar)	Dissolved salts, metals, organic pollutants	PSU, PES (support layers), CA, polyamide	Seawater desalination, drinking water production
ED	Ion-selective (no pores)	Electric Potential (10–30 V)	Dissolved ions (cations and anions)	Ion exchange membranes (cation exchange membranes, anion exchange membranes)	Brackish water desalination, industrial wastewater treatment, salt recovery
Membrane Distillation	0.1–0.6	Thermal Gradient	Water vapor (salt excluded)	PP, PTFE, PVDF, zeolites, ceramics	Desalination, high-salinity wastewater treatment

bacteria, proteins, yeast cells, and major infectious agents using a physical separation method. Because of its ability to remove a wide variety of large contaminants, MF is considered a versatile membrane process. MF membranes have pores with diameters ranging from 0.1 to 5 μm. A rather open membrane structure generally retains particles whose diameter exceeds 0.1 μm [32]. Since hydrodynamic resistance is low, these membranes require minimal hydrostatic pressures to ensure high contaminant retention and solvent flow rates.

Membranes are made from various materials, with two types: polymers and ceramics. Ceramics have a longer lifespan, mechanical resistance, thermal stability, and resistance to extreme pH conditions, making them more economical and sterilizable through heating or steam passage, unlike polymers [33]. MF finds its most widespread application in the industrial field of water treatment. For over ten years now, the MF market has been dominated by wastewater treatment. This method is commonly used in combination with other traditional or membrane-based techniques [34].

The scientific community has aimed to create new membrane materials that promote innovation, economic efficiency, and environmental sustainability. El Ataoui et al. [35] have worked on the manufacture of a ceramic membrane based on Moroccan clay and sugar residues, demonstrating the durability and cost-effectiveness of this material to be used in the treatment of industrial polluted water after optimization of the parameters by surface response methodology. The improved membrane has a permeability of 1,246 L h^{-1} m^{-2} bar^{-1}, a porosity of 42.71%, and an average pore size of 0.88 μm, as well as a compressive strength of 56.10 MPa. In addition, it demonstrates exceptional resistance to chemical corrosion in basic environments. The optimum clay membrane underwent industrial wastewater filtration. As a result, removal rates of turbidity, COD, and electrical conductivity are 98.26%, 62.86%, and 46.14%, respectively. In addition, the cleaning procedure restored 88% of the membrane's original water permeability.

Similarly, a study was conducted on the preparation of a clay-based membrane, but this time incorporated with cornstarch and silica. This work will focus on the key properties of this membrane. This results from the determination of the optimum conditions, which are a mass percentage of 2.5% SiO_2, 5% starch, and 92.5% clay at a sintering temperature of 905°C, generating a ceramic membrane with a porosity of 43%, an average pore size of 1.45 μm, and a mechanical strength of 32 MPa. The resulting membrane offered a permeability of 960 h^{-1} m^{-2} bar^{-1} and was particularly effective in clarifying wastewater from the textile and tanning industries, removing 95% of colors for tanning and 99% for textiles, and over 98% of turbidity. This shows the feasibility of this membrane. What's more, it can be recycled [36].

3.2.2 UF

UF is a membrane-based, pressure-driven (2–8 bars) separation technique. This method uses a porous membrane with pore sizes ranging from 2 to 100 nm, enabling molecular selectivity of nano-sized elements [37]. More specifically, in the UF process, separation is achieved by the application of a concentration gradient combined with pressure exerted across the membrane, in addition to the porosity of the latter, enabling the separation of nanometric particles present in wastewater via a size-based filtering process [38].

UF membrane technology also demonstrates its ability to separate or concentrate macromolecules of various molecular dimensions, including certain plastics, silica, oils, nanoparticles, bacterial endotoxins, solid sediments, and various other viruses [39].

Many recent studies have focused on the application of UF in the treatment of wastewater. For example, a recent study on the 2025 horizon developed a mixed UF membrane made from polyethersulfone (PES). To prevent fouling, they incorporated tannic acid-modified molybdenum disulfide oxide. This membrane demonstrates its effectiveness in treatment with a water flow rate of 140.98 $L{\cdot}m^{-2}{\cdot}h^{-1}{\cdot}bar^{-1}$ and removes 91.31% of the macromolecular organic compounds present in rare earth wastewater, while allowing the passage of inorganic salts such as $CaCl_2$ and $MgSO_4$ (3.26% and 4.84% retention, respectively) [40]. Another project using a membrane made from a silica nanocomposite modified with cellulose acetate aims to implement UF under a pressure of one bar to treat phenol present in wastewater. Thus, this material has a water flux of 126.38 $L\ m^{-2}h^{-1}$ and an elimination rate reaching up to 74%, which suggests that UF is a promising method for the treatment and removal of various pollutants present in wastewater [41].

3.2.3 NF

NF is a separation method with a nominal pore size of around 1–2 nm and a molecular weight cutoff (MWCO) of 0.1–5 kDa. This process requires higher pressure (~6–12 bar) than MF and UF systems. It effectively eliminates viruses, humic compounds, and microorganisms. Furthermore, NF may eliminate alkalinity and water hardness, also known as softening membrane [37].

Among the advantages of NF are better rejection of divalent ions, lower operational pressure, reduced energy consumption, better rejection of divalent ions, and increased flow rate. Thanks to its characteristics, NF stands out as an innovative and promising technology, suitable for large-scale deployment for the removal of various pollutants (dyes, oils and fats, heavy metals, suspended particles, raw materials, as well as other auxiliary chemicals present in polluted waters) [42,43].

A recent study on wastewater treatment in the textile industry showed that an NF membrane was designed by depositing pure chitosan in degraded layers on a non-woven polyester support, with the aim of removing the dye Direct Blue 78 (DB78). Membrane characteristics: Surface roughness (Ra) is 25.5 nm, porosity is 45%, and average pore size is 6.8 nm, all under a pressure of 7.8 bar. With an initial concentration of 50 $mg\ L^{-1}$, the dye removal rate exceeds 99% [44]. Another NF membrane developed from polysulfone (PSU) modified with folic acid-functionalized pectin and sodium alginate for treating pharmaceutical wastewater shows excellent water permeability reaching 44.1 $L\ m^{-2}{\cdot}h$ and an almost complete removal rate (100%) of ceftriaxone and cefixime. Moreover, the NF test conducted on pharmaceutical wastewater demonstrated an effective separation of total dissolved solids, COD, and turbidity by the membrane (total dissolved solids: 52.3%; COD: 89.9%; turbidity: 100%) [45]. This allows us to conclude that UF is effective and promising for various treatments.

3.2.4 RO

The water purification technique using RO, which employs a semi-permeable membrane to remove impurities and pollutants present in the water, is commonly adopted. This technology is based on the principles of osmosis, a natural process where water moves from an area of low solute concentration to an area of high solute concentration through a semi-permeable membrane [46].

This method has a pore size of less than 1 nm, with an MWCO set at 100 Da. In addition, this technology requires a high operating pressure, between 12 and 80 bar, and facilitates the removal of all harmful ions as well as a large majority of non-ionic molecules in solution [37]. Therefore, this technology produces a very low effluent concentration but requires a significant initial investment and high operating costs [47].

The RO membrane is often made from thin-film composite, which only allows the passage of water while eliminating a large part or almost all of the pollutants (bacteria, salts, and organic molecules). Several characteristics evaluate the effectiveness of RO membranes, namely selectivity, fouling resistance, and impermeability [48].

For instance, Hui Gong et al. [49] have developed a compact wastewater treatment technique using DT-RO, achieving a water recovery rate of 90% with a flow of 35.6 liters per square meter per hour under a pressure of 25–28 bars. This method meets strict elimination standards while being energy-efficient and concentrating nutrients for potential valorization.

3.2.5 ED

ED is a separation technique based on an ion exchange membrane as a dynamic engine, selectively directing ion flows under the effect of an electric field, enabling ion separation. This technique is used to treat wastewater, such as pharmaceutical and electronic industrial waste, and to recover liquid waste [50,51].

In ED, ions move according to their charge, under the influence of an electric field. Ion exchange membranes enable specific separation of cations and anions thanks to their fixed charges: Cationic membranes contain negatively charged groups that favor cation migration, while anionic membranes have positively charged groups that facilitate anion passage. Under the influence of an electric field, cations pass through the cationic membranes and are stopped by the anionic membranes, while anions penetrate the anionic membranes and are retained by the cationic membranes. This process facilitates ion concentration, transforming the "dilute" solution into a "concentrated" one [52–54] (Figure 3.1).

A recent study used the ED technique to remove polyphenols from water discharged by the petroleum industry, using a bipolar membrane. After optimizing the parameters, the study shows that the optimum conditions were set at a current density of 41.89 mA cm^{-2} and a volume flow rate of 12.84 mL min^{-1}, and at a feed concentration of 214 ppm after a filtration time of 85.50 min, the polyphenol recovery rate reached 75.4% [55]. Another study shows that, thanks to ED, 85.4% of the cadmium present in wastewater from zinc smelting has been removed. The absence of a spacer reduces membrane fouling. Physical and acid cleaning is more effective than the reverse operation for minimizing accumulations [56].

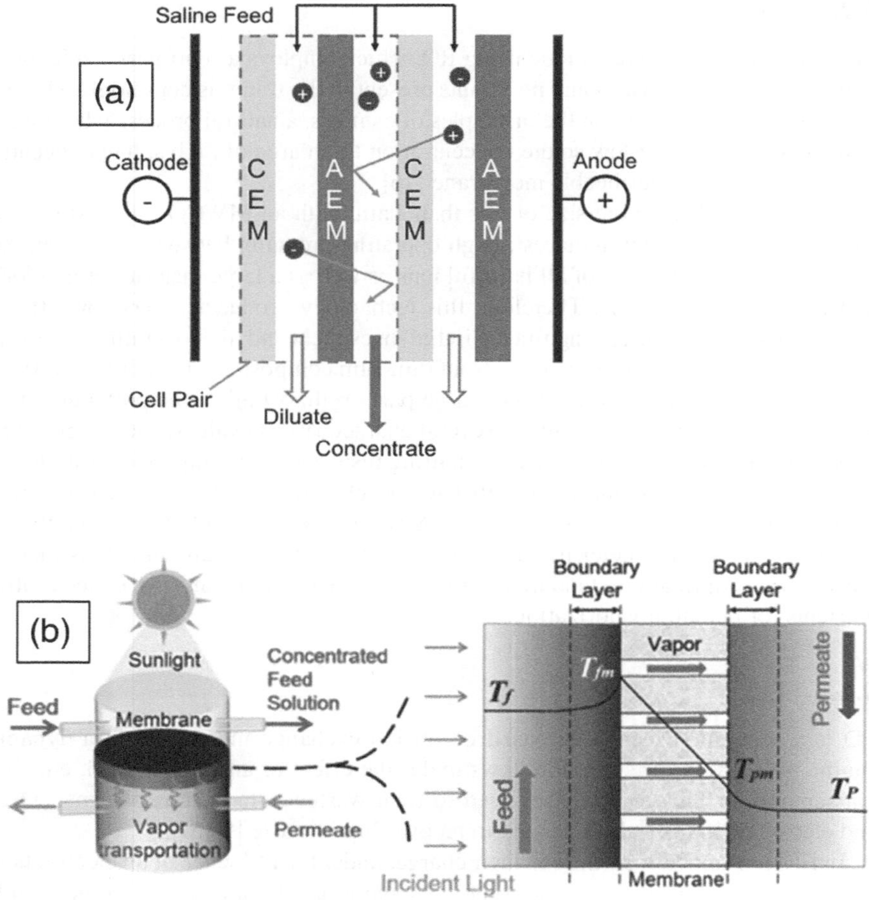

FIGURE 3.1 (a) Series of alternating cation exchange membranes and anion exchange membranes in an ED. (Reprinted with permission from [63].) (b) Diagram of photothermal membrane distillation. (Reprinted with permission from [64].)

3.2.6 Membrane Distillation

Membrane distillation is a separation technique that relies on vapor pressure and a thermal gradient created on the concentrate side, allowing only vapor to pass through a highly hydrophobic membrane with very fine pores. An increase in vapor pressure promotes the transfer of water vapor through the hydrophobic membrane from the hot side of the membrane to a cold zone. However, water molecules in the liquid state cannot cross the membrane due to the balance between the surface free energy elements and the surface structure of the membrane [57,58] (Figure 3.1).

Membrane distillation can be classified into several types [59]:

- Direct Contact Membrane Distillation: Both the feed and the permeate stream are in direct contact with the membrane on both sides.
- Air Gap Membrane Distillation: An air gap is provided between the membrane and condensing surface to minimize heat losses.
- Vacuum Membrane Distillation: The permeate side is subjected to a vacuum in order to increase the vapor flux.
- Sweeping Gas Membrane Distillation: The permeate side is swept with an inert gas to transport the vapors to an external condenser.

Several studies show the potential of membrane distillation application in wastewater treatment [60]. For example, research was carried out with membrane distillation for the treatment of highly polluted coking wastewater. A new composite PACH membrane composed of a polytétrafluoroéthylène (PTFE) membrane, cation exchange layers, anion exchange layers, and a polyamide layer was prepared by the researchers. This membrane exhibited superior resistance to 0.4 mM sodium dodecyl sulfate (bare PTFE, 0.1 mM) and the ability to concentrate NaCl up to 200 $g{\cdot}L^{-1}$. The coking wastewater concentration factor was realized ten times, and the crystallization was realized and run continuously for 168 h without fouling, without scaling, and except for the wetted [61]. Another study concerns the application of oxygen plasma-irradiated membranes (PTFE and polyvinylidene fluoride, PVDF) for dye removal from wastewater in an air gap membrane distillation process. Modified PTFE membranes demonstrated high flux (9.53 kg m^{-2} h) and over 99% dye removal; modified PVDF membranes, with higher flux (17.50 kg m^{-2} h), achieved 48.92% dye removal, suggesting wetting problems [62].

3.3 MECHANISMS OF MEMBRANE FILTRATION

Membrane filtration represents a pivotal technique in contemporary water treatment systems, characterized by the use of engineered materials and precise separation mechanisms to eliminate a wide range of contaminants. The performance of membrane processes is fundamentally governed by mechanisms such as size exclusion, adsorption phenomena, electrostatic interactions, and biological degradation [67]. These interactions not only determine the separation performance but also play a role in the ease of scaling up and running the technology.

For instance, in UF, three major mechanisms are responsible for the removal of substances: (i) physical rejection, where components that are larger than the pore size of the membrane (e.g., extracellular cyanotoxins or humic substances) are physically retained or adsorbed on the membrane surface [68,69]; (ii) chemical interactions, which are determined by membrane physicochemistry (e.g., hydrophobicity) and the specific features of substances, such as polarity or functional groups, in the water chemistry [67]; and (iii) biological degradation, where biodegradable substances are removed through membrane bioreactor (MBR) systems, e.g., activated sludge MBR or biofilm MBR combined filtration and microbial degradation [70–72].

Membranes are classified structurally into microporous or dense (nonporous) and functionally by the size of the desired solutes [73]. Surface charge also has a major influence on filtration performance, particularly in NF, which exhibits greater selectivity between divalent and some monovalent ions than MF and UF, and operates at lower pressures and energy than RO [74].

3.3.1 Porosity and Pore Size Distribution

The performance and selectivity of a membrane-based filtration system are primarily influenced by the porosity and pore size distribution of the membrane, which collectively control permeate flux, fouling tendency, and the overall filtration quality. Porosity, defined as the ratio of void volume to total membrane volume, affects hydraulic permeability, and membranes with higher porosity tend to offer higher flow rates. But porousness is a balancing act: Too much void fraction can come at the expense of selectivity and mechanical strength. Similarly, pore size distribution determines the diversity and how often certain pore diameters appear in the matrix of the membrane and directly influences the filtration precision and resistance toward fouling [75].

Based on the average pore size, membranes are typically classified into MF (0.1–10 μm), UF (2–100 nm), NF (1–10 nm), and RO (<1 nm), each of which is capable of removing different categories of contaminants, from suspended solids to micropollutants, light macromolecules, and monovalent and multivalent ions. Such a narrower and more homogeneous pore size distribution is found to benefit both the separation efficiency and fouling resistance, while the presence of the heterogeneous distributions may cause localized clogging and inconsistent rejection rates. Analytical methods such as capillary flow porometry are commonly used to measure pore size distribution and allow accurate determination of pore size and associated structural heterogeneities that are important to predict filter performance [75].

Operation affects not only the membrane structure but also its fluid dynamics. For instance, asymmetric and multibore, which, in contrast to the simply hollow membrane, may have anisotropic porosity resulting from the presence of macrovoids or layered supports, and such anisotropic porosity can change local velocity profiles and the tendency for fouling at low porosity regions. It has been shown by computational fluid dynamics that such heterogeneity can lead to preferential pathways and non-uniform contaminant retention. Moreover, the feedwater matrix plays a substantial role in porosity development over time, with natural organic matter clogging the pores 100 nm and smaller, reducing effective porosity by as much as 40%, while systems exposed to low natural organic matter groundwater retain as much as 90% of their initial porosity after years in service.

Mechanistically, membrane separation is mainly based on sieving in the pore with the membrane serving as a high-precision filter, allowing smaller solutes to pass and blocking larger pollutants [75–77]. The pore dimension, in between molecular and atomic sizes, determines the selectivity of this barrier. In NF and RO, in addition to the effects of solute size and charge, solute diffusivity through the polymer matrix and solubility in it (Knudsen diffusion) and electrostatic interactions between charged solutes and the functional groups on the membrane surface may also play a role [75,76]. NF membranes possessing surface charges normally provide higher exclusion efficiency of charged emerging contaminants through both size exclusion and electrostatic repulsion.

More specifically, for RO membranes with pore sizes in the nanometer scale, in the range between 0.1 and 1 nm, the separation process is not solely driven by size exclusion but also by the different solubility and diffusivity of the solutes through the tight polyamide film. Such an active layer is generally supported by a microporous PSU support and a mesoporous polyester backing to maintain mechanical stability and flow spreading [74]. The design and characterization of these layered structures is critical to maintaining high performance under different operation or environmental conditions. At the end, the selection of membrane materials focusing on porosity and pore size distribution is still significant for the sustainable, efficient water filtration aspect and practical application in both conventional and advanced water treatment conditions [75].

3.3.2 Filtration Principles: Size Exclusion, Adsorption, and Electrostatic Interactions

Steric, adsorption, and electrostatic interactions have been recognized as the three major mechanisms determining dynamic membrane filtration. The contribution of each mechanism is empirical and is related to the membrane's physicochemical properties, water source, and operating conditions. It is fundamental to understand such mechanisms and how they are related in order to optimize the membrane performance, specifically when applied to wastewater treatment and the selective removal of target pollutants.

Steric exclusion, or sieving, is the simplest mechanism where solutes that are bigger than the membrane pore size are physically trapped [76]. This phenomenon is more important in MF and UF processes that are commonly applied for the removal of particles, bacteria, and macromolecules. In NF and RO, the exclusion takes place at the molecular level, and thus, efficient retention of the organic micropollutants, e.g., sulfaguanidine (214 Da), with rejection rates higher than 95% is possible. But for small neutral molecules, such as N-nitrosodimethylamine, this mechanism alone is not enough, and therefore, additional retention pathways are required [77–79].

Adsorption is involved in the removal of the solutes eluted without being captured by size exclusion. It is characterized by the clicking and/or the sticking of contaminants either to the membrane surface or to the pore walls by means of the Van der Waals attraction, hydrophobic interactions, or some specific chemical bonds. Membrane functionalization, compared to, e.g., incorporation of Fe/Pd nanoparticles, can improve adsorption capacity up to ten-fold for specific compounds like trichloroethylene. In addition, the hydrophobic substances, e.g., albendazole, show a strong adsorption tendency toward NF membranes, which can accelerate adsorption for exclusion. However, adsorption can result in significant membrane fouling and is detrimental to long-range performance.

Electrostatic interactions are crucial, especially in NF or in RO, in which charged membranes interact with ionic species through Coulombic forces. For instance, negatively charged polyamide membranes reject efficiently divalent ions like sulfate (SO_4^{2-}) by Donnan exclusion with rejections higher than 99%. On the contrary, positively charged ceramic membranes functionalized with alumina nanoparticles exhibit high retention of anionic species like phosphate (up to 85%). These interactions are particularly significant at certain pH levels; the influence on PES membrane zeta potential, for instance, from −25 (pH 3) to −45 mV (pH 10) influences their performance toward ionic species [78–80].

In addition to these basic processes, membrane separation is similarly controlled by molecular diffusion, based on the membrane concentration gradient. This is particularly the case for systems like ED or capacitive deionization, where an external electric field drives the selective migration of ions through ion exchange membranes. The efficacy of this separation process depends on the pore size and MWCO. For example, compared to MF (0.1–10 μm), UF (0.001–0.1 μm) membranes have a higher performance to remove viruses, which is much larger than that for other substances. In contrast, NF and RO membranes can remove salts, small organic compounds, and some heavy metals [77–79].

The performance of these processes for eliminating disinfection by-products (DBPs) indicates the complementarity between them [81]. Another general conclusion is the dominant role of steric exclusion for the larger (<260 Da, e.g., haloacetic acids, HAAs) DBPs, where even at low pH, where we expect to minimize other interacting forces, such as electrostatic effects, retention higher than 95% can be found in the case of RO membranes [82]. Electrostatic interaction is critical for low molecular weight DBPs, as indicated by the NTR7410 NF membrane, which has a high MWCO of 20,000 Da and can remove greater than 80% of HAA5 [83,84]. Finally, adsorption has a strong contribution in the early filtration stage, which permits the efficient removal of 61.6%–93.7% of halomethanes by the RO membranes [84,85]. However, such behavior is less effective when the adsorption sites are saturated [86,87].

3.4 MEMBRANE MATERIALS AND CHARACTERISTICS

The choice of type of membrane materials significantly influences separation performance, operation stability, and antifouling and chemical stability. Membranes can be classified as either organic (polymeric) or inorganic (ceramic or metallic) (Figure 3.2). A great progress toward improving the selectivity, mechanical strength, and antifouling ability of the membranes has been made by modifying the membrane surface as well as constructing new membrane materials, expanding their application scope in water treatment and wastewater treatment.

3.4.1 Organic Membranes (Polymeric Membranes)

There are widespread applications of polymeric membranes in separation processes due to their cost-effectiveness, ease of preparation, and chemical tunability. Frequently used polymers are PVDF, PSU, PES, polyamide, and cellulose acetate. These are amenable to phase inversion, electrospun, and interfacial polymerization techniques, enabling fine-tuning of porosity and surface properties of the membranes [74,88].

By default, polymeric membranes offer processing versatility, easy and low-cost manufacture, and chemical modifiability. Nevertheless, they also have disadvantages, including weaker thermal and chemical stability than inorganic membranes

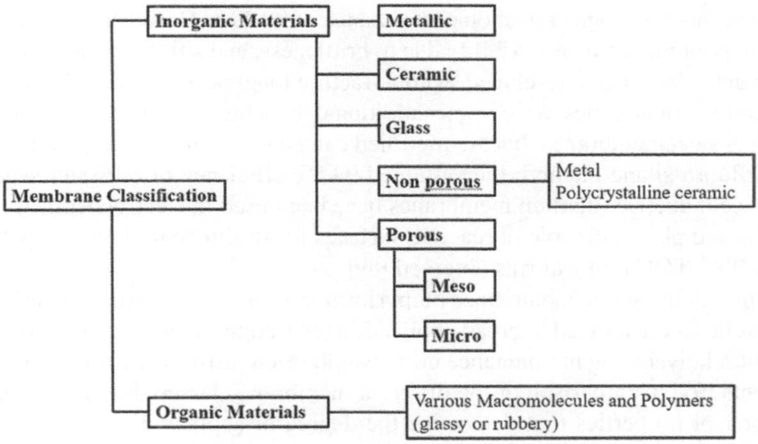

FIGURE 3.2 Classification of membranes based on materials used in fabrication.

and greater fouling propensity. Especially, the wetting property on the surface is responsible for the organic fouling and performance degradation. This drawback has been recently alleviated by improvements such as the inclusion of nanocomposites, the incorporation of copolymerization, and surface functionalization, which further increased the hydrophilicity and structural stability of the membranes.

Novel changes, such as the incorporation of silica-coated carbon nanotubes into PVDF matrixes, have shown up to 25% increase in water flux and pretty good anti-fouling properties. Apparently, inclusion of graphene oxide in polyamide membranes also promotes chlorine resistance and improves mechanical strength. PTFE membranes, even if inherently resistant to fouling, are in fact formulated with hydrophilic additives, such as polyvinylpyrrolidone, in order to work efficiently in aqueous environments. In processes like UF, selecting MWCO activates a compromise between permeability and selectivity, where larger MWCO membranes have higher water permeance at the detriment of the contaminant rejection [89].

Porous polymer films are also frequently used as an underlying support in electrochemical systems, which provide a mechanical framework for conductive deposits. Although cost-effective and low risk of secondary contamination, their conductivity is still much lower than that of metallic counterparts, and hybrid structures by integrating other conductors are needed [90–95].

3.4.2 Inorganic Membranes (Ceramic and Metallic Membranes)

Inorganic membranes made of alumina, zirconia, titania, silica, and stainless steel are well acknowledged for their thermal stability and chemical inertness. These characteristics make them highly useful for harsh operating environments where solvents are unfriendly or high temperatures exist [74,96].

Ceramic membranes demonstrate the best performance under severe conditions, operating frequently at more than 400°C, or under acidic or alkaline conditions. In the field of potable water, for instance, asymmetric α-alumina membranes with a 6 nm pore size show pathogen rejection performance of 99.99%. However, ceramic materials have poor mechanical flexibility due to brittleness, and silica-zirconia composites, for instance, have been developed, with a fracture toughness increased by 30%.

Metallic membranes have some additional benefits in particular applications. Stainless steel membranes that are modified can be made superhydrophobic by using the perfluoro-silane method; they attained 99.5% efficiency of oil/water separation. Electric conductive titanium membranes have been used in electrochemical reaction systems and played the role of reactive surfaces for in situ degradation of pollutants, and ca 90% TOC removal was obtained [89].

Although inorganic membranes outperform in terms of properties, their high cost and fragile character lead to broad application restrictions. However, their long useful life and relatively low maintenance costs would often justify their use in harsh environments [97]. Comparison of the different membrane classes shows the fingerprint character of properties that determine the degree of applicability in separation. In Table 3.2, a comprehensive assessment of the polymeric, inorganic, and mixed matrix membranes, including successful features and potential limitations, is summarized. This categorization system makes it easier for users to find the optimal membrane

TABLE 3.2
Comparative Analysis of Various Membrane Categories and Their Properties

Membrane Classification	Beneficial Attributes	Limiting Factors
Polymeric membranes	• Straightforward synthesis methodology and fabrication processes • Economically viable production • Appreciable mechanical durability • Conducive to industrial scale-up and modular configuration modifications • Principal separation mechanism: Solution-diffusion transport	• Inadequate resistance to chemical agents and thermal conditions • Susceptibility to plasticization phenomena • Challenges in precise pore dimension regulation • Inherent performance constraints regarding permeability-selectivity correlation
Inorganic membranes	• Exceptional chemical, mechanical, and thermal resistance properties • Capacity for pore architecture modification • Alleviation of the permeability-selectivity compromise • Operational viability under extreme environmental conditions • Multiple separation mechanisms: Molecular sieving (<6 Å), surface diffusion (<10–20 Å), capillary condensation • (<30 Å), and Knudsen diffusion (<0.1 μm)	• Structural fragility concerns • Significant capital investment requirements • Technical challenges in mass production methodologies

(Continued)

TABLE 3.2 (*Continued*)
Comparative Analysis of Various Membrane Categories and Their Properties

Membrane Classification	Beneficial Attributes	Limiting Factors
Mixed matrix membranes	• Enhanced mechanical integrity and thermal stability characteristics • Mitigated plasticization effects • Reduced operational energy consumption parameters • Structural integrity maintenance under high-pressure conditions • Transcendence of conventional permeability-selectivity limitations • Superior separation efficiency compared to conventional polymeric substrates • Operational principles incorporating synergistic polymeric-inorganic membrane mechanisms	• Structural brittleness at elevated filler concentrations within the polymeric matrix • Chemical and thermal stability contingent upon the polymeric matrix composition

combination for their operating conditions, environmental demands, and separation purposes. As clearly deduced from the comparative analysis, although polymeric membranes provide economic and fabrication benefits, their inorganic analogues are more resistant, and on the other hand, there are mixed matrix membranes that are effectively the best candidate by taking the advantages of both elements.

3.4.3 Membrane Surface Modifications and Functionalization

Surface engineering is essential to improving membrane performance, including fouling resistance, selectivity, and chemical tolerance. The modification methods involve physical and chemical approaches to engineered surface properties without affecting the mechanical strength [98,99].

Hydrophilic coatings such as polyethylene glycol and polydopamine are deposited to suppress foulant attachment via hydration layer generation. Chemical grafting involves the covalent attachment of functional groups (plasma treated, UV polymerization, click chemistry) to improve selectivity and antifouling.

Nanomaterial integration involves the incorporation of materials such as graphene oxide, carbon nanotubes, AgNPs, and metal-organic frameworks in the membrane matrix to enhance its hydrophilicity, mechanical strength, and filtration efficiency. TiO_2-covered membranes, for example, are known to possess a photocatalytic self-cleaning property under UV irradiation. Zwitterionic and charged surfaces are also known for their good antifouling ability through the mechanism of hydrated and electrostatic repulsion.

Advanced approaches such as hydrophilic/hydrophobic patterning have proved useful to improve oil–water separation performance (decreasing transmembrane pressure and improving separation efficiency). Biomimetic functionalization, such as aquaporin incorporation, achieves water flux rates that are far higher than those of the RO membrane, but protein can be degraded, although protein degradation remains a limiting factor for commercialization.

Further methods, including polymer brush growth, polyelectrolyte layer-by-layer deposition, and copolymerization, offer additional vectors for the introduction of functional groups, adjustment of surface roughness, and longer operating life [96,100–103]. Importantly, the surface chemistry can be tailored to meet specific requirements in some applications, for example, by promoting ammonium rejection by inducing a positive charge or by reducing fouling in MBRs and membrane distillation applications. A comparison of studies suggests that omniphobic surfaces may manifest a long-term fouling resistance, whereas hydrophilic coatings may be better suited for short-term antifouling (long-term antifouling) performance [104,105].

3.5 APPLICATIONS OF MEMBRANE TECHNOLOGIES IN WASTEWATER TREATMENT

Membrane technologies are widely employed in the advanced treatment of wastewater owing to their high treatment efficiency and capability to remove different contaminants. These processes employ semi-permeable membranes that retain pathogens, suspended particles, and different inorganic and organic pollutants, while allowing the passage of water molecules [106]. Membrane Technologies are widely used/employed, and a variety of separation capacities are available. However, for resource recovery as well as cost-efficient water recovery from complex waste streams, novel techniques like forward osmosis (FO) [107], distillation using membranes [108], and capacitive deionization [109] are currently under investigation [110]. The applicability of the membranes for the treatment of persistent pollutants, such as dyes, drugs, and endocrine disrupters, could be strengthened due to the compatibility of membranes with biological systems, improved oxidation, and adsorption processes [111]. Membrane technologies are at the cutting edge of wastewater treatment innovation, offering scalable and reliable alternatives for residential, industrial, and ecological uses in response to the growing global need for ecologically responsible water usage.

3.5.1 Municipal Wastewater Treatment

The aim of municipal wastewater treatment, a crucial component of urban infrastructure, is to eliminate contaminants from sewage generated by homes, companies, and commercial establishments [112]. The most prevalent kind of wastewater that belongs to the low-strength substance group is municipal wastewater [113]. The main goal is to create an effluent that can be recycled or released into the environment in a way that protects water quality and human health [114]. Membrane technologies are becoming more prevalent in municipal systems to increase efficiency, reduce environmental impact, and promote water reuse programs [115].

3.5.1.1 MBRs

MBR technology has been growing rapidly in the treatment of wastewater in recent years, greatly enhancing the effectiveness of biological treatment [116]. The MBR combines membrane separation (filtration) with biological treatment for activated sludge (Figure 3.3) [117]. The waste chemicals' biodegradation is carried out by the biological unit; however, the treated water and mixed liquor are separated via the membrane separation process. The membrane unit of the membrane biological sludge process operates instead of the conventional gravity sedimentation unit (clarifier). Its pore size is 0.01–0.1 mm, preventing particles and bacteria from being filtered along with the permeate [118]. By combining MF or UF membranes precisely throughout the biological treatment process, MBRs significantly decrease the demand for secondary clarifiers, resulting in a relatively compact system footprint and improved treatment efficiency [117]. One of the fundamental benefits of MBR systems is their capacity to sustain high biomass concentrations, which leads to better elimination of nutrients and organic material like nitrogen and phosphorus [119]. This feature assures consistent effluent quality, even under fluctuating load conditions, making MBRs especially appropriate for metropolitan locations with limited space and severe discharge criteria [120]. Furthermore, the effluent produced is almost entirely free of suspended particles and pathogens, which makes it suitable for non-potable recycling applications, including agriculture and processing in industries [121].

3.5.1.2 Anaerobic MBRs

One promising advancement in the treatment of municipal wastewater is the use of anaerobic MBRs, which bring together the advantages of membrane filtration and anaerobic digestion to provide efficient and environmentally friendly wastewater management [122]. Anaerobic MBRs make it easier for organic materials in wastewater to be transformed into methane-rich biogas that may be utilized to produce energy (Figure 3.4). Furthermore, the low organic content of municipal wastewater prevents sufficient biogas generation [123]. In addition to lowering wastewater treatment's energy footprint, this procedure helps produce renewable energy [124]. The membrane

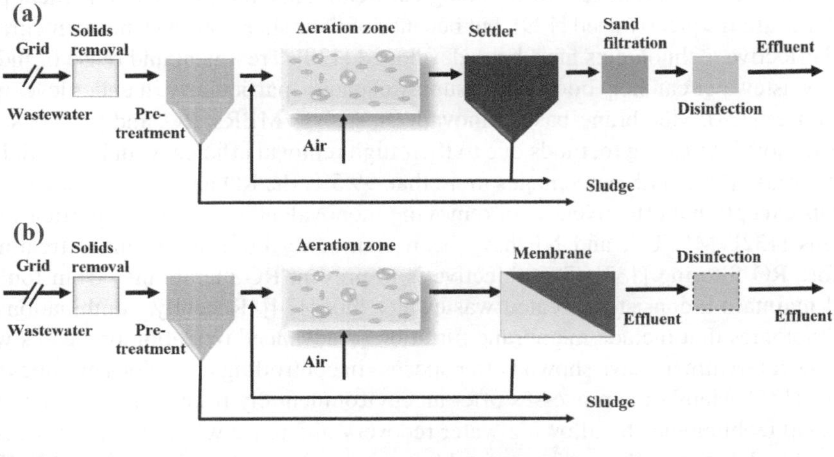

FIGURE 3.3 (a) The process of activated sludge. (b) The MBR process.

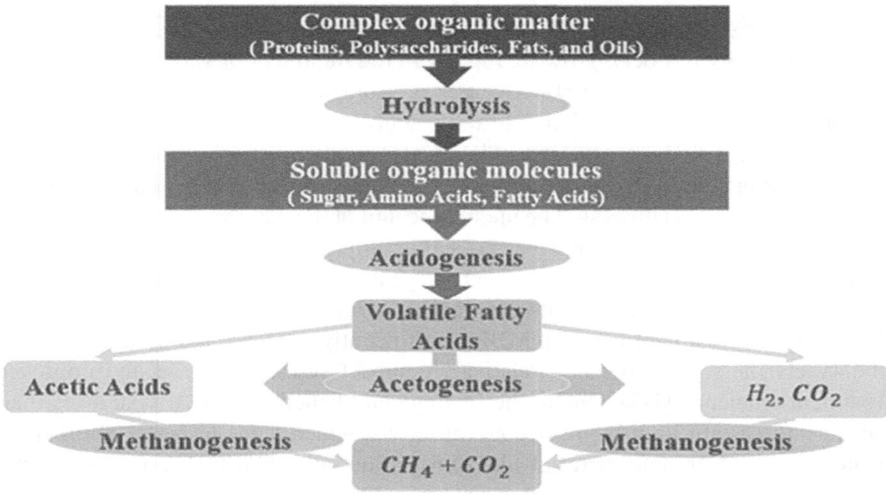

FIGURE 3.4 Anaerobic treatment phases.

filtration element removes floating debris and pathogens, producing discharge that passes strict discharge criteria and is acceptable for recycling applications [125]. In contrast with traditional aerobic processes, anaerobic MBRs produce less extra sludge, cutting sludge handling, and disposal expenses [126]. Also, they contain nutrient recovery, such as phosphorus and nitrogen, which enables their recovery and potential recycling as fertilizers, matching with ecological principles [127].

3.5.2 Industrial Wastewater Treatment

Discharging industrial effluent (such as metallurgy, petrochemicals, pharmaceuticals, food processing, and textiles) into surface and groundwater resources harms the environment significantly [128], requiring more sophisticated treatment technology than traditional approaches. Both biological treatments and physical–chemical procedures are frequently used [129], but because of their shortcomings, more integrated and effective technologies have been developed [130]. Treatment and reuse of industrial wastewater can help boost water supply capacity, particularly in underdeveloped countries [131]. Membrane-based innovations, such as MBRs, RO, and NF, are some of the most interesting methods due to their high removal efficiency and potential for water reuse [66]. With percentages more than 99.5%, the RO has been acknowledged for its exceptional effectiveness in removing monovalent ions and tiny particles like germs [132]. MF, UF, and NF have been commonly used as an initial treatment before RO filtering [133]. The objective is to prevent RO membranes from fouling and maintain a consistent treated wastewater flux [134]. Recently, combinations of technologies that include membrane filtration or advanced oxidation processes with biological treatment have shown better success in controlling hazardous organics and COD [135]. Membrane processes offer an environmentally friendly substitute to traditional technologies by allowing water recovery and reuse within the process, considerably decreasing the environmental footprint and operational expenses [136,137].

3.5.3 Desalination and Water Reuse

Desalination is a procedure that removes salts from the water and produces purified water [138]. Desalination reduces water stress in coastal and inland locations by producing purified water from the sea, salty groundwater, drainage water, and purified sewage [19]. Desalination technologies provide the possibility to address the freshwater shortage, especially in coastal locations [139]. A pair of techniques of FO desalination exists: direct and indirect (Figure 3.5).

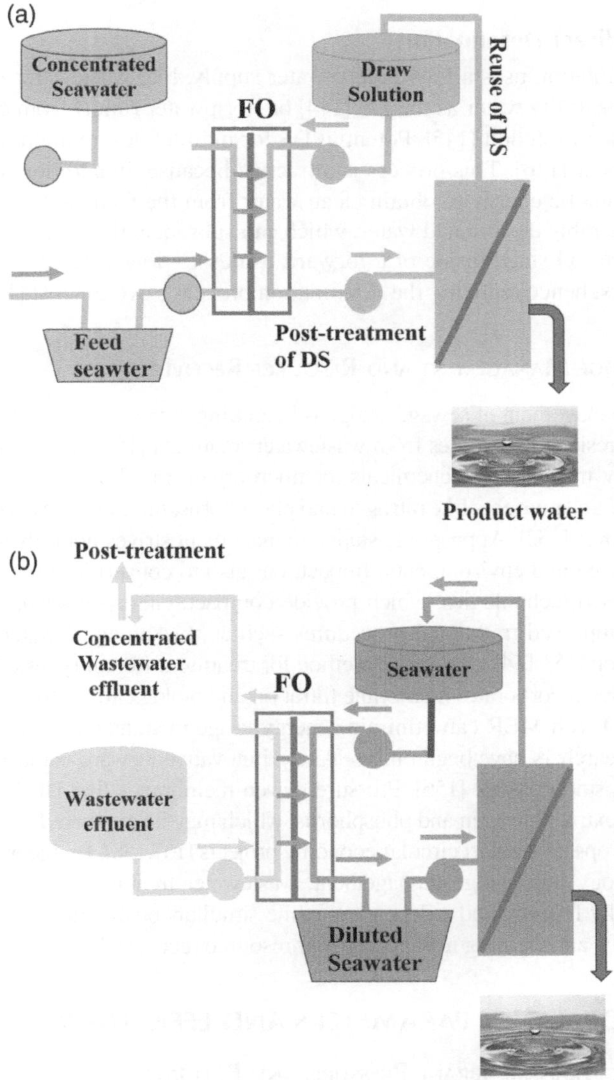

FIGURE 3.5 (a) Direct and (b) indirect FO processes for water desalination. (Reprinted with permission from [150].)

3.5.3.1 Direct Desalination

Direct desalination shares similarities with other traditional membrane-based desalination processes that directly extract freshwater from salty water, like RO and NF [140]. The feed solution for direct FO desalination is saltwater, whereas the draw solution (DS) is an osmotic reagent that can be a non-volatile salt [141]. To recover fresh water in this procedure, the DS must be separated from the remaining diluted DS during the drawing liquid recovery step [142]. Similar to RO desalination, direct FO desalination may necessitate a pretreatment of the feed seawater [143].

3.5.3.2 Indirect Desalination

Indirect desalination uses a low-quality water supply, like wastewater effluent, along with elevated salinity water as the DS [144] or stormwater runoff from cities, which is utilized as a feed solution [145]. Potential DS for indirect desalination includes brackish and saltwater [146]. This process is attractive because, in addition to the free DS, it uses free osmotic energy to obtain clean water from the feed [147], creating diluted DS, or considerably desalinated water, which can subsequently be fully desalinated by a low-pressure RO stage inside of a forward osmosis – low pressure reverse osmosis hybrid process, hence reducing the desalination process's overall cost [148,149].

3.5.4 Sludge Management and Resource Recovery

Sustainable management of sewage sludge is becoming increasingly important because it is a mud-like residue that comes from wastewater treatment [151]. It may contain impurities like heavy metals, other chemicals, or microorganisms. It also includes significant organic matter and nutrients like nitrogen and phosphorus, making it an excellent fertilizer and soil improver [152]. Appropriate sludge management strives not only to decrease the amount produced and environmental impact, but also to convert it into a resource [114]. Membrane-based technologies, which provide compact, energy-efficient solutions, have increasingly improved traditional procedures such as thickening, dewatering, digesting, and incineration [153,154]. A popular method for treating different types of wastewater is the MBR, which incorporates membrane filtration and biologically activated sludge [155]. Integrating FO with MBR can minimize energy usage in standard MBRs. In the recent five years, researchers have begun investigating innovative forward osmosis – membrane bioreactor or osmotic MBR [156]. Pressure-driven membranes like UF and NF are also being used to extract nitrogen and phosphorus, which may be recovered and used again as fertilizer for crops, to support circular economy projects [157,158]. By incorporating membrane technology into sludge management, wastewater treatment plants become more environmentally friendly and independent while simultaneously improving environmental performance and creating new avenues for resource recovery [159].

3.6 PERFORMANCE PARAMETERS AND EFFICIENCY

3.6.1 Flux, Transmembrane Pressure, and Fouling

The efficacy and efficiency of membrane-based wastewater treatment are heavily influenced by three key parameters: flux, transmembrane pressure, and membrane fouling [160]. Flux signifies the pace at which the purified water passes the membrane, which has a direct impact on system productivity [160]. Nevertheless, persistent high fluxes frequently result in increased transmembrane pressure, the pressure difference required to force the liquid through the membrane, which is a sensitive measure of how well the system is functioning [161]. An increasing transmembrane pressure frequently indicates the beginning of fouling, the buildup of organic, inorganic, or biological contaminants on the membrane surface, eventually lowering flux and raising energy demand [162]. Reduced flow, decreased efficiency, and highlighting their function in performance optimization and membrane lifespan extension are some of the basic ways fouling affects membrane performance [163].

3.6.2 Membrane Lifetime and Durability

In membrane-based wastewater treatment methods, membrane lifetime and durability are critical performance metrics [98]. The effectiveness of membranes can be considerably decreased over time due to a variety of physical, chemical, and biological stressors, including fouling, scaling, and chemical degradation [134,164]. The authors investigate RO membranes, focusing on a qualitative and quantitative diagnostic study of RO. They found an estimate of a five-year average lifespan for these RO components, which is two years greater than the membrane manufacturers' typical three-year warranty. Although the short age of the majority of installations limits the data available, research indicates that the 15–20 year membrane life typically guaranteed by the majority of ceramic membrane product vendors is reasonable [165]. Their extended lifespan is due to their robust structure, which is often composed of materials like aluminum oxide, which provides excellent resistance to mechanical wear, heat stress, and chemical deterioration [166]. The most recent polymeric membrane modules may last up to eight years [167]. Depending on the kind, feeding materials, and process circumstances, membranes gradually lose their effectiveness. Fouling is the biggest problem for membrane systems. Therefore, the most popular technique for fixing this problem is cleaning [168]. For the chemical agents to react with the accumulated components, the cleaning process must be long enough. Flux recovery is improved by extending the cleaning duration at the start of the cleaning procedure. However, due to the chemicals' limited ability to dissolve the created layer, the efficiency eventually declines [169]. Chemical cleaning also impacts membrane longevity and may partially degrade the mechanical strength of flat-sheet polymeric membranes [170]. Ceramic membranes are a sustainable choice for long-term water treatment because of their lengthy service life, which reduces operating costs and environmental effects by minimizing waste and replacements [171].

3.6.3 Cost-Effectiveness and Energy Consumption

Membrane technology offers financial sustainability, reduced chemical consumption, eco-friendliness, and simple use [66]. Therefore, cost-effectiveness and energy consumption are important characteristics that define the economic feasibility of membrane-based systems in wastewater treatment [172]. MF eliminates micrometer-sized particles with 0.1–10 μm pore diameters [173]. The typical operating pressure range for MF operations is 0.1–8 bars, which is marginally lower than that of UF [174]. The typical energy usage of an MF system is between 0.1 and 0.3 kWh m^{-3}, while there are considerable variations [175]. Because of its cheap operating costs, MF is an affordable main therapy option [176]. UF membranes have pores that are smaller than those of MF (0.01–0.1 μm) [177]. UF operates at pressures below 1 bar and consumes more energy [67]. The energy consumption of UF is higher, at about 0.2 kWh m^{-3} [178]. Because of its exceptional water quality, UF is appropriate for secondary or tertiary treatment phases [179]. The size of the pores in NF membranes varies between 0.001 and 0.01 μm [180]. NF membranes work at pressures between 10 and 30 bar and can reject organic molecules with different molecular weights [181]. The energy consumption for NF ranges between 1.20 and 1.30 kWh m^{-3} [182]. RO employs semi-permeable membranes to remove nearly all dissolved contaminants, including monovalent salts, nitrates, and organic compounds [183], and the pressures vary between 15 and 50 bar or 50–100 bar (for seawater desalination) [184]. RO systems consume around 2–5 kWh m^{-3} [185]. Each kind of membrane technology has unique advantages and disadvantages concerning cost-effectiveness and energy usage. Economic concerns, water quality regulations, and specific treatment aims influence the choice of membrane technology [137].

3.7 CHALLENGES AND LIMITATIONS OF MEMBRANE TECHNOLOGIES

3.7.1 Membrane Fouling and Scaling

Membrane fouling and scaling represent critical challenges in wastewater treatment applications. Fouling refers to the accumulation of unwanted substances on the membrane surface or within its pores, which hinders performance. There are several types of fouling: particulate fouling, caused by suspended solids; organic fouling, caused by dissolved organics such as oils and grease; biofouling, which arises from microbial growth; and scaling, resulting from the precipitation of sparingly soluble salts. The mechanisms of fouling include adsorption of molecules onto the membrane surface, pore blocking, and cake layer formation, each contributing to reduced permeability and increased energy demands [134]. Fouling is influenced by multiple factors such as feedwater quality, operating pressure, membrane material, and system hydrodynamics. For example, membranes with hydrophilic surfaces are less prone to fouling due to reduced organic and microbial adhesion. Mitigation strategies include pretreatment techniques like coagulation, filtration, and pH adjustment, as well as regular chemical cleaning using acids, alkalis, or disinfectants. Moreover, the development of antifouling coatings and surface-modified membranes has emerged as a promising solution to prolong membrane life and performance. In RO systems, membrane surface fouling generally evolves through three primary pathways: the accumulation of suspended matter (cake layer formation), mineral crystallization (scaling), and microbial colonization (biofouling) [186]. Cake fouling develops when fine particles, whether of organic, inorganic, or biological origin, are retained at the membrane interface, forming a dense layer that impedes water passage [187]. Scaling, on the other hand, results from the precipitation and subsequent crystallization of sparingly soluble salts directly onto the membrane surface, a process often intensified by local supersaturation conditions. Biofouling emerges when microorganisms initially adhering to the membrane begin to multiply and establish complex microbial communities that form structured biofilms [187]. These fouling phenomena can manifest both on the membrane's surface and within its pores, as illustrated in Figure 3.6.

3.7.2 Maintenance and Operational Issues

Membrane systems, although effective, often demand high levels of maintenance. Frequent cleaning cycles are required to manage fouling and scaling, which in turn increases chemical consumption and operational downtime. The use of harsh cleaning agents can lead to degradation of membrane materials, necessitating regular module replacement, which is both costly and labor-intensive. Mechanical failures such as seal leaks, fiber breakage, and pump malfunctions can also occur, affecting system reliability. Another operational concern is energy demand, particularly in pressure-driven systems like RO. High transmembrane pressures can lead to elevated energy costs, making energy optimization essential. Pressure drop across the membrane modules must be carefully managed to avoid efficiency losses. Operator expertise is also critical; systems require skilled personnel for routine monitoring, maintenance, and troubleshooting. Additionally, automation and control systems add complexity but can improve reliability and allow for early detection of anomalies [189].

3.7.3 Economic and Environmental Considerations

Economic evaluation of membrane systems must account for both capital expenditures and operational expenditures. While initial investment costs for membrane modules and infrastructure are high, operational costs related to cleaning chemicals, energy, and replacement modules can surpass capital expenditures over the

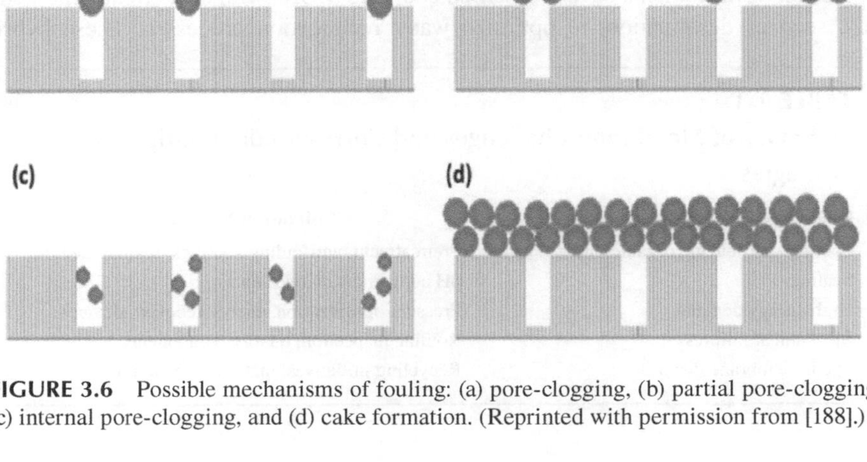

FIGURE 3.6 Possible mechanisms of fouling: (a) pore-clogging, (b) partial pore-clogging, (c) internal pore-clogging, and (d) cake formation. (Reprinted with permission from [188].)

system's lifespan. Proper lifecycle cost analysis is thus essential to understand the true financial impact. Environmental impacts are also a concern. The disposal of spent membranes, which are often non-biodegradable, poses a solid waste challenge. Moreover, chemicals used in membrane cleaning can generate hazardous waste if not properly neutralized. In comparison to conventional treatment systems, membranes may reduce sludge generation, but the use of energy-intensive processes may offset this benefit. Therefore, energy efficiency and chemical minimization strategies are necessary for sustainable operation (Table 3.3).

3.8 RECENT INNOVATIONS AND ADVANCEMENTS

3.8.1 Novel Membrane Materials and Coatings

Recent years have seen significant progress in the development of advanced membrane materials to combat fouling and enhance selectivity. Polymeric membranes have been modified using nanomaterials to increase their hydrophilicity and antimicrobial properties [190]. Ceramic membranes, while more costly, offer superior thermal and chemical stability, making them suitable for harsh wastewater environments [191]. Notably, graphene oxide and aquaporin-based membranes mimic natural water channels and exhibit exceptional permeability and selectivity. Surface functionalization through plasma treatment or grafting hydrophilic groups has also been used to create antifouling surfaces. These modifications reduce the adhesion of foulants and enhance cleaning efficiency. Bioinspired membranes, replicating properties found in nature, such as lotus-leaf effects, are also being explored.

3.8.2 Hybrid Systems and MBRs

Hybrid systems, particularly MBRs, have emerged as advanced solutions in wastewater treatment by integrating biological degradation with membrane filtration. MBRs combine the activated sludge process with membrane technologies like MF or UF, effectively separating solids from treated water [192]. This integration results in high-quality effluent, reduced footprint, and enhanced removal of contaminants [193].

Recent advancements have focused on coupling MBRs with other treatment methods, such as desalination, to optimize water reclamation processes. These hybrid

TABLE 3.3
Summary of Membrane Challenges and Corresponding Mitigation Strategies

Challenge	Mitigation Strategy
Fouling (particulate, organic, biofouling)	Pretreatment, antifouling coatings, regular cleaning
Scaling	pH adjustment, antiscalants
High energy demand	Pressure optimization, energy recovery devices
Mechanical failures	Routine inspection, quality components
Spent membrane disposal	Recycling initiatives, material innovation

configurations aim to address challenges like membrane fouling and energy consumption while improving overall treatment efficiency. MBRs are particularly beneficial in scenarios requiring stringent effluent standards, limited space, or specific contaminant removal, making them suitable for both municipal and industrial applications [194].

3.8.3 Smart Membranes and Sensors in Wastewater Treatment

Smart membranes represent a new frontier in wastewater treatment, characterized by their ability to respond to environmental stimuli such as pH, temperature, or specific chemicals. These membranes can adjust their permeability or selectivity in real time, enhancing treatment efficiency and adaptability [195]. Incorporating sensors into membrane systems allows for real-time monitoring of water quality parameters, enabling proactive adjustments to treatment processes. This integration facilitates early detection of membrane fouling, system malfunctions, or contaminant breakthroughs, ensuring consistent effluent quality. The development of smart membranes and sensor-integrated systems is driven by the need for more resilient, efficient, and adaptive wastewater treatment solutions, especially in the face of varying influent characteristics and regulatory requirements [195].

3.9 CONCLUSION

In conclusion, membrane technologies represent an essential and constantly evolving solution to the growing challenges associated with water management and wastewater treatment. Thanks to their capacity for efficient separation, their adaptability to various contexts (municipal, industrial, desalination), and recent innovations in materials and intelligent systems, these technologies are emerging as key tools for guaranteeing water quality and promoting resource reuse. However, despite their many advantages, obstacles remain, including clogging, energy costs, and membrane durability. To maximize their impact, efforts must continue in research, process optimization, and integration of hybrid solutions to make these systems more efficient, accessible, and ecologically viable on a large scale.

REFERENCES

[1] N. El Messaoudi, M. El Khomri, A. El Mouden, A. Bouich, A. Jada, A. Lacherai, H.M.N. Iqbal, S.I. Mulla, V. Kumar, J.H.P. Américo-Pinheiro, Regeneration and reusability of non-conventional low-cost adsorbents to remove dyes from wastewaters in multiple consecutive adsorption–desorption cycles: A review, *Biomass Convers. Biorefinery.* 14 (2024) 11739–11756. https://doi.org/10.1007/s13399-022-03604-9.

[2] M. Kasbaji, M. Mennani, M. Oubenali, A. Ait Benhamou, A. Boussetta, E.H. Ablouh, M. Mbarki, N. Grimi, M. El Achaby, A. Moubarik, Bio-based functionalized adsorptive polymers for sustainable water decontamination: A systematic review of challenges and real-world implementation, *Environ. Pollut.* 335 (2023) 122349. https://doi.org/10.1016/j.envpol.2023.122349.

[3] N. Rashmi, S.A. Salmataj, P.S. Kumar, P. Bhat, Exploring chemically and physically modified plant-based fiber biomass for biosorption in wastewater treatment: A concise review, *J. Water Process Eng.* 67 (2024). https://doi.org/10.1016/j.jwpe.2024.106245.

[4] K.K. Kesari, R. Soni, Q.M.S. Jamal, P. Tripathi, J.A. Lal, N.K. Jha, M.H. Siddiqui, P. Kumar, V. Tripathi, J. Ruokolainen, Wastewater treatment and reuse: A review of its applications and health implications, *Water. Air. Soil Pollut.* 232 (2021). https://doi.org/10.1007/s11270-021-05154-8.

[5] P.R. Yaashikaa, P.S. Kumar, A. Saravanan, D.V.N. Vo, Advances in biosorbents for removal of environmental pollutants: A review on pretreatment, removal mechanism and future outlook, *J. Hazard. Mater.* 420 (2021). https://doi.org/10.1016/j.jhazmat.2021.126596.

[6] C.A. Martínez-Huitle, M.A. Rodrigo, I. Sirés, O. Scialdone, A critical review on latest innovations and future challenges of electrochemical technology for the abatement of organics in water, *Appl. Catal. B Environ.* 328 (2023). https://doi.org/10.1016/j.apcatb.2023.122430.

[7] K.M. Al-Qahtani, Water purification using different waste fruit cortexes for the removal of heavy metals, *J. Taibah Univ. Sci.* 10 (2016) 700–708. https://doi.org/10.1016/j.jtusci.2015.09.001.

[8] K.S. Bharathi, S.T. Ramesh, Removal of dyes using agricultural waste as low-cost adsorbents: A review, *Appl. Water Sci.* 3 (2013) 773–790.

[9] S. Walia, C. Murleedharn, J. Band, M. Kanwar, A. Kumar, Quantitation of antibiotic resistance genes pollution in hospital waste water effluent and Urban Clinton River Water, Michigan, USA, *Curr. Med. Res. Pract.* 6 (2016) 149–151. https://doi.org/10.1016/j.cmrp.2016.07.005.

[10] W. Tan, L. Zhang, F. Fu, S. Bowman, P. Wang, Y. Li, Y. Zhang, The heavy metal adsorption and plant cultivation performance of grafting modified plant medium made with recycled fibers, *J. Clean. Prod.* 329 (2021) 129788. https://doi.org/10.1016/j.jclepro.2021.129788.

[11] M. Zaynab, R. Al-Yahyai, A. Ameen, Y. Sharif, L. Ali, M. Fatima, K.A. Khan, S. Li, Health and environmental effects of heavy metals, *J. King Saud Univ. Sci.* 34 (2022) 101653. https://doi.org/10.1016/j.jksus.2021.101653.

[12] A. Tolun, Z. Altintas, Optimizing antioxidant and thermal stability in olive mill wastewater through hydroxypropyl-β-cyclodextrin complexation: A comprehensive study, *Carbohydr. Polym. Technol. Appl.* 9 (2025) 100631. https://doi.org/10.1016/j.carpta.2024.100631.

[13] X. Lin, K. Chan, K. Kingkhambang, H. Hayashi, A. Zinchenko, Hydrothermal preparation of pharmaceuticals adsorbents from chitin and chitosan: Optimization and mechanism, *Bioresour. Technol.* 414 (2024). https://doi.org/10.1016/j.biortech.2024.131583.

[14] S. Arita, D. Kristianti, L.N. Komariah, Effectiveness of biomass-based fly ash in pulp and paper liquid waste treatment, *South African J. Chem. Eng.* 41 (2022) 79–84. https://doi.org/10.1016/j.sajce.2022.05.004.

[15] A. Omer, N.A. Elagib, M. Zhuguo, F. Saleem, A. Mohammed, Water scarcity in the Yellow River Basin under future climate change and human activities, *Sci. Total Environ.* 749 (2020) 141446. https://doi.org/10.1016/j.scitotenv.2020.141446.

[16] S. Raychaudhuri, M. Raychaudhuri, R.K. Mohanty, S.K. Rautaray, Effect of treated wastewater on heavy metal accumulation, fish growth and associated implications on consumer health, *J. Taiwan Inst. Chem. Eng.* 166 (2024) 105652. https://doi.org/10.1016/j.jtice.2024.105652.

[17] S. Raj, H. Singh, J. Bhattacharya, Treatment of textile industry wastewater based on coagulation-flocculation aided sedimentation followed by adsorption: Process studies in an industrial ecology concept, *Sci. Total Environ.* 857 (2023) 159464. https://doi.org/10.1016/j.scitotenv.2022.159464.

[18] R. Changotra, H. Rajput, A. Dhir, Treatment of real pharmaceutical wastewater using combined approach of Fenton applications and aerobic biological treatment, *J. Photochem. Photobiol. A Chem.* 376 (2019) 175–184. https://doi.org/10.1016/j.jphotochem.2019.02.029.

[19] C.A. Quist-Jensen, F. Macedonio, E. Drioli, Membrane technology for water production in agriculture: Desalination and wastewater reuse, *Desalination.* 364 (2015) 17–32. https://doi.org/10.1016/j.desal.2015.03.001.

[20] M.W. Li, W.H. Zhang, H. Meng, W. Gu, X. Zhang, H. Guo, Acid-resistant triazine-amine-based polyamide-sulfonamide nanofiltration membranes for efficient acidic wastewater treatment, *Sep. Purif. Technol.* 366 (2025) 132762. https://doi.org/10.1016/j.seppur.2025.132762.

[21] R. Liu, J. Du, Q. Chen, R. Zhao, F. Yu, K. Simoens, A. Volodin, P. Luis, W. Ye, R. Dewil, M. Xie, J. Lin, B. Van der Bruggen, Facile construction of a tight polypeptide-based composite ultrafiltration membrane with antimicrobial activity for sustainable treatment of highly saline organic wastewater, *J. Memb. Sci.* 720 (2025) 123796. https://doi.org/10.1016/j.memsci.2025.123796.

[22] R. Yuan, Y. Qin, H. Wang, X. He, Changes in water quality along the process of coal chemical reverse osmosis wastewater advanced treatment: Spectral analysis and molecular level exploration, *Sep. Purif. Technol.* 360 (2025) 131267. https://doi.org/10.1016/j.seppur.2024.131267.

[23] P.M. Ruya, M.P. Silva, G. Reyniers, G.A. Ruya, S.S. Lim, I.G. Wenten, X. Yang, Intensifying transition metal ion removal and recovery from acidic wastewater via electrodialysis (ED)-based process, *Water Res.* 280 (2025). https://doi.org/10.1016/j.watres.2025.123504.

[24] R.V. Patel, A. Yadav, V.K. Shahi, Advances in membrane distillation for wastewater treatment: Innovations, challenges, and sustainable opportunities, *Sci. Total Environ.* 969 (2025) 178749. https://doi.org/10.1016/j.scitotenv.2025.178749.

[25] S. Foorginezhad, M.M. Zerafat, A.F. Ismail, P.S. Goh, Emerging membrane technologies for sustainable water treatment: A review on recent advances, *Environ. Sci. Adv.* (2025) 530–570. https://doi.org/10.1039/d4va00378k.

[26] N. Abdullah, N. Yusof, A.F. Ismail, W.J. Lau, Insights into metal-organic frameworks-integrated membranes for desalination process: A review, *Desalination.* 500 (2021) 114867. https://doi.org/10.1016/j.desal.2020.114867.

[27] X. Sun, H. Shiraz, R. Wong, J. Zhang, J. Liu, J. Lu, N. Meng, Enhancing the performance of PVDF/GO ultrafiltration membrane via improving the dispersion of GO with homogeniser, *Membranes.* 12 (2022) 1–12. https://doi.org/10.3390/membranes12121268.

[28] P.S. Goh, A.F. Ismail, A review on inorganic membranes for desalination and wastewater treatment, *Desalination.* 434 (2018) 60–80. https://doi.org/10.1016/j.desal.2017.07.023.

[29] Y. Davoodbeygi, M. Askari, E. Salehi, S. Kheirieh, A review on hybrid membrane-adsorption systems for intensified water and wastewater treatment: Process configurations, separation targets, and materials applied, *J. Environ. Manage.* 335 (2023) 117577. https://doi.org/10.1016/j.jenvman.2023.117577.

[30] P.R.D. Rogers, Green chemistry and sustainable technology, *Green Chem. Sustain. Technol.* (2020). https://doi.org/10.1201/9780367808310.

[31] S.F. Anis, R. Hashaikeh, N. Hilal, Microfiltration membrane processes: A review of research trends over the past decade, *J. Water Process Eng.* 32 (2019). https://doi.org/10.1016/j.jwpe.2019.100941.

[32] Y. Zhang, Q. Fu, Algal fouling of microfiltration and ultrafiltration membranes and control strategies: A review, *Sep. Purif. Technol.* 203 (2018) 193–208. https://doi.org/10.1016/j.seppur.2018.04.040.

[33] I. Fooladi, P. Ghanbarizadeh, A. Azari, M. Abbasi, R. Karami, M. Akrami, Titania and zirconia ceramic nanofiltration membrane fabrication by coating method on mullite and mullite-alumina microfiltration supports for industrial wastewater treatment, *Arab. J. Chem.* 17 (2024). https://doi.org/10.1016/j.arabjc.2024.105973.

[34] J. Landaburu-Aguirre, Polymeric membranes for micronanoplastic sampling and removal from water effluents, *Polym. Sci. Peer Rev. J.* 4 (2022) 10–12. https://doi.org/10.31031/psprj.2022.04.000591.

[35] K. El ataoui, Y. El maguana, S. Chlaih, M. Harech, A. Bouazizi, M. El masloumi, H. Aouad, N. Elhadiri, Enhancing microfiltration membrane efficiency for treating industrial wastewater derived from sugar industry waste through Doehlert design optimization, *Desalin. Water Treat.* 320 (2024). https://doi.org/10.1016/j.dwt.2024.100860.

[36] H. El Moustansiri, S. El Abbadi, M. Douma, A. Bouazizi, D. El Machtani Idrissi, M. Bechelany, M. Ouammou, N. Tijani, Development of low-cost wollastonite based-membrane from clay for efficient microfiltration of textile and tannery wastewaters, *Sep. Purif. Technol.* 359 (2025). https://doi.org/10.1016/j.seppur.2024.130770.

[37] M.D. Garba, M. Usman, M.A.J. Mazumder, A. Al-Ahmed, Inamuddin, complexing agents for metal removal using ultrafiltration membranes: A review, *Environ. Chem. Lett.* 17 (2019) 1195–1208. https://doi.org/10.1007/s10311-019-00861-5.

[38] I. Kammakakam, Z. Lai, Next-generation ultrafiltration membranes: A review of material design, properties, recent progress, and challenges, *Chemosphere.* 316 (2023). https://doi.org/10.1016/j.chemosphere.2022.137669.

[39] T. Ahmad, C. Guria, S. Shekhar, Effects of inorganic salts in the casting solution on morphology of poly(vinyl chloride)/bentonite ultrafiltration membranes, *Mater. Chem. Phys.* 280 (2022) 125805. https://doi.org/10.1016/j.matchemphys.2022.125805.

[40] H. Zhong, Y. Wang, Q. Jiang, H. Tian, C. Jiang, P. Han, Y. Qin, L. Li, S. Gui, Preparation of TA-O-MoS_2/PES mixed matrix ultrafiltration membrane for rare earth wastewater treatment, *J. Environ. Chem. Eng.* 12 (2024). https://doi.org/10.1016/j.jece.2024.114523.

[41] H. Waleed, B. Hudaib, M. Al-Harahsheh, M. Allawzi, Synthesis and characterization of a modified silica/cellulose acetate nanocomposite ultrafiltration membrane for phenol removal from wastewater, *Sep. Purif. Technol.* 353 (2025). https://doi.org/10.1016/j.seppur.2024.128391.

[42] Z. Samavati, A. Samavati, P.S. Goh, A. Fauzi Ismail, M. Sohaimi Abdullah, A comprehensive review of recent advances in nanofiltration membranes for heavy metal removal from wastewater, *Chem. Eng. Res. Des.* 189 (2023) 530–571. https://doi.org/10.1016/j.cherd.2022.11.042.

[43] B.A.M. Al-Rashdi, D.J. Johnson, N. Hilal, Removal of heavy metal ions by nanofiltration, *Desalination.* 315 (2013) 2–17. https://doi.org/10.1016/j.desal.2012.05.022.

[44] A. Eteba, M. Bassyouni, M.H. Elzahar, M.Z. El-shekhiby, Development and optimization of biodegradable chitosan-based nanofiltration membranes for effective textile wastewater treatment, *Sep. Purif. Technol.* 367 (2025). https://doi.org/10.1016/j.seppur.2025.132852.

[45] G. Moradi, S. Zinadini, M. Rahimi, F. Gholami, Folic acid functionalized pectin-co-sodium alginate surface modified nanofiltration membranes for treatment of wastewater containing pharmaceuticals, *J. Water Process Eng.* 73 (2025). https://doi.org/10.1016/j.jwpe.2025.107638.

[46] S.A. Moezzi, S. Rastgar, M. Faghani, Z. Ghiasvand, A. Javanshir Khoei, Optimization of carbon membrane performance in reverse osmosis systems for reducing salinity, nitrates, phosphates, and ammonia in aquaculture wastewater, *Chemosphere.* 376 (2025). https://doi.org/10.1016/j.chemosphere.2025.144304.

[47] M. Sarai Atab, A.J. Smallbone, A.P. Roskilly, An operational and economic study of a reverse osmosis desalination system for potable water and land irrigation, *Desalination.* 397 (2016) 174–184. https://doi.org/10.1016/j.desal.2016.06.020.

[48] G. Blandin, A.R.D. Verliefde, J. Comas, I. Rodriguez-Roda, P. Le-Clech, Efficiently combining water reuse and desalination through forward osmosis-reverse osmosis (FO-RO) hybrids: A critical review, *Membranes*. 6 (2016). https://doi.org/10.3390/membranes6030037.

[49] H. Gong, Z. Li, Z. Jin, H. Xu, K. Wang, Evaluation of direct water reclamation from municipal wastewater by disk tube reverse osmosis membrane (DT-RO) based compact process, *Desalin. Water Treat.* 129 (2018) 194–200. https://doi.org/10.5004/dwt.2018.23117.

[50] P.S. Senanayake, A. Lugo, M.F. Ahmed, Z. Stoll, N.E. Moe, J. Barber, W. Shane Walker, P. Xu, H. Wang, Electrodialysis modeling for desalination and resource recovery, *Curr. Opin. Chem. Eng.* 47 (2025). https://doi.org/10.1016/j.coche.2024.101081.

[51] A. Campione, L. Gurreri, M. Ciofalo, G. Micale, A. Tamburini, A. Cipollina, Electrodialysis for water desalination: A critical assessment of recent developments on process fundamentals, models and applications, *Desalination*. 434 (2018) 121–160. https://doi.org/10.1016/j.desal.2017.12.044.

[52] Y. Mei, C.Y. Tang, Recent developments and future perspectives of reverse electrodialysis technology: A review, *Desalination*. 425 (2018) 156–174. https://doi.org/10.1016/j.desal.2017.10.021.

[53] G.J. Doornbusch, M. Tedesco, J.W. Post, Z. Borneman, K. Nijmeijer, Experimental investigation of multistage electrodialysis for seawater desalination, *Desalination*. 464 (2019) 105–114. https://doi.org/10.1016/j.desal.2019.04.025.

[54] S.K.A. Al-Amshawee, M.Y.B.M. Yunus, Electrodialysis membrane with concentration polarization – A review, *Chem. Eng. Res. Des.* 201 (2024) 645–678. https://doi.org/10.1016/j.cherd.2023.10.060.

[55] A. Beiranvand, M.R. Omidkhah, M. Moharrami, S. Davari, H. Mahdavi, Characterization and optimization of electrodialysis with bipolar membranes with improved alkaline stability for phenol recovery from petroleum wastewater, *Chem. Eng. J. Adv.* 22 (2025) 100756. https://doi.org/10.1016/j.ceja.2025.100756.

[56] K.J. Min, H.J. An, K.Y. Park, Cadmium-treatment efficiency and membrane fouling during electrodialysis of wastewater discharged from zinc smelting, *Chemosphere*. 332 (2023). https://doi.org/10.1016/j.chemosphere.2023.138881.

[57] M. Jiang, Z. Fang, Z. Liu, X. Huang, H. Wei, C.Y. Yu, Application of membrane distillation for purification of radioactive liquid, *Clean. Eng. Technol.* 12 (2023). https://doi.org/10.1016/j.clet.2022.100589.

[58] M.I. Siyal, C.K. Lee, C. Park, A.A. Khan, J.O. Kim, A review of membrane development in membrane distillation for emulsified industrial or shale gas wastewater treatments with feed containing hybrid impurities, *J. Environ. Manage.* 243 (2019) 45–66. https://doi.org/10.1016/j.jenvman.2019.04.105.

[59] M.M.A. Shirazi, A. Kargari, A review on applications of membrane distillation (MD) process for wastewater treatment, *J. Membr. Sci. Res.* 1 (2015) 101–112.

[60] P. Biniaz, N.T. Ardekani, M.A. Makarem, M.R. Rahimpour, Water and wastewater treatment systems by novel integrated membrane distillation (MD), *ChemEngineering*. 3 (2019) 1–36. https://doi.org/10.3390/chemengineering3010008.

[61] R. Yang, R. Zhao, Y. Liu, Q. Wu, Z. Zhong, H. Liu, C. Liu, K. Zuo, A novel composite membrane for zero liquid discharge of coking wastewater in membrane distillation, *Chem. Eng. J.* 512 (2025). https://doi.org/10.1016/j.cej.2025.162521.

[62] H. Susane, Y.F. Wang, S.J. You, Enhanced performance of air gap membrane distillation for azo dye wastewater treatment using oxygen-plasma-modified PVDF and PTFE membranes, *J. Taiwan Inst. Chem. Eng.* 170 (2025). https://doi.org/10.1016/j.jtice.2025.106006.

[63] K.G. Nayar, P. Sundararaman, C.L. O'Connor, J.D. Schacherl, M.L. Heath, M.O. Gabriel, S.R. Shah, N.C. Wright, A.G. Winter, V, Feasibility study of an electrodialysis system for in-home water desalination in urban India, *Dev. Eng.* 2 (2016) 38–46. https://doi.org/10.1016/j.deveng.2016.12.001.

[64] A.G. Razaqpur, Y. Wang, X. Liao, Y. Liao, R. Wang, Progress of photothermal membrane distillation for decentralized desalination: A review, *Water Res.* 201 (2021). https://doi.org/10.1016/j.watres.2021.117299.

[65] M. M. Sabee, A. Rusli, N. Ahmad, Z. A. Hamid, Materials and applications for functional polymer membranes. *Adv. Funct. Membr. Mater. Appl.* 120 (2022) 72–110.

[66] E.O. Ezugbe, S. Rathilal, Membrane technologies in wastewater treatment: A review, *Membranes.* 10 (2020). https://doi.org/10.3390/membranes10050089.

[67] W. Gao, H. Liang, J. Ma, M. Han, Z.-L. Chen, Z.-S. Han, G.-B. Li, Membrane fouling control in ultrafiltration technology for drinking water production: A review, *Desalination.* 272 (2011) 1–8. https://doi.org/10.1016/j.desal.2011.01.051.

[68] I. Sutzkover-Gutman, D. Hasson, R. Semiat, Humic substances fouling in ultrafiltration processes, *Desalination.* 261 (2010) 218–231. https://doi.org/10.1016/j.desal.2010.05.008.

[69] J. Lee, H.W. Walker, Mechanisms and factors influencing the removal of microcystin-LR by ultrafiltration membranes, *J. Memb. Sci.* 320 (2008) 240–247. https://doi.org/10.1016/j.memsci.2008.04.007.

[70] O. Sagbo, Y. Sun, A. Hao, P. Gu, Effect of PAC addition on MBR process for drinking water treatment, *Sep. Purif. Technol.* 58 (2008) 320–327. https://doi.org/10.1016/j.seppur.2007.05.003.

[71] J. yu Tian, H. Liang, Y. ling Yang, S. Tian, G. bai Li, Membrane adsorption bioreactor (MABR) for treating slightly polluted surface water supplies: As compared to membrane bioreactor (MBR), *J. Memb. Sci.* 325 (2008) 262–270. https://doi.org/10.1016/j.memsci.2008.07.047.

[72] J. yu Tian, H. Liang, X. Li, S. jie You, S. Tian, G. bai Li, Membrane coagulation bioreactor (MCBR) for drinking water treatment, *Water Res.* 42 (2008) 3910–3920. https://doi.org/10.1016/j.watres.2008.05.025.

[73] H. Wang, H. Zhou, Understand the basics of membrane filtration, *Chem. Eng. Prog.* 109 (2013) 33–40.

[74] C.A. Buckner, R.M. Lafrenie, J.A. Dénommée, J.M. Caswell, D.A. Want, G.G. Gan, Y.C. Leong, P.C. Bee, E. Chin, A.K.H. Teh, S. Picco, L. Villegas, F. Tonelli, M. Merlo, J. Rigau, D. Diaz, M. Masuelli, S. Korrapati, P. Kurra, S. Puttugunta, S. Picco, L. Villegas, F. Tonelli, M. Merlo, J. Rigau, D. Diaz, M. Masuelli, M. Tascilar, F.A. de Jong, J. Verweij, R.H.J. Mathijssen, We are IntechOpen, the world's leading publisher of Open Access books Built by scientists, for scientists TOP 1%, *Intech.* 11 (2016) 13. https://www.intechopen.com/books/advanced-biometric-technologies/liveness-detection-in-biometrics.

[75] M.B. Tanis-Kanbur, R.I. Peinador, J.I. Calvo, A. Hernández, J.W. Chew, Porosimetric membrane characterization techniques: A review, *J. Memb. Sci.* 619 (2021) 637459. https://doi.org/10.1016/j.memsci.2020.118750.

[76] L. Rizzo, W. Gernjak, P. Krzeminski, S. Malato, C.S. McArdell, J.A.S. Perez, H. Schaar, D. Fatta-Kassinos, Best available technologies and treatment trains to address current challenges in urban wastewater reuse for irrigation of crops in EU countries, *Sci. Total Environ.* 710 (2020) 136312. https://doi.org/10.1016/j.scitotenv.2019.136312.

[77] B.T. Akinyemi, O.D. Ogundele, A.B. Afolabi, Advancements in sustainable membrane technologies for enhanced remediation and wastewater treatment: A comprehensive review, *Acadlore Trans. Geosci.* 2 (2023) 196–207. https://doi.org/10.56578/atg020402.

[78] M. Frappa, F. Macedonio, E. Drioli, Progress of membrane engineering for water treatment, *J. Membr. Sci. Res.* 6 (2020) 269–279. https://doi.org/10.22079/JMSR.2019.108451.1265.

[79] N. Kim, J. Lee, S. Kim, S.P. Hong, C. Lee, J. Yoon, C. Kim, Short review of multichannel membrane capacitive deionization: Principle, current status, and future prospect, *Appl. Sci.* 10 (2020). https://doi.org/10.3390/app10020683.
[80] Š. Schlosser, Membrane filtration, *Eng. Asp. Food Biotechnol.* (2013) 112–132. https://doi.org/10.1201/b15426.
[81] S. Xia, M. Liu, H. Yu, D. Zou, Pressure–driven membrane filtration technology for terminal control of organic DBPs: A review, *Sci. Total Environ.* 904 (2023) 166751. https://doi.org/10.1016/j.scitotenv.2023.166751.
[82] L. Yang, Q. She, M.P. Wan, R. Wang, V.W.C. Chang, C.Y. Tang, Removal of haloacetic acids from swimming pool water by reverse osmosis and nanofiltration, *Water Res.* 116 (2017) 116–125. https://doi.org/10.1016/j.watres.2017.03.025.
[83] R. Chalatip, R. Chawalit, R. Nopawan, Removal of haloacetic acids by nanofiltration, *J. Environ. Sci.* 21 (2009) 96–100. https://doi.org/10.1016/S1001-0742(09)60017-6.
[84] B. Chen, C. Zhang, L. Wang, J. Yang, Y. Sun, Removal of disinfection byproducts in drinking water by flexible reverse osmosis: Efficiency comparison, fates, influencing factors, and mechanisms, *J. Hazard. Mater.* 401 (2021) 123408. https://doi.org/10.1016/j.jhazmat.2020.123408.
[85] F.-X. Kong, X.-M. Wang, H.-W. Yang, J.-F. Chen, C.-M. Guo, T. Zhang, Y.F. Xie, The role of solubility on the rejection of trace organics by nanofiltration membrane: Exemplified with disinfection by-products, *Environ. Sci. Pollut. Res.* 24 (2017) 18400–18409. https://doi.org/10.1007/s11356-017-9282-0.
[86] P. Xu, J.E. Drewes, T.U. Kim, C. Bellona, G. Amy, Effect of membrane fouling on transport of organic contaminants in NF/RO membrane applications, *J. Memb. Sci.* 279 (2006) 165–175. https://doi.org/10.1016/j.memsci.2005.12.001.
[87] M.G. Shin, W. Choi, S.J. Park, S. Jeon, S. Hong, J.H. Lee, Critical review and comprehensive analysis of trace organic compound (TOrC) removal with polyamide RO/NF membranes: Mechanisms and materials, *Chem. Eng. J.* 427 (2022) 130957. https://doi.org/10.1016/j.cej.2021.130957.
[88] M.G. Buonomenna, Nano-enhanced reverse osmosis membranes, *Desalination* 314 (2013) 73–88.
[89] Y. Qi, D. Li, S. Zhang, F. Li, T. Hua, Electrochemical filtration for drinking water purification: A review on membrane materials, mechanisms and roles, *J. Environ. Sci.*. 141 (2024) 102–128. https://doi.org/10.1016/j.jes.2023.06.033.
[90] B. Khorshidi, J. Hajinasiri, G. Ma, S. Bhattacharjee, M. Sadrzadeh, Thermally resistant and electrically conductive PES/ITO nanocomposite membrane, *J. Memb. Sci.* 500 (2016) 151–160. https://doi.org/10.1016/j.memsci.2015.11.015.
[91] X. Wang, H. Wang, F. Li, X. Hu, Z. Xie, T. Hua, Activation of peroxymonosulfate in an electrochemical filter by $MnFe_2O_4$-rGO electro-assisted catalytic membrane for the degradation of oxytetracycline, *J. Environ. Chem. Eng.* 10 (2022) 107008. https://doi.org/10.1016/j.jece.2021.107008.
[92] Z.K. Li, W.Z. Lang, W. Miao, X. Yan, Y.J. Guo, Preparation and properties of PVDF/SiO_2@GO nanohybrid membranes via thermally induced phase separation method, *J. Memb. Sci.* 511 (2016) 151–161. https://doi.org/10.1016/j.memsci.2016.03.048.
[93] N. Li, W. Wang, C. Ma, L. Zhu, X. Chen, B. Zhang, C. Zhong, A novel conductive rGO/ZnO/PSF membrane with superior water flux for electrocatalytic degradation of organic pollutants, *J. Memb. Sci.* 641 (2022) 300387. https://doi.org/10.1016/j.memsci.2021.119901.
[94] N. Li, W. Wang, L. Zhu, W. Cui, X. Chen, B. Zhang, Z. Zhang, A novel electro-cleanable PAN-ZnO nanofiber membrane with superior water flux and electrocatalytic properties for organic pollutant degradation, *Chem. Eng. J.* 421 (2021) 127857. https://doi.org/10.1016/j.cej.2020.127857.

[95] W.L. Jiang, X. Xia, J.L. Han, Y.C. Ding, M.R. Haider, A.J. Wang, Graphene modified electro-Fenton catalytic membrane for in situ degradation of antibiotic florfenicol, *Environ. Sci. Technol.* 52 (2018) 9972–9982. https://doi.org/10.1021/acs.est.8b01894.
[96] Y. Mansourpanah, F. Emamian, Membrane and bioseparation, *Adv. Membr. Technol.* (2020) 1–34. https://doi.org/10.5772/intechopen.86954.
[97] B.A.R. Hassan, Z.B.M. Yusoff, M.A. Hassali, S. B. Othman, Supportive and palliative care in solid cancer patients. In *Cancer treatment-conventional and innovative approaches*. IntechOpen, 2013. https://doi.org/10.5772/55358.
[98] N. Shehata, D. Egirani, A.G. Olabi, A. Inayat, M.A. Abdelkareem, K.J. Chae, E.T. Sayed, Membrane-based water and wastewater treatment technologies: Issues, current trends, challenges, and role in achieving sustainable development goals, and circular economy, *Chemosphere*. 320 (2023) 137993. https://doi.org/10.1016/j.chemosphere.2023.137993.
[99] J. Landaburu-Aguirre, R. García-Pacheco, S. Molina, L. Rodríguez-Sáez, J. Rabadán, E. García-Calvo, Fouling prevention, preparing for re-use and membrane recycling. Towards circular economy in RO desalination, *Desalination*. 393 (2016) 16–30. https://doi.org/10.1016/j.desal.2016.04.002.
[100] S. Zinadini, V. Vatanpour, A.A. Zinatizadeh, M. Rahimi, Z. Rahimi, M. Kian, Preparation and characterization of antifouling graphene oxide/polyethersulfone ultrafiltration membrane: Application in MBR for dairy wastewater treatment, *J. Water Process Eng.* 7 (2015) 280–294. https://doi.org/10.1016/j.jwpe.2015.07.005.
[101] N.N.R. Ahmad, W.L. Ang, C.P. Leo, A.W. Mohammad, N. Hilal, Current advances in membrane technologies for saline wastewater treatment: A comprehensive review, *Desalination*. 517 (2021) 115170. https://doi.org/10.1016/j.desal.2021.115170.
[102] F. Galiano, I. Friha, S.A. Deowan, J. Hoinkis, Y. Xiaoyun, D. Johnson, R. Mancuso, N. Hilal, B. Gabriele, S. Sayadi, A. Figoli, Novel low-fouling membranes from lab to pilot application in textile wastewater treatment, *J. Colloid Interface Sci.* 515 (2018) 208–220. https://doi.org/10.1016/j.jcis.2018.01.009.
[103] S. Mansour, A. Giwa, S.W. Hasan, Novel graphene nanoplatelets-coated polyethylene membrane for the treatment of reject brine by pilot-scale direct contact membrane distillation: An optimization study, *Desalination*. 441 (2018) 9–20. https://doi.org/10.1016/j.desal.2018.04.026.
[104] Y.C. Woo, Y. Kim, M. Yao, L.D. Tijing, J.S. Choi, S. Lee, S.H. Kim, H.K. Shon, Hierarchical composite membranes with robust omniphobic surface using layer-by-layer assembly technique, *Environ. Sci. Technol.* 52 (2018) 2186–2196. https://doi.org/10.1021/acs.est.7b05450.
[105] X. Du, Z. Zhang, K.H. Carlson, J. Lee, T. Tong, Membrane fouling and reusability in membrane distillation of shale oil and gas produced water: Effects of membrane surface wettability, *J. Memb. Sci.* 567 (2018) 199–208. https://doi.org/10.1016/j.memsci.2018.09.036.
[106] M.A. Shannon, P.W. Bohn, M. Elimelech, J.G. Georgiadis, B.J. Mar̃ias, A.M. Mayes, Science and technology for water purification in the coming decades, *Nature*. 452 (2008) 301–310. https://doi.org/10.1038/nature06599.
[107] T.S. Chung, X. Li, R.C. Ong, Q. Ge, H. Wang, G. Han, Emerging forward osmosis (FO) technologies and challenges ahead for clean water and clean energy applications, *Curr. Opin. Chem. Eng.* 1 (2012) 246–257. https://doi.org/10.1016/j.coche.2012.07.004.
[108] E. Drioli, A. Ali, F. Macedonio, Membrane distillation: Recent developments and perspectives, *Desalination*. 356 (2015) 56–84. https://doi.org/10.1016/j.desal.2014.10.028.
[109] Y. Oren, Capacitive deionization (CDI) for desalination and water treatment - past, present and future (a review), *Desalination*. 228 (2008) 10–29. https://doi.org/10.1016/j.desal.2007.08.005.

[110] T.Y. Cath, A.E. Childress, M. Elimelech, Forward osmosis: Principles, applications, and recent developments, *J. Memb. Sci.* 281 (2006) 70–87. https://doi.org/10.1016/j.memsci.2006.05.048.

[111] A.M. Comerton, R.C. Andrews, D.M. Bagley, P. Yang, Membrane adsorption of endocrine disrupting compounds and pharmaceutically active compounds, *J. Memb. Sci.* 303 (2007) 267–277. https://doi.org/10.1016/j.memsci.2007.07.025.

[112] B.J. Singh, A. Chakraborty, R. Sehgal, A systematic review of industrial wastewater management: Evaluating challenges and enablers, *J. Environ. Manage.* 348 (2023) 119230. https://doi.org/10.1016/j.jenvman.2023.119230.

[113] M.L. Sikosana, K. Sikhwivhilu, R. Moutloali, D.M. Madyira, Municipal wastewater treatment technologies: A review, *Procedia Manuf.* 35 (2019) 1018–1024. https://doi.org/10.1016/j.promfg.2019.06.051.

[114] K. Obaideen, N. Shehata, E.T. Sayed, M.A. Abdelkareem, M.S. Mahmoud, A.G. Olabi, The role of wastewater treatment in achieving sustainable development goals (SDGs) and sustainability guideline, *Energy Nexus.* 7 (2022) 100112. https://doi.org/10.1016/j.nexus.2022.100112.

[115] J.A. Silva, Wastewater treatment and reuse for sustainable water resources management: A systematic literature review, *Sustain.* 15 (2023). https://doi.org/10.3390/su151410940.

[116] F. Meng, F. Yang, J. Xiao, H. Zhang, Z. Gong, A new insight into membrane fouling mechanism during membrane filtration of bulking and normal sludge suspension, *J. Memb. Sci.* 285 (2006) 159–165. https://doi.org/10.1016/j.memsci.2006.08.020.

[117] T.U. Rahman, H. Roy, M.R. Islam, M. Tahmid, A. Fariha, A. Mazumder, N. Tasnim, M.N. Pervez, Y. Cai, V. Naddeo, M.S. Islam, The advancement in Membrane Bioreactor (MBR) technology toward sustainable industrial wastewater management, *Membranes.* 13 (2023). https://doi.org/10.3390/membranes13020181.

[118] J. Hoinkis, S.A. Deowan, V. Panten, A. Figoli, R.R. Huang, E. Drioli, Membrane bioreactor (MBR) technology - A promising approach for industrial water reuse, *Procedia Eng.* 33 (2012) 234–241. https://doi.org/10.1016/j.proeng.2012.01.1199.

[119] F.Y. Sun, X.M. Wang, X.Y. Li, An innovative membrane bioreactor (MBR) system for simultaneous nitrogen and phosphorus removal, *Process Biochem.* 48 (2013) 1749–1756. https://doi.org/10.1016/j.procbio.2013.08.009.

[120] J. He, Z. Zhang, F. Cui, X. Tan, X. Zheng, R. Cheng, Global techno-economic analysis of MBR for hospital wastewater treatment, *Sci. Total Environ.* 956 (2024) 177172. https://doi.org/10.1016/j.scitotenv.2024.177172.

[121] E.N. Ferri, L. Bolelli, Wastewater remediation treatments aimed at water reuse: Recent outcomes from pilot- and full-scale tests, *Appl. Sci.* 15 (2025). https://doi.org/10.3390/app15052448.

[122] Z. An, J. Zhu, M. Zhang, Y. Zhou, X. Su, H. Lin, F. Sun, Anaerobic membrane bioreactor for the treatment of high-strength waste/wastewater: A critical review and update, *Chem. Eng. J.* 470 (2023) 144322. https://doi.org/10.1016/j.cej.2023.144322.

[123] V. Stazi, M.C. Tomei, Enhancing anaerobic treatment of domestic wastewater: State of the art, innovative technologies and future perspectives, *Sci. Total Environ.* 635 (2018) 78–91. https://doi.org/10.1016/j.scitotenv.2018.04.071.

[124] M. Zielińska, A. Ojo, Anaerobic Membrane Bioreactors (AnMBRs) for wastewater treatment: Recovery of nutrients and energy, and management of fouling, *Energies.* 16 (2023). https://doi.org/10.3390/en16062829.

[125] J. Rihter Pikl, A. Lobnik, M. Roš, H. El Khiar, N. Uranjek, Microfibres and coliforms determination and removal from wastewater treatment effluent, *Clean. Eng. Technol.* 22 (2024). https://doi.org/10.1016/j.clet.2024.100806.

[126] Y.N. Kanafin, D. Kanafina, S. Malamis, E. Katsou, V.J. Inglezakis, S.G. Poulopoulos, E. Arkhangelsky, Anaerobic membrane bioreactors for municipal wastewater treatment: A literature review, *Membranes.* 11 (2021). https://doi.org/10.3390/membranes11120967.

[127] A. Soo, H.K. Shon, A nutrient circular economy framework for wastewater treatment plants, *Desalination*. 592 (2024) 118090. https://doi.org/10.1016/j.desal.2024.118090.
[128] M. Moudi, M. Nasrabadi, Archive of SID. ir. *J. Environ. Res. 15* (30) (2025).
[129] K.H. Hama Aziz, F.S. Mustafa, Advanced oxidation processes for the decontamination of heavy metal complexes in aquatic systems: A review, *Case Stud. Chem. Environ. Eng*. 9 (2024) 100567. https://doi.org/10.1016/j.cscee.2023.100567.
[130] F. García-Ávila, A. Zambrano-Jaramillo, C. Velecela-Garay, K. Coronel-Sánchez, L. Valdiviezo-Gonzales, Effectiveness of membrane technologies in removing emerging contaminants from wastewater: Reverse Osmosis and Nanofiltration. *Water Cycle* 6 (2025), 357–373. https://doi.org/10.1016/j.watcyc.2024.12.004.
[131] X. Zheng, Z. Zhang, D. Yu, X. Chen, R. Cheng, S. Min, J. Wang, Q. Xiao, J. Wang, Overview of membrane technology applications for industrial wastewater treatment in China to increase water supply, *Resour. Conserv. Recycl.* 105 (2015) 1–10. https://doi.org/10.1016/j.resconrec.2015.09.012.
[132] C. Muro, F. Riera, M. del Carmen Diaz, Membrane separation process in wastewater treatment of food industry, *Food Ind. Process. - Methods Equip.* (2012). https://doi.org/10.5772/31116.
[133] C.R. Bartels, M. Wilf, K. Andes, J. Iong, Design considerations for wastewater treatment by reverse osmosis, *Water Sci. Technol.* 51 (2005) 473–482. https://doi.org/10.2166/wst.2005.0670.
[134] Y.A. Tayeh, M.Y.D. Alazaiza, T.M. Alzghoul, M.J. Bashir, A comprehensive review of RO membrane fouling: Mechanisms, categories, cleaning methods and pretreatment technologies, *J. Hazard. Mater. Adv.* 18 (2025) 100684. https://doi.org/10.1016/j.hazadv.2025.100684.
[135] Y.N. Kanafin, A. Satayeva, P. Abdirova, V.J. Inglezakis, E. Arkhangelsky, S.G. Poulopoulos, Membrane bioreactor and advanced oxidation processes for combined treatment of synthetic wastewater containing naproxen, bisphenol A, and sulfamethoxazole, *J. Water Process Eng.* 55 (2023) 104250. https://doi.org/10.1016/j.jwpe.2023.104250.
[136] H.K. Shon, S. Vigneswaran, S.A. Snyder, Effluent organic matter (EfOM) in wastewater: Constituents, effects, and treatment, *Crit. Rev. Environ. Sci. Technol.* 36 (2006) 327–374. https://doi.org/10.1080/10643380600580011.
[137] A.I. Osman, Z. Chen, A.M. Elgarahy, M. Farghali, I.M.A. Mohamed, A.K. Priya, H.B. Hawash, P.S. Yap, Membrane technology for energy saving: Principles, techniques, applications, challenges, and prospects, *Adv. Energy Sustain. Res.* 5 (2024). https://doi.org/10.1002/aesr.202400011.
[138] J.E. Miller, L.R. Evans, Forward osmosis: A new approach to water purification and desalination, *Sandia Rep.* (2006) 51 p. https://doi.org/10.2172/893156.
[139] R. Valladares Linares, Z. Li, S. Sarp, S.S. Bucs, G. Amy, J.S. Vrouwenvelder, Forward osmosis niches in seawater desalination and wastewater reuse, *Water Res.* 66 (2014) 122–139. https://doi.org/10.1016/j.watres.2014.08.021.
[140] L. Chekli, S. Phuntsho, H.K. Shon, S. Vigneswaran, J. Kandasamy, A. Chanan, A review of draw solutes in forward osmosis process and their use in modern applications, *Desalin. Water Treat.* 43 (2012) 167–184. https://doi.org/10.1080/19443994.2012.672168.
[141] Q. Ge, M. Ling, T.S. Chung, Draw solutions for forward osmosis processes: Developments, challenges, and prospects for the future, *J. Memb. Sci.* 442 (2013) 225–237. https://doi.org/10.1016/j.memsci.2013.03.046.
[142] D. Li, X. Zhang, G.P. Simon, H. Wang, Forward osmosis desalination using polymer hydrogels as a draw agent: Influence of draw agent, feed solution and membrane on process performance, *Water Res.* 47 (2013) 209–215. https://doi.org/10.1016/j.watres.2012.09.049.

[143] Z.Y. Li, V. Yangali-Quintanilla, R. Valladares-Linares, Q. Li, T. Zhan, G. Amy, Flux patterns and membrane fouling propensity during desalination of seawater by forward osmosis, *Water Res.* 46 (2012) 195–204. https://doi.org/10.1016/j.watres.2011.10.051.
[144] R. Valladares Linares, Z. Li, M. Abu-Ghdaib, C.H. Wei, G. Amy, J.S. Vrouwenvelder, Water harvesting from municipal wastewater via osmotic gradient: An evaluation of process performance, *J. Memb. Sci.* 447 (2013) 50–56. https://doi.org/10.1016/j.memsci.2013.07.018.
[145] Z. Li, R. Valladares Linares, M. Abu-Ghdaib, T. Zhan, V. Yangali-Quintanilla, G. Amy, Osmotically driven membrane process for the management of urban runoff in coastal regions, *Water Res.* 48 (2014) 200–209. https://doi.org/10.1016/j.watres.2013.09.028.
[146] D. Khanafer, I. Ibrahim, S. Yadav, A. Altaee, A. Hawari, J. Zhou, Brine reject dilution with treated wastewater for indirect desalination, *J. Clean. Prod.* 322 (2021) 129129. https://doi.org/10.1016/j.jclepro.2021.129129.
[147] D. Patel, A. Mudgal, V. Patel, J. Patel, K. Park, P. Davies, R.R. Alegre, Energy, exergy, economic and environment analysis of standalone forward osmosis (FO) system for domestic wastewater treatment, *Desalination.* 567 (2023). https://doi.org/10.1016/j.desal.2023.116995.
[148] T.Y. Cath, N.T. Hancock, C.D. Lundin, C. Hoppe-Jones, J.E. Drewes, A multi-barrier osmotic dilution process for simultaneous desalination and purification of impaired water, *J. Memb. Sci.* 362 (2010) 417–426. https://doi.org/10.1016/j.memsci.2010.06.056.
[149] V. Yangali-Quintanilla, Z. Li, R. Valladares, Q. Li, G. Amy, Indirect desalination of Red Sea water with forward osmosis and low pressure reverse osmosis for water reuse, *Desalination.* 280 (2011) 160–166. https://doi.org/10.1016/j.desal.2011.06.066.
[150] Z. Li, R.V. Linares, S. Sarp, G. Amy, *Direct and indirect seawater desalination by forward osmosis*, Elsevier B.V., 2018. https://doi.org/10.1016/B978-0-444-63961-5.00009-2.
[151] B.M. Cieślik, J. Namieśnik, P. Konieczka, Review of sewage sludge management: Standards, regulations and analytical methods, *J. Clean. Prod.* 90 (2015) 1–15. https://doi.org/10.1016/j.jclepro.2014.11.031.
[152] H. Kominko, K. Gorazda, Z. Wzorek, Potentiality of sewage sludge-based organo-mineral fertilizer production in Poland considering nutrient value, heavy metal content and phytotoxicity for rapeseed crops, *J. Environ. Manage.* 248 (2019) 109283. https://doi.org/10.1016/j.jenvman.2019.109283.
[153] X. Hao, Q. Chen, M.C.M. van Loosdrecht, J. Li, H. Jiang, Sustainable disposal of excess sludge: Incineration without anaerobic digestion, *Water Res.* 170 (2020). https://doi.org/10.1016/j.watres.2019.115298.
[154] W. Xue, A.P.P. Aung, S. Guerrero-Cruz, K. Xiao, Y. He, A.K. Anal, A.S. Tabucanon, An innovative application of osmotic microbial fuel cell (OsMFC) for enhanced activated sludge thickening and stabilization with bioelectricity generation, *Water Res.* 275 (2025) 123199. https://doi.org/10.1016/j.watres.2025.123199.
[155] A. Achilli, T.Y. Cath, E.A. Marchand, A.E. Childress, The forward osmosis membrane bioreactor: A low fouling alternative to MBR processes, *Desalination.* 239 (2009) 10–21. https://doi.org/10.1016/j.desal.2008.02.022.
[156] E.R. Cornelissen, D. Harmsen, K.F. de Korte, C.J. Ruiken, J.J. Qin, H. Oo, L.P. Wessels, Membrane fouling and process performance of forward osmosis membranes on activated sludge, *J. Memb. Sci.* 319 (2008) 158–168. https://doi.org/10.1016/j.memsci.2008.03.048.
[157] M. Zielińska, K. Bułkowska, Use of membrane techniques for removal and recovery of nutrients from liquid fraction of anaerobic digestate, *Membranes.* 15 (2025). https://doi.org/10.3390/membranes15020045.

[158] A. Cosenza, H. Gulhan, C.M. Maida, G. Mannina, Nutrient recovery from wastewater treatment by ultrafiltration membrane for water reuse in view of a circular economy perspective, *Bioresour. Technol.* 363 (2022) 127929. https://doi.org/10.1016/j.biortech.2022.127929.
[159] Y. Zhu, F. Gao, H. Jia, Z. Chen, B.J. Ni, J. Wang, Applications of membrane technology in the resource recovery of power lithium-ion battery precursor wastewater: A review, *Environ. Res.* 275 (2025). https://doi.org/10.1016/j.envres.2025.121372.
[160] J. Yang, M. Monnot, L. Ercolei, P. Moulin, Membrane-based processes used in municipal wastewater treatment for water reuse: State-of-the-art and performance analysis, *Membranes.* 10 (2020) 1–56. https://doi.org/10.3390/membranes10060131.
[161] S. Muthukumaran, D.A. Nguyen, K. Baskaran, Performance evaluation of different ultrafiltration membranes for the reclamation and reuse of secondary effluent, *Desalination.* 279 (2011) 383–389. https://doi.org/10.1016/j.desal.2011.06.040.
[162] K.N.O. Shannon, R. Holman, An evaluation of fouling potential and methods to control fouling of membranes, (2007) 6417–6444.
[163] X. Du, Y. Shi, V. Jegatheesan, I. Ul Haq, A review on the mechanism, impacts and control methods of membrane fouling in MBR system, *Membranes.* 10 (2) (2020) 24. https://doi.org/10.3390/membranes10020024.
[164] A. Gul, J. Hruza, F. Yalcinkaya, Fouling and chemical cleaning of microfiltration membranes: A mini-review, *Polymers.* 13 (2021). https://doi.org/10.3390/polym13060846.
[165] P. Jarvis, I. Carra, M. Jafari, S.J. Judd, Ceramic vs polymeric membrane implementation for potable water treatment, *Water Res.* 215 (2022) 118269. https://doi.org/10.1016/j.watres.2022.118269.
[166] M.H. Ben Salem, G. Soucy, D. Marceau, A. Godefroy, S. Charest, Review of the properties and degradation mechanisms of refractories in aluminum reduction cells, *Metals.* 15 (2025) 1–18. https://doi.org/10.3390/met15030278.
[167] M.B. Asif, Z. Zhang, Ceramic membrane technology for water and wastewater treatment: A critical review of performance, full-scale applications, membrane fouling and prospects, *Chem. Eng. J.* 418 (2021) 129481. https://doi.org/10.1016/j.cej.2021.129481.
[168] S.S. Madaeni, T. Mohamamdi, M.K. Moghadam, Chemical cleaning of reverse osmosis membranes, *Desalination.* 134 (2001) 77–82. https://doi.org/10.1016/S0011-9164(01)00117-5.
[169] S.S. Madaeni, H.R. Tavakolian, F. Rahimpour, Cleaning optimization of microfiltration membrane employed for milk sterilization, *Sep. Sci. Technol.* 46 (2011) 571–580. https://doi.org/10.1080/01496395.2010.534118.
[170] D.F. Ayala, V. Ferre, S.J. Judd, Membrane life estimation in full-scale immersed membrane bioreactors, *J. Memb. Sci.* 378 (2011) 95–100. https://doi.org/10.1016/j.memsci.2011.03.013.
[171] A. Samadi, L. Gao, L. Kong, Y. Orooji, S. Zhao, Waste-derived low-cost ceramic membranes for water treatment: Opportunities, challenges and future directions, *Resour. Conserv. Recycl.* 185 (2022) 106497. https://doi.org/10.1016/j.resconrec.2022.106497.
[172] S.F. Ahmed, F. Mehejabin, A. Momtahin, N. Tasannum, N.T. Faria, M. Mofijur, A.T. Hoang, D.V.N. Vo, T.M.I. Mahlia, Strategies to improve membrane performance in wastewater treatment, *Chemosphere.* 306 (2022) 135527. https://doi.org/10.1016/j.chemosphere.2022.135527.
[173] M. Moslehi, H. Mahdavi, Controlled pore size nanofibrous microfiltration membrane via multi-step interfacial polymerization: Preparation and characterization, *Sep. Purif. Technol.* 223 (2019) 96–106. https://doi.org/10.1016/j.seppur.2019.04.041.
[174] U. Merin, G. Daufin, Crossflow microfiltration in the dairy industry: state-of-the-art. *Le Lait*, 70 (4) (1990), 281–291. https://doi.org/10.1051/lait:1990421.

[175] E.W. Tow, A.L. Hartman, A. Jaworowski, I. Zucker, S. Kum, M. AzadiAghdam, E.R. Blatchley, A. Achilli, H. Gu, G.M. Urper, D.M. Warsinger, Modeling the energy consumption of potable water reuse schemes, *Water Res.* X. 13 (2021) 100126. https://doi.org/10.1016/j.wroa.2021.100126.

[176] H. El Mrabet, M. Elazhar, I. Moussaoui, J. Moumouh, F.Z. Addar, M. Tahaikt, F. Elazhar, T. Mohamed, N. El Haloui, A. Elmidaoui, Technico-economic assessment of membrane filtration as post-treatment of biological denitrification for brackish groundwater, *Desalin. Water Treat.* 322 (2025) 101160. https://doi.org/10.1016/j.dwt.2025.101160.

[177] J. Thuvander, A.S. Jönsson, Extraction of galactoglucomannan from thermomechanical pulp mill process water by microfiltration and ultrafiltration-Influence of microfiltration membrane pore size on ultrafiltration performance, *Chem. Eng. Res. Des.* 105 (2016) 171–176. https://doi.org/10.1016/j.cherd.2015.12.003.

[178] N. Tangsubkul, K. Parameshwaran, S. Lundie, A.G. Fane, T.D. Waite, Environmental life cycle assessment of the microfiltration process, *J. Memb. Sci.* 284 (2006) 214–226. https://doi.org/10.1016/j.memsci.2006.07.047.

[179] S. Munirasu, M.A. Haija, F. Banat, Use of membrane technology for oil field and refinery produced water treatment - A review, *Process Saf. Environ. Prot.* 100 (2016) 183–202. https://doi.org/10.1016/j.psep.2016.01.010.

[180] N. F. M. Tahirun, Preparation of nanofiltration hollow fiber membrane for acetic acid removal from synthetic hydrolysate (Doctoral dissertation, UMP) (2014).

[181] F.J. Benítez, J.L. Acero, A.I. Leal, M. González, The use of ultrafiltration and nanofiltration membranes for the purification of cork processing wastewater, *J. Hazard. Mater.* 162 (2009) 1438–1445. https://doi.org/10.1016/j.jhazmat.2008.06.036.

[182] A. Rivero-Falcón, Y. López-López, B. Peñate Suárez, N. Melián-Martel, Performance evaluation of nanofiltration membranes for SWRO brine valorisation: Insights from a pilot plant under real conditions, *Sep. Purif. Technol.* 359 (2025). https://doi.org/10.1016/j.seppur.2024.130774.

[183] Y.A. Tayeh, A comprehensive review of reverse osmosis desalination: Technology, water sources, membrane processes, fouling, and cleaning, *Desalin. Water Treat.* 320 (2024) 100882. https://doi.org/10.1016/j.dwt.2024.100882.

[184] B. Van Der Bruggen, L. Lejon, C. Vandecasteele, Reuse, treatment, and discharge of the concentrate of pressure-driven membrane processes, *Environ. Sci. Technol.* 37 (2003) 3733–3738. https://doi.org/10.1021/es0201754.

[185] A.A. Alsarayreh, M.A. Al-Obaidi, A.M. Al-Hroub, R. Patel, I.M. Mujtaba, Evaluation and minimisation of energy consumption in a medium-scale reverse osmosis brackish water desalination plant, *J. Clean. Prod.* 248 (2020) 119220. https://doi.org/10.1016/j.jclepro.2019.119220.

[186] Q. Lei, M. Zhang, L. Shen, R. Li, B.Q. Liao, H. Lin, A novel insight into membrane fouling mechanism regarding gel layer filtration: Flory-Huggins based filtration mechanism, *Sci. Rep.* 6 (2016) 1–9. https://doi.org/10.1038/srep33343.

[187] A. Uppu, A. Chaudhuri, S. Prasad Das, Numerical modeling of particulate fouling and cake-enhanced concentration polarization in roto-dynamic reverse osmosis filtration systems, *Desalination.* 468 (2019) 114053. https://doi.org/10.1016/j.desal.2019.06.019.

[188] L.I. Qrenawi, A.A. Abuhabib, A review on sources, types, mechanisms, characteristics, impacts and control strategies of fouling in ro membrane systems, *Desalin. Water Treat.* 208 (2020) 43–69. https://doi.org/10.5004/dwt.2020.26448.

[189] A. Penezić, B. Gašparović, V. Cuculić, S. Strmečki, T. Djakovac, M. Mlakar, Dissolved trace metals and organic matter distribution in the Northern Adriatic, an increasingly oligotrophic Shallow Sea, *Water.* 14 (2022). https://doi.org/10.3390/w14030349.

[190] Z. Zhou, S. Rajabzadeh, A.R. Shaikh, Y. Kakihana, W. Ma, H. Matsuyama, Effect of surface properties on antifouling performance of poly(vinyl chloride-co-poly(ethylene glycol)methyl ether methacrylate)/PVC blend membrane, *J. Memb. Sci.* 514 (2016) 537–546. https://doi.org/10.1016/j.memsci.2016.05.008.
[191] S. Judd, Membrane technology. In *Membranes for Industrial wastewater recovery and re-use* (pp. 13–74). Elsevier Science, 2003. https://doi.org/10.1016/b978-185617389-6/50003-5 .
[192] S. Al-Asheh, M. Bagheri, A. Aidan, Membrane bioreactor for wastewater treatment: A review, *Case Stud. Chem. Environ. Eng.* 4 (2021). https://doi.org/10.1016/j.cscee.2021.100109.
[193] J. Kim, B. Wu, S. Jeong, S. Jeong, M. Kim, Recent advances of membrane-based hybrid membrane bioreactors for wastewater reclamation, *Front. Membr. Sci. Technol.* 3 (2024) 1–14. https://doi.org/10.3389/frmst.2024.1361433.
[194] J. Kim, B. Wu, S. Jeong, S. Jeong, M. Kim, Recent advances of membrane-based hybrid membrane bioreactors for wastewater reclamation. *Front. Membr. Sci. Technol.* 3 (2024) 1361433.https://doi.org/10.3389/frmst.2024.1361433.
[195] X. Liu, G. Zhang, K.B. Al Mohawes, N.M. Khashab, Smart membranes for separation and sensing, *Chem. Sci.* (2024) 18772–18788. https://doi.org/10.1039/d4sc04793a.

4 Innovative Adsorption Techniques for Efficient Liquid Waste Treatment

Materials, Mechanisms, and Applications

Meriem Kasbaji, Mohamed Amine Kasbaji, Mohamed M'Barki, Ahmed Gamouh, Hafid Abderrafia, Mohamed Berkani, Mounir El Achaby, Catia Algieri, and Mustapha Oubenali

4.1 INTRODUCTION

Water is widely recognized as the most vital element within the cosmos. In the human organism and other mammalian species, the mean proportion of water constitutes approximately 60%, rendering it indispensable for survival. Moreover, water plays a significant role in the cultivation of food and fibers necessary for both human and animal consumption. More critically, water acts as the immediate habitat for aquatic organisms, offering shelter, dissolved oxygen, essential minerals, and organic matter as nutrients to facilitate their growth. Thus, water can be regarded as an integral component of the ecological system (Amenaghawon et al., 2023; Kasbaji et al., 2024; Vardhan et al., 2019).

Despite the relative abundance of water in the universe, it is predominantly found as saltwater in the oceans and seas, which possess salinity levels that render them unsuitable for human consumption. Only about 3% of the total water available in the universe exists as freshwater, with nearly 80% of this freshwater being trapped in the polar ice caps. However, the majority of accessible freshwater is consistently subjected to pollution, primarily as a result of anthropogenic activities (Crini and Lichtfouse, 2019). Therefore, a continuous supply of potable water is essential for our daily endeavors. Additionally, the paper, plastic, and metal industries depend on clean water for their production processes. Consequently, an increased awareness regarding the conservation of water through the understanding of the usage and recycling of water-intensive products is necessary. The escalating demand for clean water is driving the trend of employing wastewater and greywater in agricultural and

DOI: 10.1201/9781003624905-4

aquacultural practices. Enhanced water treatment technologies and frameworks are imperative to meet this demand, as current systems are approaching their capacity limits in providing adequate water treatment to satisfy the everyday needs of both human beings and the environment. Water is typically contaminated as a result of industrial discharges or inadequate sanitation infrastructure. Irresponsible human actions, such as the disposal of waste into water bodies, including rivers and lakes, exacerbate the issue of water pollution. Traditional methods of water treatment, distribution, and discharge are increasingly unsustainable due to their extensive dependence on centralized infrastructure. Industrialization serves as the primary catalyst for anthropogenic activities, significantly influenced by population growth and urban expansion (Kasbaji et al., 2023b; Zhang et al., 2016). Industrial effluents laden with organic and inorganic contaminants are frequently released into the environment and aquatic ecosystems, thereby jeopardizing both aquatic organisms and human health. For instance, phenolic compounds originating from pharmaceutical or textile industry effluents are commonly toxic and hazardous, even at minuscule concentrations (Kasbaji et al., 2023a). They possess the potential to induce severe skin and eye irritation upon exposure, while prolonged exposure may lead to liver and heart failure. Other toxic organic pollutants encompass polyaromatic hydrocarbons, pesticides, insecticides, and chlorinated hydrocarbons, whereas toxic inorganic pollutants include mercury, arsenic, cadmium, and lead (Kabir et al., 2014; Kasbaji et al., 2023c; Wang et al., 2019). Consequently, it is essential to develop methodologies for the removal of these toxic contaminants from wastewater to ensure the maintenance of a healthy lifestyle and environment.

Various approaches to industrial effluent treatment exist in practice. The most prominent include reverse osmosis, ion exchange, electrodialysis, electrolysis, and adsorption (Crini and Lichtfouse, 2019; Kabir et al., 2014; Peramune et al., 2022; Russo et al., 2021). Adsorption distinguishes itself as particularly efficient, economical, and accessible for wastewater processing (Kasbaji et al., 2023b; Kausar et al., 2022; Roa et al., 2021; Rodoshi et al., 2024). The financial investment for wastewater treatment through adsorption ranges approximately from 10 to 200 USD per million liters (AlAqad et al., 2024). The adaptability of adsorption technology allows its application across nearly all wastewater contaminant types, whether dissolved or undissolved, organic or inorganic, or biological in nature, with removal rates exceeding 90% (Rajendran et al., 2022). The adsorption mechanism functions as a surface interaction requiring an effective adsorbent material with strong attraction (either physical or chemical) to wastewater pollutants. Ideal adsorbents possess several key characteristics: non-toxicity, environmental compatibility, substantial adsorption capacity, high selectivity for target contaminants, and straightforward recyclability. Notably, nanomaterials including carbon nanotubes, graphene, and metal oxides exhibit these desirable adsorbent properties and have demonstrated exceptional performance as adsorption materials (Ding et al., 2025; Kasbaji et al., 2023c; Peng et al., 2025).

Aquatic environments become contaminated with heavy metal ions such as Pb^{2+}, Cd^{2+}, Cr^{3+}, Cr^{6+}, Ni^{2+}, Hg^{2+}, Cu^{2+}, Zn^{2+}, and Co^{2+} through natural sources like parent rocks and minerals, as well as human activities including agriculture, energy production, sewage disposal, and metallurgical operations (AlAqad et al., 2024; Peng et al., 2025). These metals can enter human biological systems through contaminated drinking water and food chains, potentially causing serious health complications (Mitra et al., 2022; Singh and Kostova, 2024; Singh et al., 2024; Zaynab et al., 2022). The cosmetics, leather, paper, pharmaceutical, textile, rubber, and food industries regularly utilize dyes, discharging substantial amounts of harmful synthetic colorants in their wastewater. These discharges contribute to significant health issues in humans and environmental degradation. Most dyes and their breakdown products exhibit carcinogenic and highly toxic properties (Kasbaji et al., 2023a). Therefore, purifying dye-contaminated water before environmental release is crucial for reducing toxicity, preventing water pollution, and protecting ecological systems (Farhan Hanafi and Sapawe, 2020; Kapanga et al., 2024). The removal of heavy metals and dyes through adsorption techniques, alongside photocatalytic breakdown of dyes, offers high efficiency, cost-effectiveness, straightforward operational requirements, and regeneration capability suitable for large-scale water and wastewater treatment implementation. Employing cost-effective yet efficient adsorbents and photocatalysts can significantly enhance these processes' economic feasibility (Akpomie and Conradie, 2020; Kasbaji et al., 2022a; Upadhyay et al., 2021). The development of economically viable yet highly efficient adsorbents and bioadsorbents has garnered significant attention as a strategic pathway to enhance the cost-effectiveness of water treatment technologies. In recent years, a growing number of studies have emphasized the adoption of adsorption processes as a sustainable and high-performance approach. This method not only offers operational simplicity and low energy requirements but also demonstrates strong potential for the removal of a wide range of contaminants, thereby making it a compelling solution for advanced water purification systems (Figure 4.1).

This chapter provides a comprehensive overview of adsorption techniques in liquid waste treatment, focusing on the various types of adsorbent materials, adsorption mechanisms, and their practical applications. It delves into the integration of different adsorbents, such as AC, bioadsorbents, and inorganic materials, and their role in addressing contaminants in wastewater. The discussion highlights real-world case studies and applications across industries, while also exploring the challenges faced in scaling up these technologies. The chapter concludes with an analysis of the environmental and economic impacts of adsorption-based treatment methods and outlines potential future advancements in the field.

4.2 TYPES OF ADSORBENT MATERIALS

Adsorbent materials are key components in the adsorption process, where contaminants from liquid waste adhere to their surface. The selection of adsorbent materials depends on various factors, including the type of contaminants, the physical and chemical properties of the adsorbent, and the specific application requirements. Various materials have been explored for their adsorptive properties, with distinct advantages and limitations for wastewater treatment. These materials can be broadly classified into several categories, including AC, inorganic materials, bioadsorbents, and composite materials (Chen et al., 2022; Farhan Hanafi and Sapawe, 2020; Kapanga et al., 2024; Perez-Calderon et al., 2023; Rajendran et al., 2022; Salem et al., 2022; Singh et al., 2024).

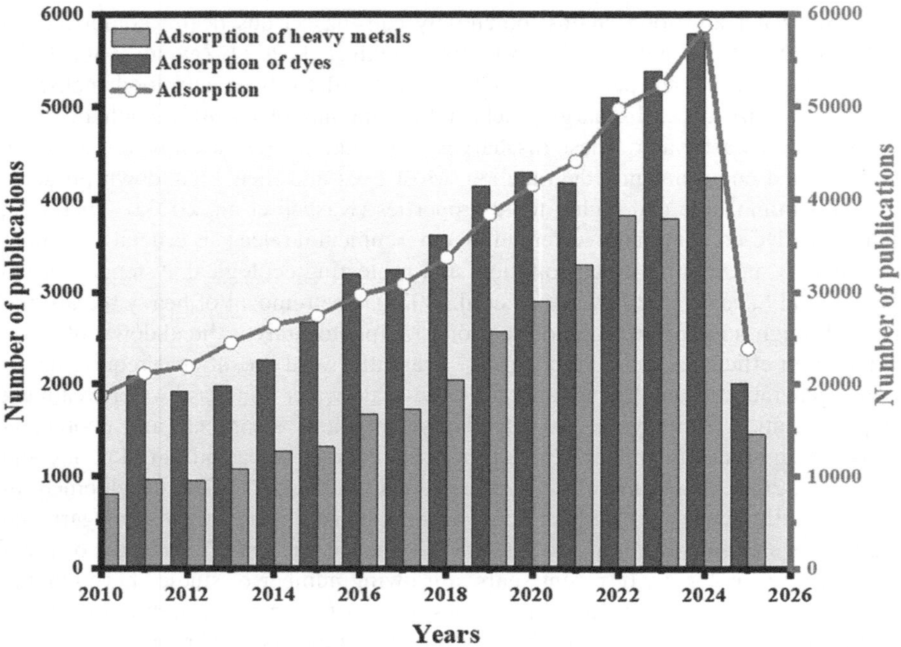

FIGURE 4.1 Illustration of the number of scientific publications by years of adsorption process, adsorption of dyes, and adsorption of heavy metals using the search terms "adsorption, adsorption of dyes, and adsorption of heavy metals." Data analysis completed using Scopus search system on April 15, 2025.

4.2.1 AC

AC is characterized as an amorphous substance, notable for its exceedingly porous architecture and a diverse array of pore dimensions. Acclaimed for its microporous nature, extensive surface area, and significant surface reactivity, AC finds widespread utilization in environmental contexts, particularly in the treatment of wastewater, attributable to its oxidized surface carbon atoms that significantly enhance adsorption efficacy (Ani et al., 2020; Azam et al., 2022; Kapanga et al., 2024; Mohan and Pittman, 2006). Figure 4.2 delineates the principal attributes of AC. The surface chemistry and pore architecture of AC are primarily influenced by the nature of its precursor material, the methodology employed for synthesis, and the conditions under which activation occurs.

The fabrication of AC is chiefly accomplished through either physical or chemical activation methods. Both strategies confer porosity to carbonaceous materials initially characterized by low surface area, transforming them into structures encompassing micropores, mesopores, and ultra-micropores, achieving surface areas of up to 2,000 m^2g^{-1} (Ani et al., 2020; Azam et al., 2022; Kapanga et al., 2024; Mohan and Pittman, 2006; Yi et al., 2021). Furthermore, AC can be categorized according to its physical manifestation: granular, pelletized, fibrous, or fabric-based (Carvalho et al.,

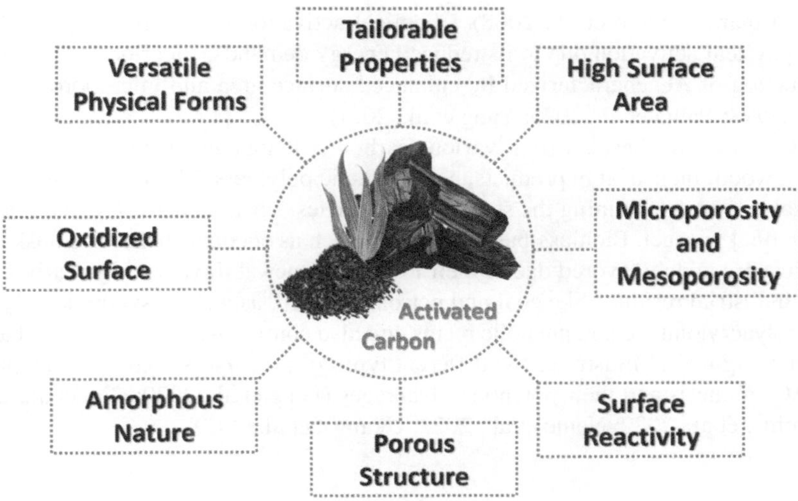

FIGURE 4.2 Main characteristics of AC.

2006; Dias et al., 2007; Huang et al., 2021; Nik et al., 2006; Onyeji and Aboje, 2011; Phiri et al., 2023; Wang et al., 2023).

The physical activation technique encompasses two principal phases: carbonization and activation. Initially, the precursor undergoes carbonization to produce char, a substance characterized by diminished volatile content and augmented fixed carbon (Pallarés et al., 2018; Pet et al., 2024; Yi et al., 2021). During this phase, the reorganization of carbon atoms into a semi-graphitic configuration initiates the development of porosity (Daud et al., 2000; Feng et al., 2020; Phiri et al., 2023). Subsequently, this char is subjected to activation at approximately 700°C utilizing steam, CO_2, or a combination of both, resulting in enhanced porosity through the selective elimination of reactive carbon atoms (Al-Qodah and Shawabkah, 2009). It is noteworthy that CO_2 activation yields more uniform pore architectures in comparison to steam activation. The conditions of activation, including temperature and gas composition, exert a significant influence on the resultant structural integrity and porosity of the AC (Ani et al., 2020; Azam et al., 2022; Girijan et al., 2021; Lee et al., 2022).

Conversely, chemical activation, often referred to as wet-oxidation, involves the impregnation of the precursor with a chemical reagent, followed by thermal treatment in an inert atmosphere at comparatively lower temperatures (300°C–800°C) (Feng et al., 2020; Gao et al., 2020). This methodology fosters dehydration and oxidation of the biomass, thus facilitating pyrolytic decomposition and augmenting carbon yield while concurrently minimizing the formation of tar and ash. Commonly employed activating agents include $ZnCl_2$, H_3PO_4, NaOH, KOH, K_2S, KCNS, H_2SO_4, HNO_3, H_2O_2, and K_2CO_3. These agents also engender surface oxygen functional groups, thereby enhancing the material's adsorption capabilities (Gao et al., 2020; Gayathiri et al., 2022; Neme et al., 2022; Ukanwa et al., 2019). Following impregnation, the resultant material is subjected to washing with acidic or alkaline solutions to eliminate residual chemicals, thereby augmenting porosity (Ahmadpour and Do,

1997; Linares-Solano et al., 2008). Chemical activation is preferentially employed over physical activation due to its reduced energy demands, superior yields, and the production of AC characterized by enhanced surface area and microporosity (Tsai et al., 1998; Yahia et al., 2025; Yang et al., 2021).

AC can be synthesized from various carbon-rich sources, including agricultural waste, wood, industrial byproducts, and synthetic polymers. These precursors play a crucial role in determining the structural properties, surface chemistry, and porosity of the final product. Biomass-based materials such as coconut shells, rice husks, and sawdust are widely favored due to their low cost, renewability, and high carbon content. Industrial residues like coal and petroleum pitch, as well as synthetic polymers like polyacrylonitrile and phenolic resins, are also commonly used for tailored applications. Figure 4.3 illustrates the different types of precursors commonly employed for AC synthesis and their potential advantages (Feng et al., 2020; Gao et al., 2020; Gayathiri et al., 2022; Neme et al., 2022; Ukanwa et al., 2019).

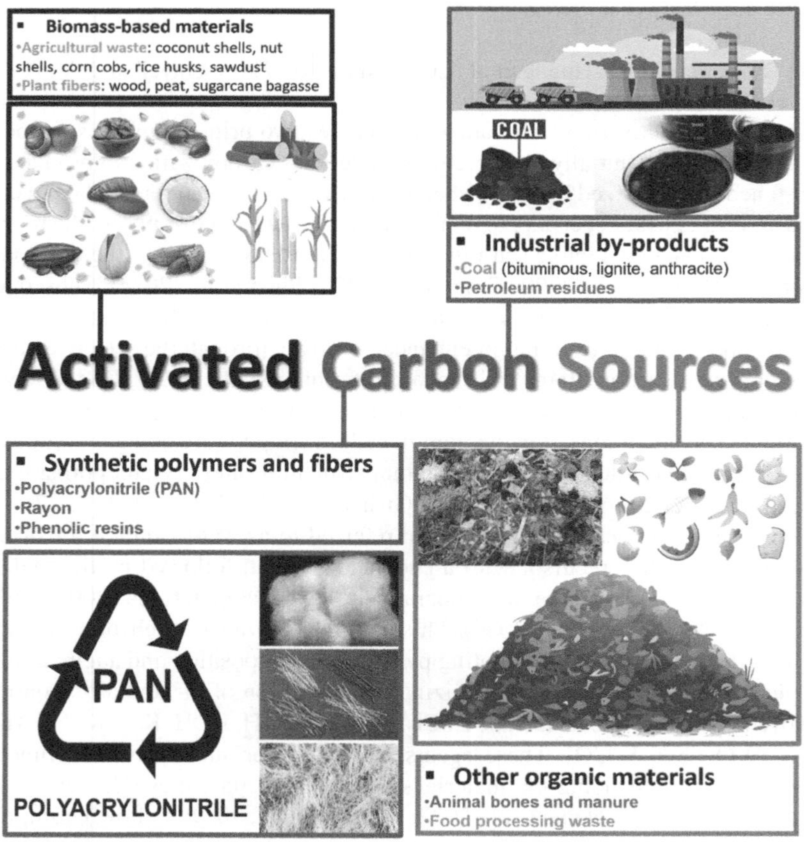

FIGURE 4.3 Raw material sources for AC production.

4.2.2 Inorganic Adsorbents

Inorganic materials such as zeolites, clay minerals, and metal oxides are also widely employed as adsorbents for wastewater treatment. These materials offer high stability, ease of regeneration, and the ability to adsorb heavy metals, phosphates, and other inorganic pollutants (Bahmanzadegan and Ghaemi, 2025; Dehmani et al., 2024; Ewis et al., 2022; Iqbal et al., 2022; Kasbaji et al., 2023a).

Zeolites, for example, are crystalline aluminosilicates with a well-defined porous structure that makes them highly efficient in ion exchange and selective adsorption (Bahmanzadegan and Ghaemi, 2025; Pérez-Botella et al., 2022). Zeolites are classified as inorganic polymers composed of a tetrahedral framework, where each unit is represented as TO_4. In this configuration, "T" may correspond to trivalent elements such as aluminum (Al), boron (B), or gallium (Ga); tetravalent elements like silicon (Si) or germanium (Ge); or pentavalent elements such as phosphorus (P). These TO_4 units are interconnected through shared oxygen atoms, forming robust three-dimensional networks of $[SiO_4]^{4-}$ and $[AlO_4]^{4-}$ tetrahedra (Bahmanzadegan and Ghaemi, 2024). The negative charge of these frameworks attracts various cations, which reside within the rings of the zeolite structure. The type, quantity, and distribution of these cations and their associated active sites play a crucial role in determining the adsorption and ion exchange properties of zeolites. Figure 4.4a illustrates several zeolite structures in ball-and-stick models. Due to their distinctive physicochemical properties, zeolites have found wide-ranging applications, including in wastewater treatment, agriculture (Kordala and Wyszkowski, 2024), fertilizers (Conversa et al., 2024), animal feed (El-Nile et al., 2023), gas separation (Yue et al., 2022), solar adsorption refrigeration (Tchernev, 2001), deodorization, and gas pollutant removal (Garofalo et al., 2023), and as solid electrolytes in construction materials (Chi et al., 2021). Moreover, zeolites are appreciated for their high thermal stability, their capacity to adsorb and incorporate aluminum ions, and their economic viability (Choi et al., 2009; Sayari et al., 2011). Figure 4.4b provides an overview of four commonly used synthesis methods for producing zeolite structures, each depicted with distinct framework representations surrounding a central structure. These methods, microwave irradiation synthesis, hydrothermal synthesis, template-assisted synthesis, and seed-assisted synthesis, present different strategies for tailoring zeolites' morphology, porosity, and structural characteristics to suit specific applications (Bahmanzadegan and Ghaemi, 2025).

Clay minerals, including bentonite and montmorillonite, are inexpensive and abundant adsorbents used to remove heavy metals and organic contaminants due to their high cation-exchange capacity (Xie et al., 2024; Yang et al., 2024). Natural mineral materials are abundantly found in nature and are characterized by their low cost, extensive reserves, and renewability (Hacıosmanoğlu et al., 2022). Clay minerals primarily consist of magnesium (Mg), aluminum (Al), and other hydrated silicate minerals, with their fundamental structural units being silica tetrahedra and aluminum oxide octahedra (Abu-Danso et al., 2020; Xie et al., 2024; Yang et al., 2024). Depending on the structural arrangement of the unit crystal layers, these minerals can form various framework structures, such as rings or cages (Moraes et al., 2017). Among these, montmorillonite, illite, vermiculite, attapulgite, sepiolite,

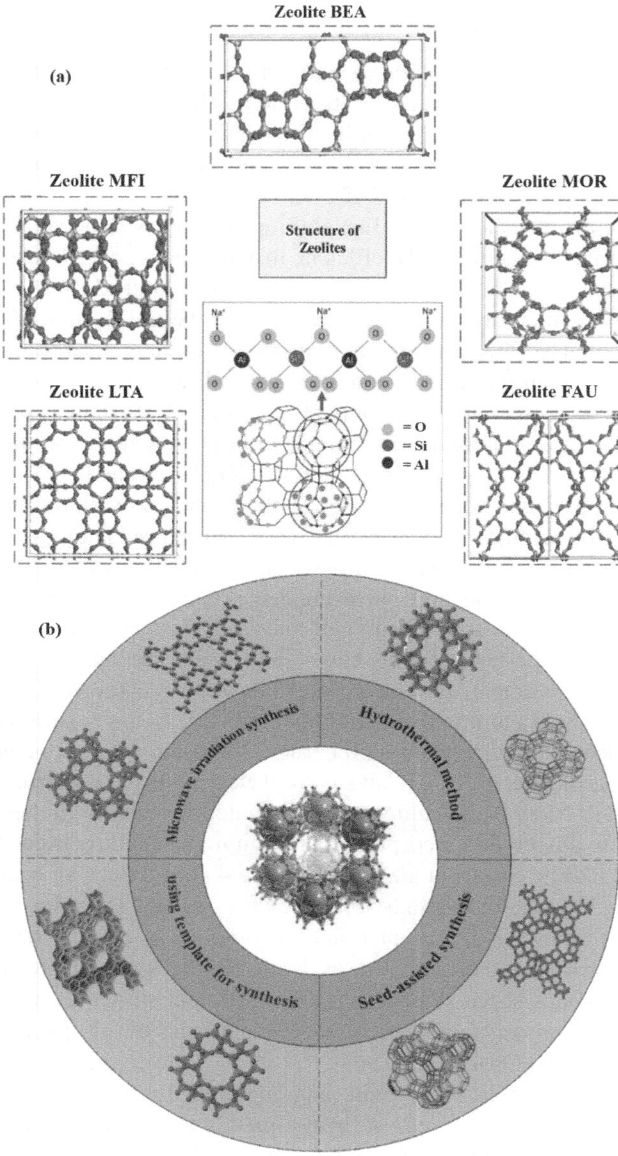

FIGURE 4.4 (a) Zeolite structures and their ball-and-stick representation, and (b) overview of synthesis methods for customizing zeolite structures. (Reprinted with permission from Bahmanzadegan and Ghaemi, 2025.)

and glauconite are examples of minerals with a 2:1 interlayer structure. In contrast, kaolinite and halloysite exhibit a typical 1:1 interstratified structure. Zeolite is an example of a framework structure (Liu et al., 2020; Zhang et al., 2011). A summary of the chemical formulas for different clay minerals is presented in Table 4.1.

TABLE 4.1
Molecular Chemical Formulas of Different Clay Minerals

Clay Mineral Species	Molecular Formula	References
Montmorillonite	$(Na,Ca)_{0\cdot 33}(Al,Mg)_2\underline{Si_4O}_{102}\cdot nH_2O$	Al Kausor et al. (2022)
Illite	$K_{1-1\cdot 5}Al_4\underline{Si}_{6\cdot 5-7}Al_{1-1\cdot 5}O_{204}$	Konan et al. (2012)
Vermiculite	$(Mg,Fe,Al)_8(Si,Al)_4O_{10}(OH)_2\cdot{}_4H_2O$	Ren et al. (2022)
Attapulgite	$(Mg,Al,Fe)_5Si_8O_{20}(OH)_2(OH_2)_4\cdot{}_4H_2O$	Liu et al. (2017)
Sepiolite	$Mg_8Si_{12}O_{30}(OH)_4(H_2O)_4\cdot{}_8H_2O$	Padilla-Ortega et al. (2011)
Glauconite	$(K,Na,Ca)(Fe,Al,Mg,Mn)_2(Si,Al)_4O_{10}(OH)_2$	Singla et al. (2020)
Kaolinite	$Al_2\underline{Si_2O}_{54}$	Cheng et al. (2012)
Halloysite	$Al_2(OH)_4Si_2O_5\cdot nH_2O$	Fu et al. (2017)
Zeolite	$Ax/q[(AlO_2)x(SiO_2)y]\cdot nH_2O$	Johnson and Arshad (2014)

Source: Xie et al. (2024).

4.2.3 Bioadsorbents

Bioadsorbents, derived from natural sources such as algae, fungi, agricultural waste, and plant-based materials, have gained attention due to their sustainability and cost-effectiveness. These materials are particularly advantageous because they are biodegradable, non-toxic, and renewable (Jabli et al., 2023; Kasbaji et al., 2023c; Subash et al., 2023; Yaashikaa et al., 2022).

For instance, algae and fungal biomass can effectively remove heavy metals and organic pollutants from wastewater through biosorption processes (Kasbaji et al., 2022a; Sarma et al., 2024). Algae represent a highly promising resource for wastewater remediation owing to their remarkable ability to adsorb and assimilate various pollutants, including organic substances and heavy metals, from contaminated water streams (Abdelfattah et al., 2023). This bioremediation process not only supports algal biomass proliferation but also offers a synergistic advantage by facilitating both environmental detoxification and sustainable bioenergy production. Salgado et al. (2023) investigated the removal of nitrogen (N) and phosphorus (P) from synthetic wastewater using *Chlorella vulgaris*, employing genetic algorithms alongside an artificial neural network model. Their findings demonstrated nutrient removal efficiencies of $92.0 \pm 0.6\%$ (6.15 ± 0.01 mg L^{-1} d^{-1}) for nitrogen and $98.2 \pm 0.2\%$ (0.92 ± 0.03 mg L^{-1} d^{-1}) for phosphorus. Likewise, Chaieb et al. (2023) reported that the halotolerant bacterium *Shewanella algae* B29, when used in conjunction with 2% cuttlebone, acted as a highly effective natural bio-sorbent for the rapid removal of textile dyes. Under optimized conditions, it achieved a decolorization efficiency of up to 91.04% for CI Reactive Red 66. These environmentally benign technologies hold substantial potential for mitigating aquatic pollution while concurrently contributing to the generation of value-added bioproducts. The principal mechanisms through which microalgae facilitate wastewater treatment involve a range of physicochemical and biological processes that collectively remove contaminants from aqueous systems. Algal systems employ a variety of sophisticated pathways to eliminate hazardous substances during wastewater remediation, including processes such as bioaccumulation, biosorption, and both extracellular and intracellular biotransformation (Bhatt et al., 2022). Figure 4.5a illustrates the diverse mechanisms and synergistic interactions involved in pollutant removal via algae–bacteria and algae–fungi consortia. These interactions are essential for enhancing the efficiency of contaminant remediation from aqueous environments (Rao et al., 2019). Organic pollutants are predominantly addressed through microbial-mediated biodegradation and metabolic transformation, while inorganic species are primarily sequestered by mechanisms such as biosorption and bioaccumulation. Figure 4.5b presents various mechanisms utilized by algal consortia to facilitate the elimination of pollutants from wastewater. The biomass-based approach primarily relies on bioaccumulation,

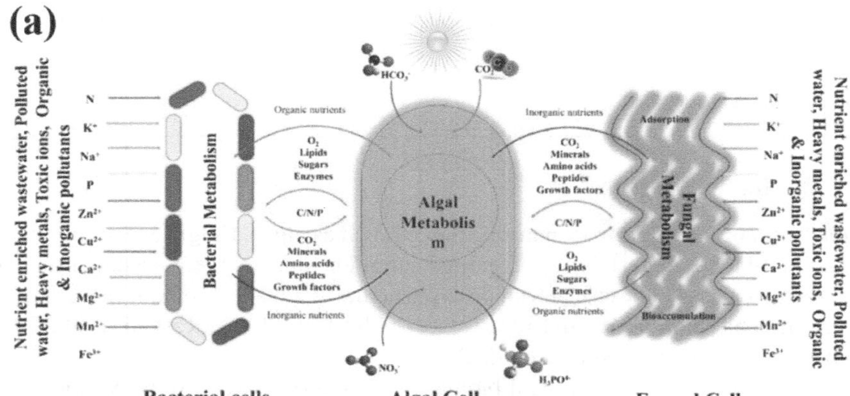

FIGURE 4.5 (a) Synergistic interactions between algae, bacteria, and fungi for enhanced pollutant remediation and (b) strategies employed by algae–fungi–bacteria consortia for efficient wastewater pollutant removal. (Reprinted with permission from Singh Mehra et al., 2025.)

enabling the effective sequestration of nitrogenous, carbonaceous, and phosphorous compounds, as well as heavy metals typically present in agricultural and industrial effluents. Adsorption operates through the physicochemical attachment of pollutants such as suspended particles and surfactants onto the surface of the algal biomass. Meanwhile, biosorption functions via passive binding mechanisms, allowing for the capture of dyes, phenolic compounds, and other organic contaminants through surface chemical interactions.

Agricultural wastes, including fruit peels, rice husks, and tea leaves, have also been investigated for their ability to adsorb a variety of contaminants (Figure 4.6). These bioadsorbents are often considered as an eco-friendly alternative to traditional adsorbents (Ishaq et al., 2023; Karić et al., 2022; Kasbaji et al., 2023b). Globally, a significant portion of agricultural residues is either incinerated or discarded without proper utilization, resulting in the loss of valuable resources and contributing to environmental degradation (Ishaq et al., 2023; Karić et al., 2022; Parthasarathy and Narayanan, 2014). These agro-wastes are composed of a wide range of organic constituents, including water, hydrocarbons, proteins, lignin, hemicellulose, lipids, and starch. The presence of diverse functional groups within these materials imparts a strong potential for adsorbing various pollutants (Kainth et al., 2024). In terms of composition, typical agricultural residues contain approximately 35–50% cellulose,

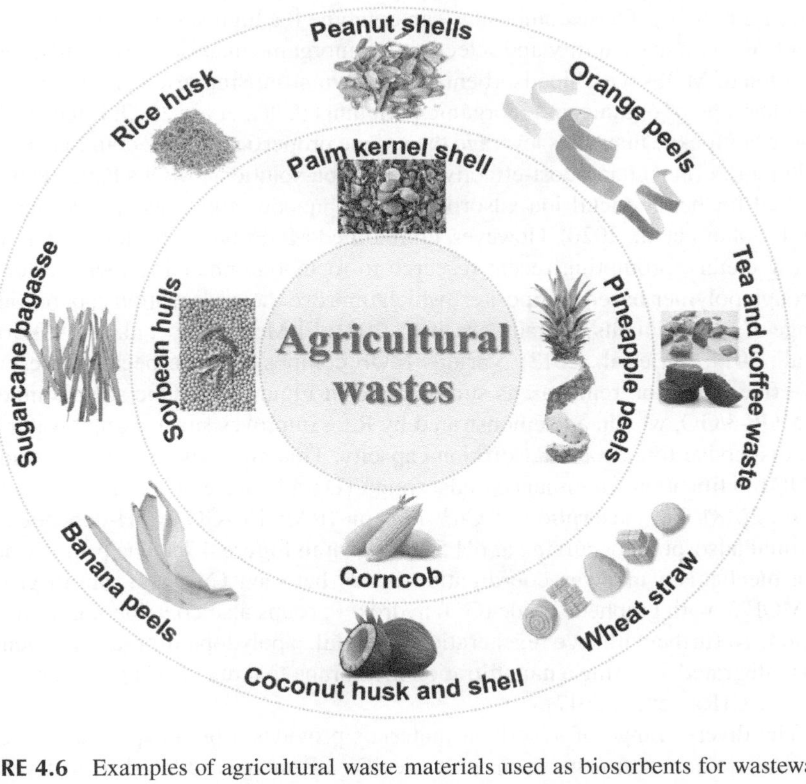

FIGURE 4.6 Examples of agricultural waste materials used as biosorbents for wastewater treatment.

20–30% lignin, and 15–30% hemicellulose, positioning lignocellulosic biomass as an abundant and renewable resource (AlAqad et al., 2024; Bortoluz et al., 2020; Gupta et al., 2021; Kasbaji et al., 2025, 2023c). Cellulose, a linear syndiotactic polymer of glucose, is formed through β-1,4-glycosidic linkages. Hemicellulose, on the other hand, is a branched heteropolymer comprising several uronic acid derivatives, such as D-glucuronic acid, D-galactose, D-mannose, L-arabinose, D-xylose, and D-glucose. Lignin is a highly crosslinked aromatic polymer composed primarily of coniferyl and p-coumaryl alcohols. The valorization of agricultural residues often begins with the extraction of cellulose, a process that necessitates the removal of both lignin and hemicellulose (Kasbaji et al., 2023c). Furthermore, bioadsorbents can be modified or functionalized to enhance their adsorption capacity, such as through chemical treatment or immobilization of functional groups (Kainth et al., 2024; Kasbaji et al., 2023c).

4.2.4 Composite Adsorbents

Composite adsorbents are materials formed by combining two or more types of adsorbents to enhance the overall adsorption efficiency. For example, the incorporation of AC with clays or biochar has been explored to improve the adsorptive properties and reduce regeneration costs (AlAqad et al., 2024; Kasbaji et al., 2025; Satyam and Patra, 2024). These composites can combine the high surface area and reusability of AC with the stability and selectivity of inorganic materials. Similarly, the combination of MOFs with bioadsorbents has shown promising results for the removal of a wide range of organic and inorganic pollutants (Chen et al., 2023; Liu et al., 2023). These composite materials leverage the unique properties of their components to provide more efficient and cost-effective adsorption solutions. MOFs have been widely studied for heavy metal ion adsorption from aqueous solutions (Daradmare et al., 2021; Soltani et al., 2020). However, unmodified MOFs have limitations in removing heavy metals, prompting recent research to focus on enhancing their performance through polymer-based composites, which improve their adsorption capabilities for a range of contaminants (Daradmare et al., 2021; El-Mehalmey et al., 2018; Furukawa et al., 2014; Siu et al., 2013). Various MOF composites have been synthesized for effective metal ion removal, as summarized in Figure 4.7a. One such composite is IRMOF-3/GO, which, as demonstrated by Rao, improves surface area, water stability, dispersive forces, and adsorption capacity. This aligns with previous studies on MOF modification for enhanced adsorption (Daradmare et al., 2021; El-Mehalmey et al., 2018). The adsorption of Cu^{2+} ions on IRMOF-3/GO is pH-dependent, with optimal adsorption occurring at pH 5, as shown in Figure 4.7b. The primary adsorption mechanism involves coordination bonds between Cu^{2+} and amino groups in IRMOF-3, with graphene oxide (GO) hydroxyl groups also contributing to this interaction. To further enhance regeneration potential, a polydopamine-coated membrane was integrated, creating a nanofiltration membrane that rejects Cu particles, facilitating reuse (Rao et al., 2017).

The diverse range of adsorbent materials provides a broad spectrum of options for wastewater treatment, each with specific advantages tailored to different contaminants. While AC remains one of the most popular adsorbents, inorganic

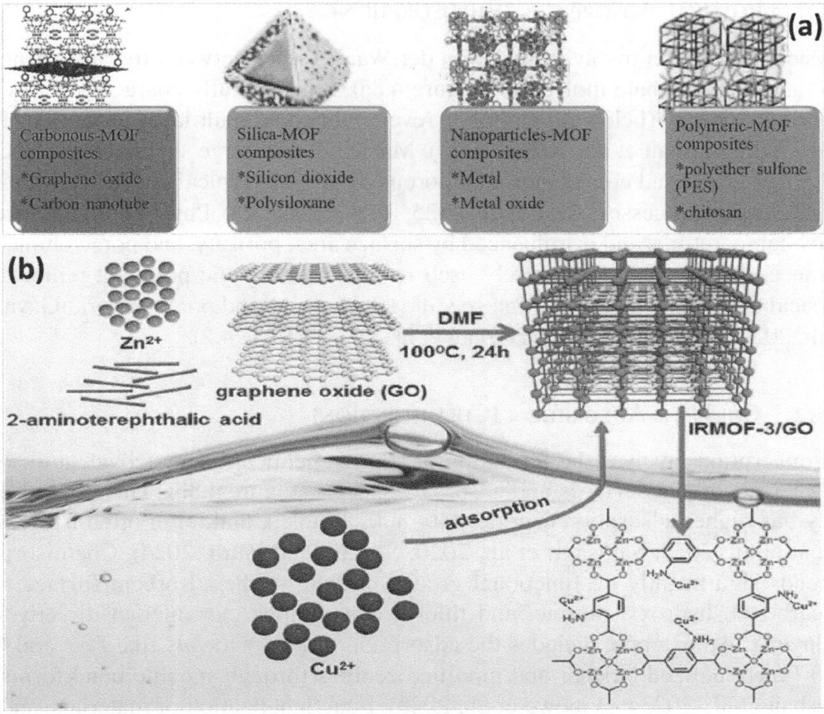

FIGURE 4.7 (a) Overview of MOF-based composites reported for heavy metal adsorption and (b) illustration of IRMOF-3/GO synthesis and its Cu^{2+} removal mechanism. (Reprinted with permission from Liu et al., 2023.)

materials, bioadsorbents, and composite adsorbents offer viable alternatives, particularly for sustainable and cost-effective treatment solutions. Ongoing research continues to explore ways to improve the performance, reusability, and environmental impact of these materials, contributing to the advancement of wastewater treatment technologies.

4.3 ADSORPTION MECHANISMS

Adsorption is a fundamental process in liquid waste treatment, playing a crucial role in the removal of pollutants through the interaction between adsorbates and the surfaces of adsorbents. Understanding the mechanisms behind adsorption is essential for the development of efficient materials and techniques for wastewater purification (Bahmanzadegan and Ghaemi, 2025; Pet et al., 2024; Satyam and Patra, 2024). Adsorption mechanisms are broadly classified into physical adsorption (physisorption), chemical adsorption (chemisorption), ion exchange, and complexation or coordination interactions, each depending on the nature of the adsorbent, the adsorbate, and the environmental conditions (Foo and Hameed, 2010; Molina-Calderón et al., 2022; Satyam and Patra, 2024; Wang and Guo, 2020).

4.3.1 Physical Adsorption (Physisorption)

Physical adsorption involves weak van der Waals forces between the adsorbent surface and the adsorbate molecules (Figure 4.8a). It is generally characterized by low adsorption energy (below 40 kJ mol^{-1}), reversibility, and multilayer formation (Feng et al., 2010; Kasbaji et al., 2024, 2022b). Materials with large surface areas, such as AC, graphene-based composites, and porous MOFs, are typically effective in physical adsorption processes (Arya et al., 2025; Rout et al., 2023). Physisorption is favored at low temperatures and is influenced by surface area, porosity, and pore volume. For instance, AC has been extensively used for dye and organic pollutant removal via physical adsorption due to its high specific surface area and pore network (Gayathiri et al., 2022; Nizam et al., 2021; Ozcan et al., 2024) (Table 4.2).

4.3.2 Chemical Adsorption (Chemisorption)

Chemisorption involves the formation of strong chemical bonds (covalent or ionic) between adsorbent surfaces and adsorbate species (Figure 4.8b). This process typically has higher adsorption energies (40–800 kJ mol^{-1}), making it often irreversible (Adjal et al., 2023; Samsami et al., 2020; Satyam and Patra, 2024). Chemisorption depends significantly on functional groups present on the adsorbent surface, such as carboxyl, hydroxyl, amine, and thiol groups, which can interact directly with pollutants. An example includes the adsorption of heavy metals like Pb^{2+} and Cu^{2+} onto functionalized biochar and modified zeolites through specific bond formation (Kasbaji et al., 2023c; Wibowo et al., 2025). Functionalization of materials, such as

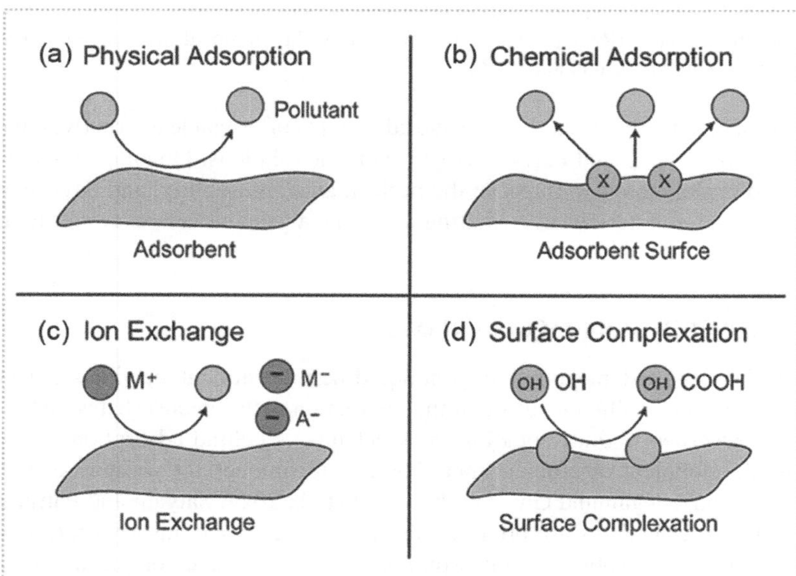

FIGURE 4.8 Different mechanisms involved in the adsorption process: (a) physisorption, (b) chemisorption, (c) ion exchange, and (d) surface complexation.

TABLE 4.2
Adsorption Mechanisms with Characteristics and Material Examples

Mechanism	Characteristics	Examples of Materials	References
Physisorption	Weak van der Waals forces, reversible	AC, MOFs, graphene	Arya et al. (2025), Rout et al. (2023)
Chemisorption	Strong chemical bonds, often irreversible	Functionalized biochar, amine-modified GO	Wibowo et al. (2025), Wu et al. (2022)
Ion exchange	Replacement of surface ions with pollutant ions	Zeolites, resins, modified clays	Wang and Peng (2010), Wibowo et al. (2025)
Surface complexation	Formation of coordinate covalent bonds	GO composites, modified MOFs	Shahabi et al. (2024) Zheng et al. (2022)

introducing amine groups onto GO or biochar, enhances chemisorption by increasing the number of active sites available for binding (Samuel Olugbenga et al., 2024) (Table 4.2).

4.3.3 Ion Exchange

Ion exchange is a specific type of chemisorption where ions from the solution are exchanged with counter-ions on the surface of the adsorbent (Figure 4.8c). Materials such as zeolites, ion exchange resins, and some modified clays operate primarily through this mechanism (Pérez-Botella et al., 2022; Wang and Peng, 2010). This process is particularly important for the removal of heavy metal ions and ammonium from aqueous solutions. For example, natural zeolites have shown efficient removal of heavy metals like Cd^{2+} and Zn^{2+} by exchanging cations present in their structure with metal ions in wastewater (Bahmanzadegan and Ghaemi, 2025; Wang and Xu, 2023) (Table 4.2).

4.3.4 Surface Complexation and Coordination

Surface complexation involves the formation of complexes between adsorbate molecules and functional groups on the adsorbent surface, often through coordinate covalent bonds (Figure 4.8d). This mechanism is significant for the adsorption of heavy metals and organic molecules containing functional groups capable of complexation, such as carboxylic acids and amines (Hiemstra and van Riemsdijk, 1996). For instance, in the case of GO/IRMOF-3 composites, Cu^{2+} ions coordinate with amino groups present on the adsorbent, enhancing the adsorption process (Shahabi et al., 2024; Zheng et al., 2022) (Rao et al., 2017). This mechanism often leads to specific, selective adsorption, depending on the chemical affinity between the pollutant and the active sites on the adsorbent (Ezzeddine et al., 2025) (Table 4.2).

4.3.5 Multiple Mechanisms Working Simultaneously

In many innovative adsorbents, multiple adsorption mechanisms can coexist, leading to synergistic effects that improve overall adsorption performance. For example, in biomass-derived ACs functionalized with nitrogen or sulfur groups, both physisorption (via pore filling) and chemisorption (via functional groups) contribute to contaminant removal (Gao et al., 2022b; Tan et al., 2015). Similarly, nanocomposite materials such as MOFs integrated with GO or polymers often leverage both physical and chemical interactions, allowing for higher adsorption capacities and selectivity (Arya et al., 2025; Lin et al., 2023).

4.3.6 Factors Influencing Adsorption Mechanisms

The adsorption process is governed by several interconnected factors that directly impact the efficiency, capacity, and mechanism of pollutant removal in liquid waste treatment. A thorough understanding of these factors is crucial for optimizing material performance and tailoring adsorption systems for specific applications (Kasbaji et al., 2024).

Surface area and porosity are fundamental to adsorption performance. Materials with high specific surface areas and well-developed porous structures, such as ACs and MOFs, provide more active sites for adsorption, leading to enhanced pollutant uptake. Additionally, microporous and mesoporous structures facilitate the diffusion and retention of contaminants, improving the adsorption rate and capacity (Arya et al., 2025; Foo and Hameed, 2010; Wang and Guo, 2020).

Surface functional groups also play a significant role by enabling specific interactions, such as electrostatic attraction, hydrogen bonding, or complexation with pollutants. For example, oxygen-containing groups on biochar or hydroxyl groups on GO enhance interactions with heavy metals and organic molecules (Kasbaji et al., 2023c; Tan et al., 2015).

pH of the solution profoundly affects adsorption by altering both the ionization state of the pollutants and the surface charge of the adsorbent. At lower pH levels, competition with protons can hinder the adsorption of cationic contaminants, while at higher pH, deprotonation of functional groups can enhance binding capacity. For instance, the removal of heavy metals like Pb^{2+} and Cu^{2+} by chitosan-based adsorbents is significantly pH-dependent (Crini and Badot, 2008; Kasbaji et al., 2023b).

Temperature influences adsorption kinetics and thermodynamics. In endothermic adsorption processes, such as dye uptake by GO composites, increasing the temperature typically enhances the adsorption capacity and diffusion rates (Zhao et al., 2011). However, for exothermic systems, higher temperatures may reduce adsorption efficiency (Kasbaji et al., 2022a; Zhao et al., 2011).

Initial pollutant concentration affects the mass transfer driving force across the interface between the solution and the adsorbent. Higher initial concentrations often lead to greater adsorption rates due to an increased gradient, but can also cause quicker saturation of active sites if the adsorbent capacity is limited (Aurich et al., 2017; Kasbaji et al., 2022b).

Contact time determines whether adsorption equilibrium can be achieved. Fast adsorption typically indicates physical adsorption or surface reactions, while longer times suggest intra-particle diffusion or chemisorption mechanisms (Ding et al., 2002; Kasbaji et al., 2023a).

TABLE 4.3
Factors Influencing Adsorption Mechanisms

Factor	Effect on Adsorption	Example Material	References
Surface area and porosity	Increased active sites and diffusion pathways	AC, MOFs	Foo and Hameed (2010)
Surface functional groups	Specific binding via electrostatic forces, hydrogen bonding, or complexation	Biochar, GO	Tan et al. (2015)
pH of the solution	Alters the adsorbent surface charge and pollutant speciation	Chitosan for metal removal	Crini and Badot (2008)
Temperature	Enhances or reduces adsorption depending on the system	GO for dye adsorption	Zhao et al. (2011)
Initial pollutant concentration	Affects the mass transfer driving force and saturation	Zeolites for ammonium adsorption	Ho and McKay (1999)
Contact time	Governs equilibrium and kinetic rate	AC for organic compounds	Ding et al. (2002) Kasbaji et al. (2023a)
Competing ions	Reduces the availability of active sites	Chitosan-based adsorbents for multimetal systems	Crini and Badot (2008)

Finally, the presence of competing ions in the solution can significantly impact adsorption efficiency. In multi-component systems, competitive adsorption can occur, where different ions vie for the same binding sites. This effect has been widely observed, for example, during the adsorption of heavy metals by chitosan and modified clays (Crini and Badot, 2008; Kasbaji et al., 2024). A detailed summary of these factors with examples and references is provided in Table 4.3.

Advances in material science, particularly the functionalization of bio-based materials, hybrid nanocomposites, and smart materials, are opening new pathways to design adsorbents tailored for specific mechanisms. Stimuli-responsive adsorbents that change surface properties upon external triggers (e.g., pH, light, magnetic fields) are particularly promising for enhancing selectivity and regeneration. Moreover, computational modeling and molecular simulations are increasingly used to predict adsorption behaviors and optimize adsorbent designs for targeted applications (AlAqad et al., 2024; Bonev and Alexandrov, 1993; Ezzeddine et al., 2025; Parthasarathy and Narayanan, 2014; Pet et al., 2024).

4.4 APPLICATIONS IN LIQUID WASTE TREATMENT

The application of adsorption techniques for liquid waste treatment has gained significant attention due to their simplicity, efficiency, and cost-effectiveness. With the rising demand for sustainable water management, innovative adsorbent materials and optimized mechanisms are continuously developed to target diverse contaminants, including heavy metals, dyes, pharmaceuticals, and organic pollutants (Crini and Lichtfouse, 2019; Donkadokula et al., 2020; Gupta and Nayak, 2012). Adsorption technology offers advantages such as high removal efficiency, operational flexibility, and the possibility of adsorbent regeneration and reuse. Recent advancements include the development of novel materials such as MOFs, biochar composites, graphene derivatives, and polymer-functionalized surfaces (Gatou et al., 2021; Selvasembian et al., 2024).

4.4.1 Removal of Heavy Metals

Heavy metals like lead (Pb^{2+}), cadmium (Cd^{2+}), chromium (Cr^{6+}), and arsenic (As^{3+}) are among the most toxic water pollutants. Their persistence and bioaccumulation necessitate efficient removal techniques (El-Gendy et al., 2017; Kasbaji et al., 2023c). Biochar-based adsorbents functionalized with amine or thiol groups have shown remarkable affinity toward heavy metals due to the formation of surface complexes and ion exchange mechanisms (Koyuncu and Kul, 2020; Li et al., 2022). Similarly, MOFs like ZIF-8 and UiO-66, due to their high surface area and tunable chemistry, demonstrate superior performance for metal ion capture (Than-ardna et al., 2025; Zango et al., 2021; Zhai et al., 2023) (Table 4.4).

4.4.2 Removal of Organic Dyes

Dyes such as methylene blue (MB), rhodamine B, and malachite green are widely found in industrial effluents. Their strong color and toxicity make their removal a priority. AC remains a benchmark material due to its high surface area and porous structure (Al-Asadi et al., 2025; Gebre Meskel et al., 2024). Recent developments focus on graphene-based materials and magnetic nanocomposites that allow easy recovery after adsorption (Behroozi et al., 2025; Saurel et al., 2018; Vaz-Ramos et al., 2025). For example, polydopamine and polyethyleneimine co-functionalized GO aerogels exhibit high adsorption capacities for anionic dyes through hydrogen bonding interactions with nitrogen-containing functional groups, demonstrating rapid synthesis and effective removal performance (Xu et al., 2020). A case study was used to evaluate the performance of AC produced from a 1:1 mixture of corn

TABLE 4.4
Summarizes Selected Materials for Heavy Metal Removal

Material	Target Pollutant	Mechanism	References
Aminated biochar	Cr^{6+}	Surface complexation, ion exchange	El-Nemr et al. (2021)
UiO-66 MOF	Ni^{2+}, Fe^{3+} and Cr^{6+}	Chemical adsorption/ electrostatic interaction	Zhai et al. (2023)
GO-modified chitosan	Pb^{2+}, Cd^{2+} and Ni^{2+}	Electrostatic interaction	Peng et al. (2024)
Zeolite	Cd^{2+}	Cationic exchange	Bai et al. (2025)

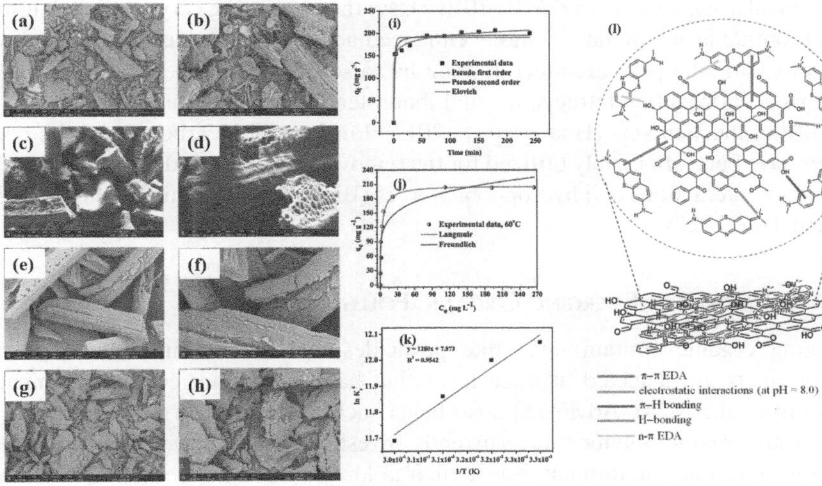

FIGURE 4.9 SEM images of AC1:0-PC (a, b), AC1:1-PC (c,d), AC1:0-C (e,f), and AC1:1-C (g,h), (i) adsorption kinetics for the MB-AC1:1-C system at concentrations of 1 g /L; pH = 8.0; temperature = 25°C, (j) adsorption isotherms for the MB-AC1:1-C system at temperatures of 60°C and fits of Langmuir and Freundlich models (dosage = 1.0 g L 1, pH = 8.0, stirring time = 240 min), (k) Van't Hoff plot for the MB-AC1:1-C system, and (l) adsorption mechanism schematic of the MB-AC1:1-C system. (Reprinted with permission from Andrade et al., 2025.)

husk and sucrose using H_2SO_4 activation. The resulting material, AC1:1-C, exhibited a high monolayer adsorption capacity of 199.4 mg g^{-1} for MB and a strong ISBET-Y index (18,864), indicating both high yield and surface area. Figure 4.9a–h shows the morphological structure via scanning electron microscopy (SEM): microfibers from corn husk and smooth-surfaced fragments from sucrose were observed, confirming their combined contribution to the porous structure. The adsorption performance was significantly affected by pH, with optimal results achieved at pH 8 (Figure 4.9i), where the AC surface became negatively charged, promoting electrostatic interaction with positively charged MB molecules. As illustrated in Figure 4.9j, fast adsorption kinetics were achieved, reaching equilibrium within 60 minutes, and higher dye concentrations led to increased adsorption. Thermodynamic analysis (Figure 4.9k) confirmed that the adsorption process was spontaneous and exothermic. Figure 4.9l depicts the proposed adsorption mechanism, including electrostatic interactions, π–π stacking, n–π electron donor–acceptor, and hydrogen bonding, all of which contributed to effective monolayer dye removal.

4.4.3 Treatment of Pharmaceutical Wastewater

The presence of pharmaceuticals like antibiotics and hormones in water has raised environmental concerns due to their endocrine-disrupting effects. MOFs and molecularly imprinted polymers have emerged as promising materials for the selective removal of pharmaceuticals (Martins et al., 2024; Zare et al., 2022). For example, hierarchically mesostructured MIL-101(Cr), synthesized using cetyltrimethylammonium bromide as a template, exhibits enhanced adsorption kinetics for dye removal due to its trimodal pore architecture, and has also demonstrated significant uptake of pharmaceuticals like tetracycline and ibuprofen via hydrophobic interactions and pore-filling mechanisms (Huang et al., 2012). Furthermore, carbon nanotube composites have been effectively utilized for the removal of sulfonamide antibiotics, relying on π–π interactions and hydrophobic forces (Biswal and Balasubramanian, 2022; Wu and Tsai, 2022).

4.4.4 Removal of Emerging Organic Pollutants

Emerging organic contaminants like pesticides, flame retardants, and personal care products are detected at trace levels but have significant ecological impacts (F. Wang et al., 2024). Advanced adsorbents such as covalent organic frameworks and porous organic polymers are currently investigated for the adsorption of these low-concentration contaminants due to their tailored pore structures and strong binding affinities (Ajay Rakkesh et al., 2024; Mohan et al., 2025).

4.4.5 Comparison of Adsorbent Performance

The selection of an appropriate adsorbent for wastewater treatment requires a comprehensive evaluation of several critical factors. These include not only the adsorption capacity and affinity toward specific pollutants but also broader considerations such as cost-effectiveness, regeneration potential, selectivity for targeted contaminants, and environmental sustainability (Table 4.5). Adsorbents derived from natural materials, such as biochars, clay minerals, and biomass-based composites, are often favored for their low cost, abundant availability, and eco-friendly nature (Kasbaji et al., 2023a). However, their performance can be limited by lower surface areas and weaker interaction mechanisms compared to engineered alternatives. In contrast, advanced materials such as functionalized graphene-based composites, magnetic nanomaterials, and MOFs typically exhibit superior adsorption efficiencies, rapid kinetics, and high selectivity due to tailored surface functionalities and porous structures. Yet, their high synthesis cost and potential environmental toxicity raise concerns for large-scale applications. Regeneration and reuse are also crucial aspects. Adsorbents with robust structural stability and chemical resilience under desorption conditions contribute significantly to sustainable and cost-efficient treatment cycles. For instance, magnetic adsorbents allow for easy recovery using external magnetic fields, while hydrogels and MOFs with tunable porosity can be regenerated through mild solvent washing or pH adjustment. Ultimately, the choice of adsorbent must strike a balance between adsorption performance, operational feasibility, environmental compatibility, and economic viability. Comparative studies under realistic conditions are essential to determine the most suitable material for specific industrial or municipal wastewater treatment scenarios (Arya et al., 2025; Chen et al., 2023; Gao et al., 2022a; Raj et al., 2024; Zhang et al., 2021).

TABLE 4.5
Comparison of Material Performance in Liquid Waste Treatment

Adsorbent Type	Target Pollutants	Advantages	Limitations	References
AC	Dyes, metals, organics	High efficiency, commercial availability	Expensive regeneration	Gayathiri et al. (2022), Koyuncu and Kul (2020), Li et al. (2020), Peng et al. (2025)
Biochar	Metals, organics	Low cost, eco-friendly	Lower capacity than AC	Alsawy et al. (2022), El-Nemr et al. (2021), Li et al. (2022), Wibowo et al. (2025), Zhou et al. (2017)
MOFs	Metals, pharmaceuticals	High selectivity, large surface area	Stability issues in water	Arya et al. (2025), Daradmare et al. (2021), Yan and Ge (2023), Zango et al. (2021), Zheng et al. (2022)
Graphene derivatives	Dyes, metals	High adsorption rates	Aggregation in water	Hu et al. (2020), Liu et al. (2023), Rao et al. (2017)
Nanocomposites (magnetic)	Dyes, pharmaceuticals	Easy separation, multifunctionality	Potential nanotoxicity risks	Daradmare et al. (2021), Dehmani et al. (2022), Rao et al. (2017), Yan and Ge (2023)

4.5 CASE STUDIES AND REAL-WORLD APPLICATIONS

The practical deployment of adsorption technologies has significantly evolved over the past few decades, driven by the urgent need for sustainable and efficient water treatment solutions (Kasbaji et al., 2022b). Numerous case studies worldwide illustrate the success of adsorption methods in removing contaminants from diverse liquid waste sources. These applications span industrial, municipal, agricultural, and even domestic sectors, underscoring the versatility and scalability of adsorption as a water purification strategy.

One of the most prominent materials utilized in real-world settings is AC, which has proven particularly effective for eliminating organic pollutants such as phenols, dyes, and pesticides. For example, the use of granular AC in a textile wastewater treatment plant in India demonstrated removal efficiencies above 64%–65% for reactive dyes and aromatic compounds (Figure 4.10a). The system was implemented in a modular format, facilitating rapid deployment and replacement of saturated carbon beds, and resulted in significant improvements in effluent quality (Sirianuntapiboon and Sansak, 2008; Walker and Weatherley, 2000). Biochar, a carbonaceous material derived from the pyrolysis of biomass, has also gained attention as a low-cost and eco-friendly adsorbent (Samuel Olugbenga et al., 2024; Wibowo et al., 2025). A study in Medellín, Colombia, used filters made from banana peel waste-derived adsorbents to remove heavy metals (Fe^{2+}, Zn^{2+}, Ni^{2+}, Cr^{2+}, Cu^{2+}) from electroplating wastewater. Both unfunctionalized biochar and commercial AC showed the highest adsorption capacities, with functionalization found unnecessary despite its higher energy and reagent costs (Figure 4.10b). The complex metal mixture required normalization of adsorption data to compare performance accurately (Arriola-Villaseñor et al., 2023).

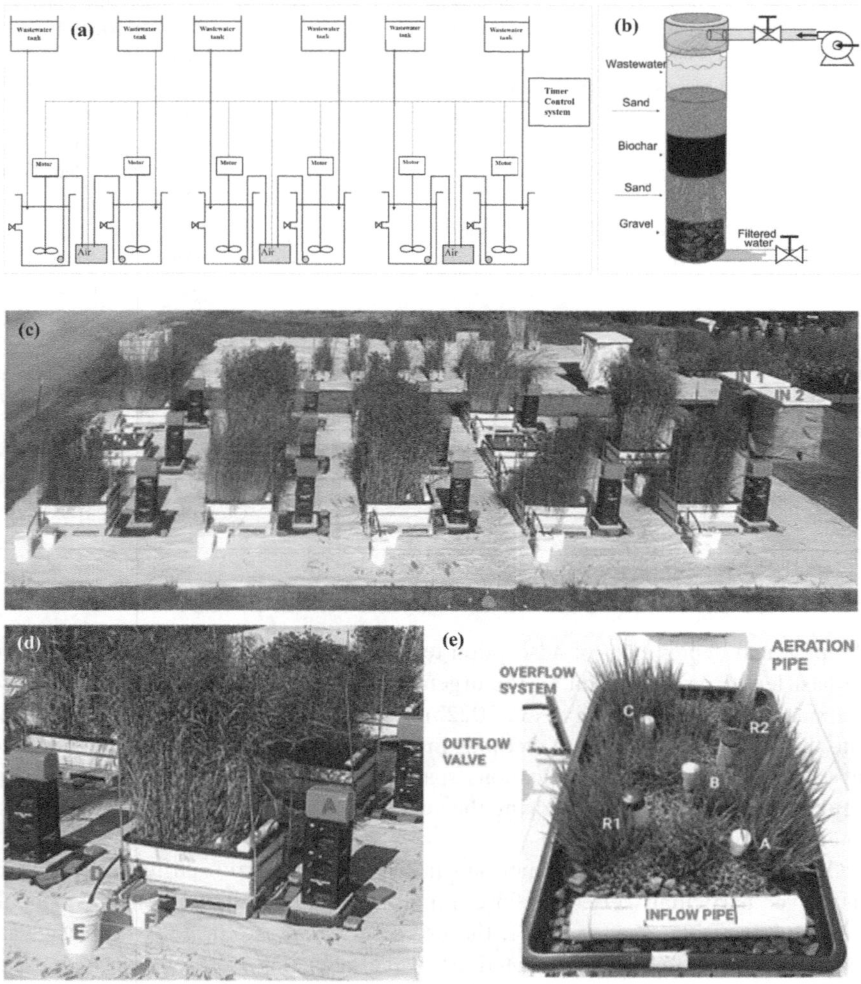

FIGURE 4.10 (a) SBR system setup and conditions: flow diagram showing the impeller (60 rpm), aeration (EK-8000, 6.0 W), and 7.5 L working volume. Operational details in text; (b) filtration system for adsorption: diagram of the custom-built filtration setup used in adsorption experiments; (c) experimental setup overview: aerial view (IN1, IN2: inflow tanks); (d) mesocosm details (A–F: components for inflow, sampling, overflow, and outflow); and (e) top view (A–C: piezometers; R1, R2: rhizotrons). (Reprinted with permissions respectively from Abas et al., 2022; Arriola-Villaseñor et al., 2023; Sirianuntapiboon and Sansak, 2008.)

This case exemplifies the potential of agricultural waste valorization in sustainable wastewater management.

In developed regions, such as Europe and North America, zeolites have been widely adopted for ammonia and heavy metal removal in municipal and aquaculture systems (Dehmani et al., 2024; Tchernev, 2001; Wang and Peng, 2010; Yue et al., 2022). A recent review highlights the growing importance of remediated water reuse for sustainable water management, especially in agriculture and industry. While innovative treatments targeting biological contaminants, persistent organic pollutants, and contaminants of emerging concerns have shown promise at the lab and pilot scales, only a few have advanced to full-scale application. Tailored approaches based on wastewater origin and use have enabled the production of high-quality treated water, though challenges remain in scaling up and fully implementing these solutions (Ferri and Bolelli, 2025). Similarly, in aquaculture facilities across Southeast Asia, clinoptilolite-type zeolites are used in recirculating systems to manage nitrogen levels and maintain water quality for fish production (Ghasemi et al., 2018). GO and other nanostructured materials represent the frontier of high-performance adsorption technologies. Although still in the early stages of field deployment, several promising pilot studies highlight their potential. A notable example is the application of a nitrogen-doped GO aerogel with a 3D network structure, synthesized via a one-step hydrothermal self-assembly and eco-friendly freeze-drying process, which achieved a high Cr(VI) adsorption capacity of 408.48 mg g^{-1} at pH 2.0. The adsorption process followed pseudo-second-order kinetics with a multi-step diffusion mechanism and demonstrated strong regeneration and recovery performance, underscoring its practical potential for wastewater treatment (Liang et al., 2018).

Clay minerals, especially montmorillonite and bentonite, are utilized extensively in low-cost household water filters in rural India and Africa. A collaborative project between non-governmental organizations and local communities in Ethiopia demonstrated the use of bentonite-ceramic composite filters for fluoride removal. These filters reduced fluoride levels from 10 mg L^{-1} to below 1.5 mg L^{-1}, the maximum allowable concentration under WHO guidelines, with an operational lifespan of over six months. This method highlights the importance of community-centered innovation in water purification (Coetzee et al., 2004; Kut et al., 2016). Industrial-scale adsorption systems are also common in the treatment of pharmaceutical and agrochemical manufacturing wastewater. For instance, a facility in the United States treating effluent from a nonsteroidal anti-inflammatory drug production line implemented a dual-stage adsorption unit using modified biochar and polymeric resins. The treatment achieved 95% removal of diclofenac and ibuprofen, reducing ecotoxicity in downstream aquatic systems (de Souza et al., 2021; Mansouri et al., 2021). In the agricultural sector, adsorption has proven effective in managing runoff containing fertilizers and pesticides. A study on subsurface flow treatment wetlands demonstrated that adding biochar to the substrate significantly enhanced the removal of chlorantraniliprole, an emerging insecticide in agricultural runoff. The outdoor experiment took place at the Montréal Botanical Garden (Canada) using water-saturated subsurface flow treatment wetland mesocosms arranged in a randomized block a posteriori design (Figure 4.10c and d). Over a one-month period, biochar-amended mesocosms achieved 90%–99% chlorantraniliprole removal, while planted systems without biochar showed poor performance (<13%). Among the macrophytes tested, Scirpus and Phragmites provided better buffering effects due to higher evapotranspiration rates. These findings highlight biochar as a promising amendment for chlorantraniliprole mitigation in subsurface flow treatment wetlands, though long-term efficiency needs further evaluation (Abas et al., 2022). Sector-wise application trends reveal that adsorption is most heavily utilized in industrial effluents and municipal wastewater treatment. However, significant applications are emerging in agriculture, aquaculture, and decentralized sanitation systems. These case studies not only affirm the operational viability of adsorption but also reflect growing innovation in integrating adsorption with complementary technologies. Nevertheless, several challenges remain in the translation of laboratory-scale adsorption findings to field-level applications. These include the variability in wastewater composition, adsorbent regeneration cost, and disposal issues related to spent adsorbents. Despite these constraints, the diverse real-world applications of adsorption reflect its robustness and adaptability across geographic and economic contexts. Continued success in this domain will require interdisciplinary collaboration between material scientists, environmental engineers, policy-makers, and local stakeholders. These efforts will ensure that adsorption technologies are not only technically effective but also socially and economically viable, fulfilling the urgent global mandate for clean water access.

4.6 ENVIRONMENTAL AND ECONOMIC IMPACT

The economic sustainability and pervasive implementation of adsorbents in wastewater treatment applications are fundamentally reliant on considerations of cost-effectiveness. Despite the numerous advantages afforded by adsorption techniques, such as operational straightforwardness, purification efficiency, and contaminant selectivity in comparison to traditional treatment methodologies, economic limitations persist as a significant obstacle. The costs associated with adsorbents are predominantly influenced by the source of raw materials, the processes of fabrication, and the physicochemical properties of the surface. AC, which is widely utilized as an adsorptive medium, commands relatively elevated market prices due to the energy-intensive nature of its production processes. Economically viable alternatives are offered by adsorbent materials that are derived from agricultural byproducts or industrial waste streams, including chitosan derivatives, both natural and synthetic zeolites, as well as chemically activated clay minerals (Ighalo et al., 2022; Rashid et al., 2021). The intrinsic characteristics of the effluent matrix, which encompass pollutant concentration levels, compositional variability, and toxicological profiles, likewise affect the economic viability of adsorption practices. Effluent streams distinguished by high contaminant concentrations or intricate chemical compositions may require augmented dosages of adsorbents, consequently increasing operational costs. Moreover, the presence of hazardous substances may necessitate post-treatment regeneration protocols or specialized disposal approaches for depleted adsorbent materials, thereby imposing further financial burdens (Rashid et al., 2021).

A notable contemporary constraint impacting adsorption technologies in wastewater remediation pertains to the regulatory frameworks and environmental considerations associated with the utilization, disposal, and management of adsorbents. Regulatory challenges encompass a lack of standardization and inconsistent guidelines concerning metrics of adsorbent quality, performance criteria, and safety evaluations, in conjunction with inadequate compliance monitoring systems for wastewater treatment processes (Robbins et al., 2022). Variations in jurisdiction across different geographic locales manifest as disparate regulatory prerequisites for the approval of adsorbents, certification standards, and methodologies for performance evaluation, leading to operational uncertainties for manufacturers of materials, entities within the supply chain, and end-users. Additionally, the complexities of regulation may hinder innovation pathways and developmental initiatives for novel adsorptive materials, as approval processes often involve prolonged timelines, substantial financial commitments, and procedural complexities. Consequently, there exists a pressing need for regulatory harmonization and procedural simplification at both national and international governance levels, alongside enhanced stakeholder awareness campaigns regarding the risk-benefit profiles of adsorbents (Beaudrie et al., 2013; Ray et al., 2009). Environmental considerations encompass potential adverse effects on human health and the integrity of ecosystems throughout the lifecycle of adsorbents, spanning production stages, application contexts, post-use disposal, and recycling efforts. The manufacturing processes of adsorbents frequently entail considerable consumption of resources and energy, simultaneously resulting in atmospheric emissions and waste byproducts (Ma et al., 2021). The operational application of adsorptive materials may introduce secondary pollutants or alter the physicochemical balance of wastewater matrixes, potentially affecting subsequent treatment processes or the health of receiving aquatic ecosystems. Disposal of used adsorbents poses risks related to contaminant leaching, pollutant migration pathways, or bioaccumulation mechanisms within soil matrixes, water systems, or living organisms. Regeneration and recycling protocols for adsorbents typically necessitate additional treatment methodologies that carry associated environmental ramifications. Therefore, comprehensive methodologies for lifecycle assessment and strategies aimed at minimizing environmental impact are imperative for adsorptive materials, necessitating the integration of principles of green chemistry and sustainable engineering paradigms throughout the phases of design conceptualization and developmental implementation (Alsawy et al., 2022; Velempini et al., 2023).

4.7 CHALLENGES AND FUTURE DIRECTIONS

The domain of adsorption technology pertaining to wastewater purification anticipates significant transformation, driven by ongoing scientific inquiry and technological innovation. The transition from traditional methodologies to advanced techniques has been enabled by groundbreaking advancements within the realms of materials science and environmental engineering. Conventional approaches, primarily employing AC substrates and zeolitic materials, have established the groundwork for the emergence of advanced methodologies that incorporate novel materials such as carbon nanotubes, MOFs, and graphene-derived architectures. This evolutionary trajectory exemplifies the intricate interrelationship among efficacy parameters, economic considerations, environmental sustainability metrics, and assessments of technological feasibility. The subsequent subsection delves into the anticipated landscape of adsorption methodologies in wastewater purification, highlighting contemporary innovations and promising developmental pathways within this specialized domain.

Within the rapidly advancing field of wastewater remediation science, a plethora of methodological approaches have been devised, each meticulously engineered to target specific categories of pollutants. These methodological frameworks encompass physical, chemical, biological, and physicochemical techniques, each displaying distinct mechanistic principles and implementation scenarios. Physical methodologies incorporate processes such as filtration mechanisms, sedimentation processes, and flotation techniques. These approaches fundamentally leverage the intrinsic physical characteristics of contaminants to facilitate their separation from aqueous environments. Such methodologies are characterized by their operational simplicity and proven effectiveness for particular contaminant classifications (Satyam and Patra, 2024).

Chemical methodologies are characterized by their distinctive use of a variety of chemical agents, including coagulants, flocculants, oxidizing agents, and reducing substances. These reagents function by altering the chemical properties of pollutants, thereby augmenting their removal efficiency. Within this methodological classification, Advanced Oxidation Processes are particularly notable for their capability to generate reactive oxygen species, specifically hydroxyl radicals, which exhibit exceptional efficacy in degrading recalcitrant organic contaminants and neutralizing pathogenic microorganisms. These processes, comprising Fenton reactions, photo-Fenton systems, ozonation techniques, photocatalytic mechanisms, and electrochemical oxidation methodologies, may proceed via either homogeneous or heterogeneous reaction pathways (Dong et al., 2022).

Within biological treatment paradigms, enzymatic remediation signifies a noteworthy innovation. This approach employs specific enzymes as biological catalysts to selectively degrade pollutant substances. Enzymatic efficacy is frequently enhanced through immobilization protocols utilizing diverse substrate materials, including membrane structures, polymeric beads, or nanomaterial platforms. Specific enzymatic systems, including laccase, peroxidase, nitrilase, and lipase, have gained particular prominence in wastewater treatment applications due to their demonstrated capacity to effectively decompose complex pollutant structures. Concurrently with enzymatic treatment methodologies, bioremediation techniques are undergoing continuous refinement. These approaches utilize microorganisms or phytological systems for pollutant removal and are optimized for both in situ and ex situ implementation scenarios, with methodologies such as bioaugmentation strategies, biostimulation protocols, and phytoremediation techniques exhibiting substantial potential (Dehnavi and Ebrahimipour, 2022; Z. Wang et al., 2024).

Among these diverse methodological approaches, adsorption, classified as a physicochemical technique, distinguishes itself through its exceptional versatility and operational efficiency. The adsorption mechanism functions by accumulating contaminant molecules onto adsorbent surfaces that exhibit heightened affinity and selectivity for specific pollutants. Adsorption efficacy is attributed to the extensive diversity of available adsorbent materials, including AC substrates, zeolitic structures, clay mineral compositions, biochar derivatives, MOF architectures, and graphene-based materials. This methodology demonstrates economic advantages, operational simplicity, and remarkable versatility in removing diverse pollutant categories, including heavy metal elements, synthetic dyes, petroleum derivatives, and micropollutants. Particularly regarding emerging contaminants, including personal care product residues, pharmaceutical compounds, and endocrine-disrupting substances, adsorption technology exhibits exceptional promise. These emerging pollutants, typically present at trace concentration levels yet possessing significant ecological and human health implications, necessitate highly selective and efficient removal methodologies. Adsorption provides a viable remediation strategy through its capacity to effectively capture and concentrate these contaminants, even at extremely low-concentration thresholds. Furthermore, adsorption processes can be integrated with complementary techniques, such as biodegradation mechanisms and advanced oxidation systems, to establish hybrid treatment platforms (Rathi and Kumar, 2021; Suiyi et al., 2024).

Adsorption efficiency parameters are determined by adsorbent characteristics, contaminant properties, and wastewater matrix composition. Optimization of these multifarious parameters is essential for effective treatment implementation. This optimization challenge highlights the significance of data-driven methodologies such as machine learning and AI applications. These computational approaches are instrumental in identifying cryptic patterns and relationships within complex, multi-dimensional datasets generated from adsorption systems. Machine learning algorithms can be calibrated to comprehend the influence of diverse operational parameters on contaminant removal efficiency and predict performance outcomes under specific operational conditions. Additionally, these computational models provide mechanistic insights and facilitate optimization of adsorbent synthesis protocols. Machine learning and AI technologies contribute significantly to process monitoring capabilities, control mechanisms, and diagnostic functionalities, thereby enhancing the comprehensive efficiency and operational reliability of adsorption systems. Moreover, AI techniques, particularly artificial neural networks, demonstrate the capacity to simulate complex adsorption phenomena at molecular scales, providing invaluable insights regarding surface interaction dynamics and pore diffusion mechanisms. These computational insights profoundly influence the synthesis strategies and application methodologies for advanced adsorbent materials (Alam et al., 2022). The synergistic integration of nanotechnology applications, advanced materials science, and AI-based analytical tools presages an auspicious future for adsorptive wastewater treatment technologies. This interdisciplinary convergence facilitates the development of sustainable and intelligent technological platforms for the elimination of both conventional and emerging contaminants. However, implementation challenges, including computational resource requirements, data acquisition constraints, and model interpretability limitations, must be systematically addressed. Surmounting these challenges necessitates collaborative initiatives among materials scientists, environmental engineers, and data science specialists.

The potential societal benefits regarding clean water accessibility and environmental health protection are immeasurable and invaluable.

REFERENCES

Abas, K., Brisson, J., Amyot, M., Brodeur, J., Storck, V., Montiel-León, J.M., Duy, S.V., Sauvé, S., Kõiv-Vainik, M., 2022. Effects of plants and biochar on the performance of treatment wetlands for removal of the pesticide chlorantraniliprole from agricultural runoff. *Ecol. Eng.* 175. https://doi.org/10.1016/j.ecoleng.2021.106477

Abdelfattah, A., Ali, S.S., Ramadan, H., El-Aswar, E.I., Eltawab, R., Ho, S.H., Elsamahy, T., Li, S., El-Sheekh, M.M., Schagerl, M., Kornaros, M., Sun, J., 2023. Microalgae-based wastewater treatment: Mechanisms, challenges, recent advances, and future prospects. *Environ. Sci. Ecotechnology* 13, 100205. https://doi.org/10.1016/j.ese.2022.100205

Abu-Danso, E., Peräniemi, S., Leiviskä, T., Kim, T.Y., Tripathi, K.M., Bhatnagar, A., 2020. Synthesis of clay-cellulose biocomposite for the removal of toxic metal ions from aqueous medium. *J. Hazard. Mater.* 381, 120871. https://doi.org/10.1016/j.jhazmat.2019.120871

Adjal, C., Timón, V., Guechtouli, N., Boussassi, R., Hammoutène, D., Senent, M.L., 2023. The role of water in the adsorption of nitro-organic pollutants on activated carbon. *J. Phys. Chem. A* 127, 8146–8158. https://doi.org/10.1021/acs.jpca.3c03877

Ahmadpour, A., Do, D.D., 1997. The preparation of activated carbon from macadamia nutshell by chemical activation. *Carbon* 35, 1723–1732. https://doi.org/10.1016/S0008-6223(97)00127-9

Ajay Rakkesh, R., Naveen, T.B., Durgalakshmi, D., Balakumar, S., 2024. Covalent organic frameworks: Pioneering remediation solutions for organic pollutants. *Chemosphere* 346, 140655. https://doi.org/10.1016/j.chemosphere.2023.140655

Akpomie, K.G., Conradie, J., 2020. Advances in application of cotton-based adsorbents for heavy metals trapping, surface modifications and future perspectives. *Ecotoxicol. Environ. Saf.* 201, 110825. https://doi.org/10.1016/j.ecoenv.2020.110825

Al-Asadi, S.T., Mussa, Z.H., Al-Qaim, F.F., Kamyab, H., Al-Saedi, H.F.S., Deyab, I.F., Kadhim, N.J., 2025. A comprehensive review of methylene blue dye adsorption on activated carbon from edible fruit seeds: A case study on kinetics and adsorption models. *Carbon Trends* 20, 100507. https://doi.org/10.1016/j.cartre.2025.100507

Al-Qodah, Z., Shawabkah, R., 2009. Production and characterization of granular activated carbon from activated sludge. *Brazilian J. Chem. Eng.* 26, 127–136. https://doi.org/10.1590/S0104-66322009000100012

Al Kausor, M., Sen Gupta, S., Bhattacharyya, K.G., Chakrabortty, D., 2022. Montmorillonite and modified montmorillonite as adsorbents for removal of water soluble organic dyes: A review on current status of the art. *Inorg. Chem. Commun.* 143, 109686. https://doi.org/10.1016/j.inoche.2022.109686

Alam, G., Ihsanullah, I., Naushad, M., Sillanpää, M., 2022. Applications of artificial intelligence in water treatment for optimization and automation of adsorption processes: Recent advances and prospects. *Chem. Eng. J.* 427. https://doi.org/10.1016/j.cej.2021.130011

AlAqad, K.M., Abdelnaby, M.M., Tanimu, A., Abdulazeez, I., Elsharif, A.M., 2024. Adsorbent materials for water treatment: A review of current trends and future challenges. *Environ. Pollut. Manag.* 2, 81–96. https://doi.org/10.1016/j.epm.2024.12.003

Alsawy, T., Rashad, E., El-Qelish, M., Mohammed, R.H., 2022. A comprehensive review on the chemical regeneration of biochar adsorbent for sustainable wastewater treatment. *npj Clean Water* 5. https://doi.org/10.1038/s41545-022-00172-3

Amenaghawon, A.N., Anyalewechi, C.L., Osazuwa, O.U., Elimian, E.A., Eshiemogie, S.O., Oyefolu, P.K., Kusuma, H.S., 2023. A comprehensive review of recent advances in the synthesis and application of metal-organic frameworks (MOFs) for the adsorptive sequestration of pollutants from wastewater. *Sep. Purif. Technol.* 311, 123246. https://doi.org/10.1016/j.seppur.2023.123246

Andrade, A.A.R., Silva, M.C., Crespo, L.H.S., Silva, T.L., Cazetta, A.L., Spessato, L., Almeida, V.C., 2025. High yield activated carbon from corn husk-sucrose prepared by chemical and physicochemical activation processes: Adsorption study of methylene blue dye. *J. Mol. Liq.* 427. https://doi.org/10.1016/j.molliq.2025.127435

Ani, J.U., Akpomie, K.G., Okoro, U.C., Aneke, L.E., Onukwuli, O.D., Ujam, O.T., 2020. Potentials of activated carbon produced from biomass materials for sequestration of dyes, heavy metals, and crude oil components from aqueous environment. *Appl. Water Sci.* 10, 1–11. https://doi.org/10.1007/s13201-020-1149-8

Arriola-Villaseñor, E., Ardila A, A.N., Barrera Z, R., Hernández, J., 2023. Using banana waste biochar for simultaneous removal of heavy metals from raw real wastewater from the electroplating industry. *Desalin. Water Treat.* 314, 88–102. https://doi.org/10.5004/dwt.2023.30079

Arya, K., Kumar, A., Kataria, R., 2025. Recent advances in MOF-based composites for the detection and adsorptive removal of Pb(II) ions in aqueous phase. *Mater. Today Sustain.* 29, 101057. https://doi.org/10.1016/j.mtsust.2024.101057

Aurich, A., Hofmann, J., Oltrogge, R., Wecks, M., Gläser, R., Blömer, L., Mauersberger, S., Müller, R.A., Sicker, D., Giannis, A., 2017. Improved isolation of microbiologically produced (2R,3S)-isocitric acid by adsorption on activated carbon and recovery with methanol. *Org. Process Res. Dev.* 21, 866–870. https://doi.org/10.1021/acs.oprd.7b00090

Azam, K., Shezad, N., Shafiq, I., Akhter, P., Akhtar, F., Jamil, F., Shafique, S., Park, Y.K., Hussain, M., 2022. A review on activated carbon modifications for the treatment of wastewater containing anionic dyes. *Chemosphere* 306, 135566. https://doi.org/10.1016/j.chemosphere.2022.135566

Bahmanzadegan, F., Ghaemi, A., 2024. Modification and functionalization of zeolites to improve the efficiency of CO_2 adsorption: A review. *Case Stud. Chem. Environ. Eng.* 9, 100564. https://doi.org/10.1016/j.cscee.2023.100564

Bahmanzadegan, F., Ghaemi, A., 2025. A comprehensive review on novel zeolite-based adsorbents for environmental pollutant. *J. Hazard. Mater. Adv.* 17, 100617. https://doi.org/10.1016/j.hazadv.2025.100617

Bai, S., Chang, Z., Ren, Z., Zhao, Y., Pang, L., 2025. Adsorption of Cd^{2+} ions onto zeolites synthesized from a mixture of coal fly ash and oil shale ash in aqueous media. *RSC Adv.* 15, 11293–11300. https://doi.org/10.1039/d5ra00752f

Beaudrie, C.E.H., Satterfield, T., Kandlikar, M., Harthorn, B.H., 2013. Expert views on regulatory preparedness for managing the risks of nanotechnologies. *PLoS One* 8, 1–9. https://doi.org/10.1371/journal.pone.0080250

Behroozi, A.H., Meunier, L., Mirahsani, A., Champagne, P., Koupaie, E.H., 2025. Graphene-based materials and technologies for the treatment of PFAS in water: A review of recent developments. *J. Hazard. Mater. Adv.* 17, 100626. https://doi.org/10.1016/j.hazadv.2025.100626

Bhatt, P., Bhandari, G., Turco, R.F., Aminikhoei, Z., Bhatt, K., Simsek, H., 2022. Algae in wastewater treatment, mechanism, and application of biomass for production of value-added product. *Environ. Pollut.* 309, 119688. https://doi.org/10.1016/j.envpol.2022.119688

Biswal, B.K., Balasubramanian, R., 2022. Adsorptive removal of sulfonamides, tetracyclines and quinolones from wastewater and water using carbon-based materials: Recent developments and future directions. *J. Clean. Prod.* 349, 131421. https://doi.org/10.1016/j.jclepro.2022.131421

Bonev, A., Alexandrov, A., 1993. Chemistry of the solid-water interface: Processes at the mineral-water and particle-water interface in natural systems. *Археология* 1, 117–125.

Bortoluz, J., Ferrarini, F., Bonetto, L.R., da Silva Crespo, J., Giovanela, M., 2020. Use of low-cost natural waste from the furniture industry for the removal of methylene blue by adsorption: Isotherms, kinetics and thermodynamics. *Cellulose* 27, 6445–6466. https://doi.org/10.1007/s10570-020-03254-y

Carvalho, A.P., Mestre, A.S., Pires, J., Pinto, M.L., Rosa, M.E., 2006. Granular activated carbons from powdered samples using clays as binders for the adsorption of organic vapours. *Microporous Mesoporous Mater.* 93, 226–231. https://doi.org/10.1016/j.micromeso.2006.03.001

Chaieb, K., Kouidhi, B., Ayed, L., Bakr Hosawi, S., Abdulbaqi Abdulhakim, J., Hajri, A., Altayb, H.N., 2023. Enhanced textile dye removal from wastewater using natural biosorbent and Shewanella algae B29: Application of Box Behnken design and genomic approach. *Bioresour. Technol.* 374, 128755. https://doi.org/10.1016/j.biortech.2023.128755

Chen, B., Ding, L., Wang, Y., Zhang, Y., 2022. High efficient adsorption for thorium in aqueous solution using a novel tentacle-type chitosan-based aerogel: Adsorption behavior and mechanism. *Int. J. Biol. Macromol.* 222, 1747–1757. https://doi.org/10.1016/j.ijbiomac.2022.09.256

Chen, J.Q., Sharifzadeh, Z., Bigdeli, F., Gholizadeh, S., Li, Z., Hu, M.L., Morsali, A., 2023. MOF composites as high potential materials for hazardous organic contaminants removal in aqueous environments. *J. Environ. Chem. Eng.* 11, 109469. https://doi.org/10.1016/j.jece.2023.109469

Cheng, H., Liu, Q., Yang, J., Ma, S., Frost, R.L., 2012. The thermal behavior of kaolinite intercalation complexes-A review. *Thermochim. Acta* 545, 1–13. https://doi.org/10.1016/j.tca.2012.04.005

Chi, X., Li, M., Di, J., Bai, P., Song, L., Wang, X., Li, F., Liang, S., Xu, J., Yu, J., 2021. A highly stable and flexible zeolite electrolyte solid-state Li–air battery. *Nature* 592, 551–557. https://doi.org/10.1038/s41586-021-03410-9

Choi, S., Drese, J.H., Jones, C.W., 2009. Adsorbent materials for carbon dioxide capture from large anthropogenic point sources. *ChemSusChem* 2, 796–854. https://doi.org/10.1002/cssc.200900036

Coetzee, P.P., Haarhoff, J., Chibi, C., 2004. *Removal of Fluoride from Drinking Water with Clay-based Defluoridators*. Water Research Commission..

Conversa, G., Pacifico, S., La Rotonda, P., Lazzizera, C., Bonasia, A., Elia, A., 2024. Foliar application of natural zeolites affects the growth and productivity of processing tomato. *Eur. J. Agron.* 154, 127100. https://doi.org/10.1016/j.eja.2024.127100

Crini, G., Badot, P.M., 2008. Application of chitosan, a natural aminopolysaccharide, for dye removal from aqueous solutions by adsorption processes using batch studies: A review of recent literature. *Prog. Polym. Sci.* 33, 399–447. https://doi.org/10.1016/j.progpolymsci.2007.11.001

Crini, G., Lichtfouse, E., 2019. Advantages and disadvantages of techniques used for wastewater treatment. *Environ. Chem. Lett.* https://doi.org/10.1007/s10311-018-0785-9

Daradmare, S., Xia, M., Le, V.N., Kim, J., Park, B.J., 2021. Metal–organic frameworks/alginate composite beads as effective adsorbents for the removal of hexavalent chromium from aqueous solution. *Chemosphere* 270, 129487. https://doi.org/10.1016/j.chemosphere.2020.129487

Daud, W.M.A.W., Ali, W.S.W., Sulaiman, M.Z., 2000. Effects of carbonization temperature on pore development in palm-shell-based activated carbon. *Carbon N. Y.* 38, 1925–1932. https://doi.org/10.1016/S0008-6223(00)00028-2

de Souza, R.M., Quesada, H.B., Cusioli, L.F., Fagundes-Klen, M.R., Bergamasco, R., 2021. Adsorption of non-steroidal anti-inflammatory drug (NSAID) by agro-industrial by-product with chemical and thermal modification: Adsorption studies and mechanism. *Ind. Crops Prod.* 161. https://doi.org/10.1016/j.indcrop.2020.113200

Dehmani, Y., Ba Mohammed, B., Oukhrib, R., Dehbi, A., Lamhasni, T., Brahmi, Y., El-Kordy, A., Franco, D.S.P., Georgin, J., Lima, E.C., Alrashdi, A.A., Tijani, N., Abouarnadasse, S., 2024. Adsorption of various inorganic and organic pollutants by natural and synthetic zeolites: A critical review. *Arab. J. Chem.* 17, 105474. https://doi.org/10.1016/j.arabjc.2023.105474

Dehmani, Y., Dridi, D., Lamhasni, T., Abouarnadasse, S., Chtourou, R., Lima, E.C., 2022. Review of phenol adsorption on transition metal oxides and other adsorbents. *J. Water Process Eng.* 49, 102965. https://doi.org/10.1016/j.jwpe.2022.102965

Dehnavi, S.M., Ebrahimipour, G., 2022. Comparative remediation rate of biostimulation, bioaugmentation, and phytoremediation in hydrocarbon contaminants. *Int. J. Environ. Sci. Technol.* 19, 11561–11586. https://doi.org/10.1007/s13762-022-04343-0

Dias, J.M., Alvim-Ferraz, M.C.M., Almeida, M.F., Rivera-Utrilla, J., Sánchez-Polo, M., 2007. Waste materials for activated carbon preparation and its use in aqueous-phase treatment: A review. *J. Environ. Manage.* 85, 833–846. https://doi.org/10.1016/j.jenvman.2007.07.031

Ding, L.P., Bhatia, S.K., Liu, F., 2002. Kinetics of adsorption on activated carbon: Application of heterogeneous vacancy solution theory. *Chem. Eng. Sci.* 57, 3909–3928. https://doi.org/10.1016/S0009-2509(02)00306-8

Ding, N., Wang, Y., Shen, Z., Liu, H., Li, Y., Sun, Y., 2025. Mechanisms and application of magnetic field enhancing pollutant adsorption in water: A review. *Sep. Purif. Technol.* 356, 129938. https://doi.org/10.1016/j.seppur.2024.129938

Dong, C., Fang, W., Yi, Q., Zhang, J., 2022. A comprehensive review on reactive oxygen species (ROS) in advanced oxidation processes (AOPs). *Chemosphere* 308. https://doi.org/10.1016/j.chemosphere.2022.136205

Donkadokula, N.Y., Kola, A.K., Naz, I., Saroj, D., 2020. A review on advanced physico-chemical and biological textile dye wastewater treatment techniques. *Rev. Environ. Sci. Biotechnol.* 19, 543–560. https://doi.org/10.1007/s11157-020-09543-z

El-Gendy, M.M.A.A., Hassanein, N.M., Ibrahim, H.A.E.H., Abd El-Baky, D.H., 2017. Heavy metals biosorption from aqueous solution by endophytic Drechslera hawaiiensis of Morus alba L. derived from heavy metals habitats. *Mycobiology* 45, 73–83. https://doi.org/10.5941/MYCO.2017.45.2.73

El-Mehalmey, W.A., Ibrahim, A.H., Abugable, A.A., Hassan, M.H., Haikal, R.R., Karakalos, S.G., Zaki, O., Alkordi, M.H., 2018. Metal-organic framework@silica as a stationary phase sorbent for rapid and cost-effective removal of hexavalent chromium. *J. Mater. Chem. A* 6, 2742–2751. https://doi.org/10.1039/c7ta08281a

El-Nemr, M.A., Ismail, I.M.A., Abdelmonem, N.M., El Nemr, A., Ragab, S., 2021. Amination of biochar surface from watermelon peel for toxic chromium removal enhancement. *Chinese J. Chem. Eng.* 36, 199–222. https://doi.org/10.1016/j.cjche.2020.08.020

El-Nile, A.E., Elazab, M.A., Soltan, Y.A., Elkomy, A.E., El-Zaiat, H.M., Sallam, S.M.A., El-Azrak, K.E.D., 2023. Nano and natural zeolite feed supplements for dairy goats: Feed intake, ruminal fermentation, blood metabolites, and milk yield and fatty acids profile. *Anim. Feed Sci. Technol.* 295, 115522. https://doi.org/10.1016/j.anifeedsci.2022.115522

Ewis, D., Ba-Abbad, M.M., Benamor, A., El-Naas, M.H., 2022. Adsorption of organic water pollutants by clays and clay minerals composites: A comprehensive review. *Appl. Clay Sci.* 229, 106686. https://doi.org/10.1016/j.clay.2022.106686

Ezzeddine, Z., Batonneau-Gener, I., Ghssein, G., Pouilloux, Y., 2025. Recent advances in heavy metal adsorption via organically modified mesoporous silica: A review. *Water* 17, 1–17. https://doi.org/10.3390/w17050669

Farhan Hanafi, M., Sapawe, N., 2020. A review on the water problem associate with organic pollutants derived from phenol, methyl orange, and remazol brilliant blue dyes. *Mater. Today Proc.* 31, A141–A150. https://doi.org/10.1016/j.matpr.2021.01.258

Feng, P., Li, J., Wang, H., Xu, Z., 2020. Biomass-based activated carbon and activators: Preparation of activated carbon from corncob by chemical activation with biomass pyrolysis liquids. *ACS Omega* 5, 24064–24072. https://doi.org/10.1021/acsomega.0c03494

Feng, Y., Gong, J.L., Zeng, G.M., Niu, Q.Y., Zhang, H.Y., Niu, C.G., Deng, J.H., Yan, M., 2010. Adsorption of Cd (II) and Zn (II) from aqueous solutions using magnetic hydroxyapatite nanoparticles as adsorbents. *Chem. Eng. J.* 162, 487–494. https://doi.org/10.1016/j.cej.2010.05.049

Ferri, E.N., Bolelli, L., 2025. Wastewater remediation treatments aimed at water reuse: Recent outcomes from pilot- and full-scale tests. *Appl. Sci.* 15. https://doi.org/10.3390/app15052448

Foo, K.Y., Hameed, B.H., 2010. Insights into the modeling of adsorption isotherm systems. *Chem. Eng. J.* 156, 2–10. https://doi.org/10.1016/j.cej.2009.09.013

Fu, L., Yang, H., Tang, A., Hu, Y., 2017. Engineering a tubular mesoporous silica nanocontainer with well-preserved clay shell from natural halloysite. *Nano Res.* 10, 2782–2799. https://doi.org/10.1007/s12274-017-1482-x

Furukawa, H., Gándara, F., Zhang, Y.B., Jiang, J., Queen, W.L., Hudson, M.R., Yaghi, O.M., 2014. Water adsorption in porous metal-organic frameworks and related materials. *J. Am. Chem. Soc.* 136, 4369–4381. https://doi.org/10.1021/ja500330a

Gao, X., Xu, Z., Li, G., 2022a. MOF-driven ultrafine Co9S8 nanocrystals embedded in N, S-codoped multilayer-assembled carbon nanoplates for efficient bifunctional oxygen electrocatalysis. *Chem. Eng. J.* 431. https://doi.org/10.1016/j.cej.2021.133385

Gao, X., Yang, S., Hu, L., Cai, S., Wu, L., Kawi, S., 2022b. Carbonaceous materials as adsorbents for CO_2 capture: Synthesis and modification. *Carbon Capture Sci. Technol.* 3, 100039. https://doi.org/10.1016/j.ccst.2022.100039

Gao, Y., Yue, Q., Gao, B., Li, A., 2020. Insight into activated carbon from different kinds of chemical activating agents: A review. *Sci. Total Environ.* 746, 141094. https://doi.org/10.1016/j.scitotenv.2020.141094

Garofalo, E., Taurino, L., Di Maio, L., Neitzert, H.C., Incarnato, L., 2023. Assessment of melt compounding with zeolites as an effective deodorization strategy for mixed plastic wastes and comparison with degassing. *Polymers (Basel).* 15. https://doi.org/10.3390/polym15081858

Gatou, M.A., Bika, P., Stergiopoulos, T., Dallas, P., Pavlatou, E.A., 2021. Recent advances in covalent organic frameworks for heavy metal removal applications. *Energies* 14. https://doi.org/10.3390/en14113197

Gayathiri, M., Pulingam, T., Lee, K.T., Sudesh, K., 2022. Activated carbon from biomass waste precursors: Factors affecting production and adsorption mechanism. *Chemosphere* 294, 133764. https://doi.org/10.1016/j.chemosphere.2022.133764

Gebre Meskel, A., Kwikima, M.M., Meshesha, B.T., Habtu, N.G., Naik, S.V.C.S., Vellanki, B.P., 2024. Malachite green and methylene blue dye removal using modified bagasse fly ash: Adsorption optimization studies. *Environ. Challenges* 14, 100829. https://doi.org/10.1016/j.envc.2023.100829

Ghasemi, Z., Sourinejad, I., Kazemian, H., Rohani, S., 2018. Application of zeolites in aquaculture industry: A review. *Rev. Aquac.* 10, 75–95. https://doi.org/10.1111/raq.12148

Girijan, S., Kumar, M., Gomber, S., 2021. Starch and powdered activated carbon amended alginate-biomass beads for metronidazole and bulk organic matter removal: Synthesis, optimization, reaction kinetics and reusability. *J. Environ. Chem. Eng.* 9, 106102. https://doi.org/10.1016/j.jece.2021.106102

Gupta, A.D., Rawat, K.P., Bhadauria, V., Singh, H., 2021. Recent trends in the application of modified starch in the adsorption of heavy metals from water: A review. *Carbohydr. Polym.* 269, 117763. https://doi.org/10.1016/j.carbpol.2021.117763

Gupta, V.K., Nayak, A., 2012. Cadmium removal and recovery from aqueous solutions by novel adsorbents prepared from orange peel and Fe_2O_3 nanoparticles. *Chem. Eng. J.* 180, 81–90. https://doi.org/10.1016/j.cej.2011.11.006

Hacıosmanoğlu, G.G., Mejías, C., Martín, J., Santos, J.L., Aparicio, I., Alonso, E., 2022. Antibiotic adsorption by natural and modified clay minerals as designer adsorbents for wastewater treatment: A comprehensive review. *J. Environ. Manage.* 317. https://doi.org/10.1016/j.jenvman.2022.115397

Hiemstra, T., van Riemsdijk, W., 1996. A surface structural approach to ion adsorption: The Charge Distribution (CD) *Model. J. Colloid Interface Sci.* 508, 488–508.

Ho, Y.S., McKay, G., 1999. Pseudo-second order model for sorption processes. *Process Biochem.* 34, 451–465. https://doi.org/10.1016/S0032-9592(98)00112-5

Hu, L., Guang, C., Liu, Y., Su, Z., Gong, S., Yao, Y., Wang, Y., 2020. Adsorption behavior of dyes from an aqueous solution onto composite magnetic lignin adsorbent. *Chemosphere* 246, 125757. https://doi.org/10.1016/j.chemosphere.2019.125757

Huang, X.X., Qiu, L.G., Zhang, W., Yuan, Y.P., Jiang, X., Xie, A.J., Shen, Y.H., Zhu, J.F., 2012. Hierarchically mesostructured MIL-101 metal-organic frameworks: Supramolecular template-directed synthesis and accelerated adsorption kinetics for dye removal. *CrystEngComm* 14, 1613–1617. https://doi.org/10.1039/c1ce06138k

Huang, Y., Yu, Q., Li, M., Jin, S., Fan, J., Zhao, L., Yao, Z., 2021. Surface modification of activated carbon fiber by low-temperature oxygen plasma: Textural property, surface chemistry, and the effect of water vapor adsorption. *Chem. Eng. J.* 418, 129474. https://doi.org/10.1016/j.cej.2021.129474

Ighalo, J.O., Omoarukhe, F.O., Ojukwu, V.E., Iwuozor, K.O., Igwegbe, C.A., 2022. Cost of adsorbent preparation and usage in wastewater treatment: A review. *Clean. Chem. Eng.* 3, 100042. https://doi.org/10.1016/j.clce.2022.100042

Iqbal, Z., Tanweer, M.S., Alam, M., 2022. Recent advances in adsorptive removal of wastewater pollutants by chemically modified metal oxides: A review. *J. Water Process Eng.* 46, 102641. https://doi.org/10.1016/j.jwpe.2022.102641

Ishaq, M., RC, C., Sharma, A., Ali, G., 2023. Agro-waste as a potential source for the development of adsorbents in water and wastewater treatment. *Int. J. Agric. Environ. Sci.* 10, 10–18. https://doi.org/10.14445/23942568/ijaes-v10i6p102

Jabli, M., Sebeia, N., El-Ghoul, Y., Soury, R., Al-Ghamdi, Y.O., Saleh, T.A., 2023. Chemical modification of microcrystalline cellulose with polyethyleneimine and hydrazine: Characterization and evaluation of its adsorption power toward anionic dyes. *Int. J. Biol. Macromol.* 229, 210–223. https://doi.org/10.1016/j.ijbiomac.2022.12.309

Johnson, E.B.G., Arshad, S.E., 2014. Hydrothermally synthesized zeolites based on kaolinite: A review. *Appl. Clay Sci.* 97–98, 215–221. https://doi.org/10.1016/j.clay.2014.06.005

Kabir, M.I., Daly, E., Maggi, F., 2014. A review of ion and metal pollutants in urban green water infrastructures. *Sci. Total Environ.* 470–471, 695–706. https://doi.org/10.1016/j.scitotenv.2013.10.010

Kainth, S., Sharma, P., Pandey, O.P., 2024. Green sorbents from agricultural wastes: A review of sustainable adsorption materials. *Appl. Surf. Sci. Adv.* 19, 100562. https://doi.org/10.1016/j.apsadv.2023.100562

Kapanga, P.M., William, G., Nyakairu, A., Isalomboto, C., Ntumba, S., Raphael, L., Tshimanga, M., Shehu, Z., 2024. A review of dye effluents polluting African surface water: Sources, impacts, physicochemical properties, and treatment methods. *Discov. Water.* https://doi.org/10.1007/s43832-024-00129-2

Karić, N., Maia, A.S., Teodorović, A., Atanasova, N., Langergraber, G., Crini, G., Ribeiro, A.R.L., Đolić, M., 2022. Bio-waste valorisation: Agricultural wastes as biosorbents for removal of (in)organic pollutants in wastewater treatment. *Chem. Eng. J. Adv*. 9. https://doi.org/10.1016/j.ceja.2021.100239

Kasbaji, M., Ibrahim, I., Mennani, M., Shaaban, S., Mokhtar, M., Salama, T.M., Moneam, I.A., Mbarki, M., Moubarik, A., Oubenali, M., 2023a. Future trends in dye removal by metal oxides and their Nano/Composites : A comprehensive review. *Inorg. Chem. Commun.* 158, 111546. https://doi.org/10.1016/j.inoche.2023.111546

Kasbaji, M., Mennani, M., Barhoumi, S., Esshouba, Y., Oubenali, M., 2024. Synergy of magnetic nanoparticles and sodium alginate-coated lignin for effective pollutant remediation, simple recovery, and cost-effective regeneration. *Langmuir*. https://doi.org/10.1021/acs.langmuir.4c02734

Kasbaji, M., Mennani, M., Boussetta, A., Grimi, N., Barba, F.J., Mbarki, M., Moubarik, A., 2022a. Bio-adsorption performances of methylene blue (MB) dye on terrestrial and marine natural fibers: Effect of physicochemical properties, kinetic models and thermodynamic parameters. *Sep. Sci. Technol.* 1–20. https://doi.org/10.1080/01496395.2022.2104733

Kasbaji, M., Mennani, M., Grimi, N., Barba, F.J., Oubenali, M., Simirgiotis, M.J., Mbarki, M., Moubarik, A., 2022b. Implementation and physico-chemical characterization of new alkali-modified bio-sorbents for cadmium removal from industrial discharges: Adsorption isotherms and kinetic approaches. *Process Biochem.* 120, 213–226. https://doi.org/10.1016/j.procbio.2022.06.010

Kasbaji, M., Mennani, M., Grimi, N., Oubenali, M., Mbarki, M., EL Zakhem, H., Moubarik, A., 2023b. Adsorption of cationic and anionic dyes onto coffee grounds cellulose/sodium alginate double-network hydrogel beads: Isotherm analysis and recyclability performance. *Int. J. Biol. Macromol.* 239, 124288. https://doi.org/10.1016/j.ijbiomac.2023.124288

Kasbaji, M., Mennani, M., Oubenali, M., Ait Benhamou, A., Boussetta, A., Ablouh, E.H., Mbarki, M., Grimi, N., El Achaby, M., Moubarik, A., 2023c. Bio-based functionalized adsorptive polymers for sustainable water decontamination: A systematic review of challenges and real-world implementation. *Environ. Pollut.* 335, 122349. https://doi.org/10.1016/j.envpol.2023.122349

Kasbaji, M., Mennani, M., Oulbaz, L., Oubenali, M., Moubarik, A., Sehaqui, H., Dahbi, M., Kassab, Z., El Achaby, M., 2025. From field to function: Exploring the versatility of alfa plant and its bio-derived materials for sustainable solutions. *Sustain. Mater. Technol.* 43, e01255. https://doi.org/10.1016/j.susmat.2025.e01255

Kausar, A., Zohra, S.T., Ijaz, S., Iqbal, M., Iqbal, J., Bibi, I., Nouren, S., El Messaoudi, N., Nazir, A., 2022. Cellulose-based materials and their adsorptive removal efficiency for dyes: A review. *Int. J. Biol. Macromol.* https://doi.org/10.1016/j.ijbiomac.2022.10.220

Konan, K.L., Peyratout, C., Smith, A., Bonnet, J.P., Magnoux, P., Ayrault, P., 2012. Surface modifications of illite in concentrated lime solutions investigated by pyridine adsorption. *J. Colloid Interface Sci.* 382, 17–21. https://doi.org/10.1016/j.jcis.2012.05.039

Kordala, N., Wyszkowski, M., 2024. Zeolite properties, methods of synthesis, and selected applications. *Molecules* 29. https://doi.org/10.3390/molecules29051069

Koyuncu, H., Kul, A.R., 2020. Biosorption study for removal of methylene blue dye from aqueous solution using a novel activated carbon obtained from nonliving lichen (Pseudevernia furfuracea (L.) Zopf.). *Surfaces and Interfaces* 19. https://doi.org/10.1016/j.surfin.2020.100527

Kut, K.M.K., Sarswat, A., Srivastava, A., Pittman, C.U., Mohan, D., 2016. A review of fluoride in African groundwater and local remediation methods. *Groundw. Sustain. Dev.* 2–3, 190–212. https://doi.org/10.1016/j.gsd.2016.09.001

Lee, J.E., Shafiq, I., Hussain, M., Lam, S.S., Rhee, G.H., Park, Y.K., 2022. A review on integrated thermochemical hydrogen production from water. *Int. J. Hydrogen Energy* 47, 4346–4356. https://doi.org/10.1016/j.ijhydene.2021.11.065

Li, A., Ge, W., Liu, L., Zhang, Y., Qiu, G., 2022. Synthesis and application of amine-functionalized $MgFe_2O_4$-biochar for the adsorption and immobilization of Cd(II) and Pb(II). *Chem. Eng. J.* 439, 135785. https://doi.org/10.1016/j.cej.2022.135785

Li, Z., Hanafy, H., Zhang, L., Sellaoui, L., Schadeck Netto, M., Oliveira, M.L.S., Seliem, M.K., Luiz Dotto, G., Bonilla-Petriciolet, A., Li, Q., 2020. Adsorption of congo red and methylene blue dyes on an ashitaba waste and a walnut shell-based activated carbon from aqueous solutions: Experiments, characterization and physical interpretations. *Chem. Eng. J.* 388, 124263. https://doi.org/10.1016/j.cej.2020.124263

Liang, Q., Luo, H., Geng, J., Chen, J., 2018. Facile one-pot preparation of nitrogen-doped ultra-light graphene oxide aerogel and its prominent adsorption performance of Cr(VI). *Chem. Eng. J.* 338, 62–71. https://doi.org/10.1016/j.cej.2017.12.145

Lin, G., Zeng, B., Li, J., Wang, Z., Wang, S., Hu, T., Zhang, L., 2023. A systematic review of metal organic frameworks materials for heavy metal removal: Synthesis, applications and mechanism. *Chem. Eng. J.* 460. https://doi.org/10.1016/j.cej.2023.141710

Linares-Solano, A., Lozano-Castelló, D., Lillo-Ródenas, M.A., Cazorla-Amorós, D., 2008. Carbon activation by alkaline hydroxides preparation and reactions, porosity and performance. *Chem. Phys. Carbon* 30, 1–62. https://doi.org/10.1201/9781420042993-5

Liu, H., Chang, L., Liu, W., Xiong, Z., Zhao, Y., Zhang, J., 2020. Advances in mercury removal from coal-fired flue gas by mineral adsorbents. *Chem. Eng. J.* 379, 122263. https://doi.org/10.1016/j.cej.2019.122263

Liu, H., Yang, J., Tian, C., Zhao, Y., Zhang, J., 2017. Mercury removal from coal combustion flue gas by modified palygorskite adsorbents. *Appl. Clay Sci.* 147, 36–43. https://doi.org/10.1016/j.clay.2017.05.006

Liu, K.G., Bigdeli, F., Sharifzadeh, Z., Gholizadeh, S., Morsali, A., 2023. Role of metal-organic framework composites in removal of inorganic toxic contaminants. *J. Clean. Prod.* 404, 136709. https://doi.org/10.1016/j.jclepro.2023.136709

Ma, Y., Zhang, X., Wen, J., 2021. Study on the harm of waste activated carbon and novel regeneration technology of it. *IOP Conf. Ser. Earth Environ. Sci.* 769. https://doi.org/10.1088/1755-1315/769/2/022047

Mansouri, F., Chouchene, K., Roche, N., Ksibi, M., 2021. Removal of pharmaceuticals from water by adsorption and advanced oxidation processes: State of the art and trends. *Appl. Sci.* 11. https://doi.org/10.3390/app11146659

Martins, R.O., Batista Junior, A.C., Melo Brito, C.C.S., Rocha, Y.A., Chaves, A.R., 2024. Molecularly imprinted polymers in solid-phase microextraction: Enhancing the analysis of pharmaceutical compounds across diverse sample matrices. *J. Pharm. Biomed. Anal. Open* 4, 100037. https://doi.org/10.1016/j.jpbao.2024.100037

Mitra, S., Chakraborty, A.J., Tareq, A.M., Emran, T. Bin, Nainu, F., Khusro, A., Idris, A.M., Khandaker, M.U., Osman, H., Alhumaydhi, F.A., Simal-Gandara, J., 2022. Impact of heavy metals on the environment and human health: Novel therapeutic insights to counter the toxicity. *J. King Saud Univ. - Sci.* 34, 101865. https://doi.org/10.1016/j.jksus.2022.101865

Mohan, B., Gupta, R.K., Pandey, V., Pombeiro, A.J.L., Ren, P., 2025. Optimized cleansing techniques: Engineered covalent-organic frameworks (COF) adsorbents for mycotoxin removal from food products. *Trends Food Sci. Technol.* 159, 104936. https://doi.org/10.1016/j.tifs.2025.104936

Mohan, D., Pittman, C.U., 2006. Activated carbons and low cost adsorbents for remediation of tri- and hexavalent chromium from water. *J. Hazard. Mater.* 137, 762–811. https://doi.org/10.1016/j.jhazmat.2006.06.060

Molina-Calderón, L., Basualto-Flores, C., Paredes-García, V., Venegas-Yazigi, D., 2022. Advances of magnetic nanohydrometallurgy using superparamagnetic nanomaterials as rare earth ions adsorbents: A grand opportunity for sustainable rare earth recovery. *Sep. Purif. Technol.* 299. https://doi.org/10.1016/j.seppur.2022.121708

Moraes, J.D.D., Bertolino, S.R.A., Cuffini, S.L., Ducart, D.F., Bretzke, P.E., Leonardi, G.R., 2017. Clay minerals: Properties and applications to dermocosmetic products and perspectives of natural raw materials for therapeutic purposes—A review. *Int. J. Pharm.* 534, 213–219. https://doi.org/10.1016/j.ijpharm.2017.10.031

Neme, I., Gonfa, G., Masi, C., 2022. Activated carbon from biomass precursors using phosphoric acid: A review. *Heliyon* 8, e11940. https://doi.org/10.1016/j.heliyon.2022.e11940

Nik, W.B.W., Rahman, M.M., Yusof, a M., Ani, F.N., Adnan, C.M.C., 2006. Production of activated carbon from palm oil shell waste and its adsorption characteristics. *Proc 1st Int. Conf. Nat. Resour. Eng. Technol.* 646–654.

Nizam, N.U.M., Hanafiah, M.M., Mahmoudi, E., Halim, A.A., Mohammad, A.W., 2021. The removal of anionic and cationic dyes from an aqueous solution using biomass-based activated carbon. *Sci. Rep.* 11, 1–17. https://doi.org/10.1038/s41598-021-88084-z

Onyeji, L.I., Aboje, A.A., 2011. Removal of heavy metals from dye effluent using activated carbon produced from coconut shell. *Int. J. Eng. Sci. Technol.* 3, 8238–8246.

Ozcan, D.O., Hendekci, M.C., Ovez, B., 2024. Enhancing the adsorption capacity of organic and inorganic pollutants onto impregnated olive stone derived activated carbon. *Heliyon* 10. https://doi.org/10.1016/j.heliyon.2024.e32792

Padilla-Ortega, E., Leyva-Ramos, R., Mendoza-Barron, J., Guerrero-Coronado, R.M., Jacobo-Azuara, A., Aragon-Piña, A., 2011. Adsorption of heavy metal ions from aqueous solution onto sepiolite. *Adsorpt. Sci. Technol.* 29, 569–584. https://doi.org/10.1260/0263-6174.29.6.569

Pallarés, J., González-Cencerrado, A., Arauzo, I., 2018. Production and characterization of activated carbon from barley straw by physical activation with carbon dioxide and steam. *Biomass and Bioenergy* 115, 64–73. https://doi.org/10.1016/j.biombioe.2018.04.015

Parthasarathy, P., Narayanan, S.K., 2014. Enhanced adsorptive removal of a model acid dye bromothymol blue from aqueous solution using magnetic chitosan-bamboo sawdust composite: Batch and column studies. *Environ. Prog. Sustain. Energy* 33, 676–680. https://doi.org/10.1002/ep.12147

Peng, D., Zhou, G., An, K., Guo, W., Yang, T., Kong, W., Shen, B., Wang, X., Lyu, H., Ma, J., 2025. Critical review on the adsorption of taste and odor compounds by activated carbon: Influencing factors, modifications methods, mechanism. *Chem. Eng. J.* 507. https://doi.org/10.1016/j.cej.2025.160677

Peng, X., Yan, J., He, C., Liu, R., Liu, Y., 2024. Sustainable triethylenetetramine modified sulfonated graphene oxide/chitosan composite for enhanced adsorption of Pb(II), Cd(II), and Ni(II) ions. *Int. J. Biol. Macromol.* 261, 129741. https://doi.org/10.1016/j.ijbiomac.2024.129741

Peramune, D., Manatunga, D.C., Dassanayake, R.S., Premalal, V., Liyanage, R.N., Gunathilake, C., Abidi, N., 2022. Recent advances in biopolymer-based advanced oxidation processes for dye removal applications: A review. *Environ. Res.* 215, 114242. https://doi.org/10.1016/j.envres.2022.114242

Pérez-Botella, E., Valencia, S., Rey, F., 2022. Zeolites in adsorption processes: State of the art and future prospects. *Chem. Rev.* 122, 17647–17695. https://doi.org/10.1021/acs.chemrev.2c00140

Perez-Calderon, J., Marin-Silva, D.A., Zaritzky, N., Pinotti, A., 2023. Eco-friendly PVA-chitosan adsorbent films for the removal of azo dye Acid Orange 7: Physical cross-linking, adsorption process, and reuse of the material. *Adv. Ind. Eng. Polym. Res.* 6, 239–254. https://doi.org/10.1016/j.aiepr.2022.12.001

Pet, I., Sanad, M.N., Farouz, M., ElFaham, M.M., El-Hussein, A., El-sadek, M.S.A., Althobiti, R.A., Ioanid, A., 2024. Review: Recent developments in the implementation of activated carbon as heavy metal removal management. *Water Conserv. Sci. Eng.* 9, 1–15. https://doi.org/10.1007/s41101-024-00287-3

Phiri, J., Ahadian, H., Sandberg, M., Granström, K., Maloney, T., 2023. The influence of physical mixing and impregnation on the physicochemical properties of pine wood activated carbon produced by one-step $ZnCl_2$ activation. *Micromachines* 14. https://doi.org/10.3390/mi14030572

Raj, A., Rego, R.M., Ajeya, K. V., Jung, H.Y., Kurkuri, M.D., 2024. Basil seeds loaded with MOFs as an eco-friendly and sustainable adsorbent for efficient removal of hazardous organic pollutants from water. *Sep. Purif. Technol.* 330, 125370. https://doi.org/10.1016/j.seppur.2023.125370

Rajendran, S., Priya, A.K., Senthil Kumar, P., Hoang, T.K.A., Sekar, K., Chong, K.Y., Khoo, K.S., Ng, H.S., Show, P.L., 2022. A critical and recent developments on adsorption technique for removal of heavy metals from wastewater-A review. *Chemosphere* 303, 135146. https://doi.org/10.1016/j.chemosphere.2022.135146

Rao, P.H., Kumar, R.R., Mohan, N., 2019. Phycoremediation: Role of algae in waste management, microorganisms for sustainability. https://doi.org/10.1007/978-981-13-7904-8_3

Rao, Z., Feng, K., Tang, B., Wu, P., 2017. Surface decoration of amino-functionalized metal-organic framework/graphene oxide composite onto polydopamine-coated membrane substrate for highly efficient heavy metal removal. *ACS Appl. Mater. Interfaces* 9, 2594–2605. https://doi.org/10.1021/acsami.6b15873

Rashid, R., Shafiq, I., Akhter, P., Iqbal, M.J., Hussain, M., 2021. A state-of-the-art review on wastewater treatment techniques: The effectiveness of adsorption method. *Environ. Sci. Pollut. Res.* 28, 9050–9066. https://doi.org/10.1007/s11356-021-12395-x

Rathi, B.S., Kumar, P.S., 2021. Application of adsorption process for effective removal of emerging contaminants from water and wastewater. *Environ. Pollut.* 280, 116995. https://doi.org/10.1016/j.envpol.2021.116995

Ray, P.C., Yu, H., Fu, P.P., 2009. Toxicity and environmental risks of nanomaterials: Challenges and future needs, *J. Environ. Sci. Health - Part C Environ. Carcinog. Ecotoxicol. Rev.* https://doi.org/10.1080/10590500802708267

Ren, S., Huang, S., Liu, B., 2022. Enhanced removal of ammonia nitrogen from rare earth wastewater by NaCl modified vermiculite: Performance and mechanism. *Chemosphere* 302, 134742. https://doi.org/10.1016/j.chemosphere.2022.134742

Roa, K., Oyarce, E., Boulett, A., ALSamman, M., Oyarzún, D., Pizarro, G.D.C., Sánchez, J., 2021. Lignocellulose-based materials and their application in the removal of dyes from water: A review. *Sustain. Mater. Technol.* 29. https://doi.org/10.1016/j.susmat.2021.e00320

Robbins, C.A., Du, X., Bradley, T.H., Quinn, J.C., Bandhauer, T.M., Conrad, S.A., Carlson, K.H., Tong, T., 2022. Beyond treatment technology: Understanding motivations and barriers for wastewater treatment and reuse in unconventional energy production. *Resour. Conserv. Recycl.* 177, 106011. https://doi.org/10.1016/j.resconrec.2021.106011

Rodoshi, N., Sharmin, T., Bin, A., 2024. Exploring the versatility of aerogels: Broad applications in biomedical engineering, astronautics, energy storage, biosensing, and current progress. *Heliyon* 10, e23102. https://doi.org/10.1016/j.heliyon.2023.e23102

Rout, D.R., Jena, H.M., Baigenzhenov, O., Hosseini-Bandegharaei, A., 2023. Graphene-based materials for effective adsorption of organic and inorganic pollutants: A critical and comprehensive review. *Sci. Total Environ.* 863, 160871. https://doi.org/10.1016/j.scitotenv.2022.160871

Russo, T., Fucile, P., Giacometti, R., Sannino, F., 2021. Sustainable removal of contaminants by biopolymers: A novel approach for wastewater treatment. Current state and future perspectives. *Processes*. https://doi.org/10.3390/pr9040719

Salem, M.A.S., Khan, A.M., Manea, Y.K., 2022. A novel nano-hybrid carbon architecture as chemo sensor for natural hazards: Active adsorption of Rose Bengal dye and detection of hazard pollutants at ppb level. *J. Environ. Chem. Eng*. 10, 107032. https://doi.org/10.1016/j.jece.2021.107032

Salgado, E.M., Esteves, A.F., Gonçalves, A.L., Pires, J.C.M., 2023. Microalgal cultures for the remediation of wastewaters with different nitrogen to phosphorus ratios: Process modelling using artificial neural networks. *Environ. Res*. 231. https://doi.org/10.1016/j.envres.2023.116076

Samsami, S., Mohamadi, M., Sarrafzadeh, M.H., Rene, E.R., Firoozbahr, M., 2020. Recent advances in the treatment of dye-containing wastewater from textile industries: Overview and perspectives. *Process Saf. Environ. Prot*. 143, 138–163. https://doi.org/10.1016/j.psep.2020.05.034

Samuel Olugbenga, O., Goodness Adeleye, P., Blessing Oladipupo, S., Timothy Adeleye, A., Igenepo John, K., 2024. Biomass-derived biochar in wastewater treatment-a circular economy approach. *Waste Manag. Bull*. 1, 1–14. https://doi.org/10.1016/j.wmb.2023.07.007

Sarma, U., Hoque, M.E., Thekkangil, A., Venkatarayappa, N., Rajagopal, S., 2024. Microalgae in removing heavy metals from wastewater – An advanced green technology for urban wastewater treatment. *J. Hazard. Mater. Adv*. 15, 100444. https://doi.org/10.1016/j.hazadv.2024.100444

Satyam, S., Patra, S., 2024. Innovations and challenges in adsorption-based wastewater remediation: A comprehensive review. *Heliyon* 10, e29573. https://doi.org/10.1016/j.heliyon.2024.e29573

Saurel, D., Orayech, B., Xiao, B., Carriazo, D., Li, X., Rojo, T., 2018. From charge storage mechanism to performance: A roadmap toward high specific energy sodium-ion batteries through carbon anode optimization. *Adv. Energy Mater*. 8, 1–33. https://doi.org/10.1002/aenm.201703268

Sayari, A., Belmabkhout, Y., Serna-Guerrero, R., 2011. Flue gas treatment via CO_2 adsorption. *Chem. Eng. J*. 171, 760–774. https://doi.org/10.1016/j.cej.2011.02.007

Selvasembian, R., Mal, J., Das, S., Verma, D.K., Anastopoulos, I., (Eds.). 2024. Emerging trends in microbial electrochemical technologies for sustainable mitigation of water resources contamination. *Springer Nature*. https://doi.org/10.1007/978-3-031-74636-9.

Shahabi, M., Ahmadpour, A., Raissi, H., 2024. Assessment of water purification by IRMOF-1 based on rGO as a new nanoengineered adsorbent: Insights from adsorption mechanism. *J. Mol. Liq*. 400, 124485. https://doi.org/10.1016/j.molliq.2024.124485

Singh, A., Kostova, I., 2024. Health effects of heavy metal contaminants Vis-à-Vis microbial response in their bioremediation. *Inorganica Chim. Acta* 568, 122068. https://doi.org/10.1016/j.ica.2024.122068

Singh Mehra, K., Abrar, I., Kant Bhatia, R., Goel, V., 2025. A comprehensive review of algae consortium for wastewater bioremediation and biodiesel production. *Energy Convers. Manag*. 325. https://doi.org/10.1016/j.enconman.2024.119428

Singh, V., Ahmed, G., Vedika, S., Kumar, P., Chaturvedi, S.K., Rai, S.N., Vamanu, E., Kumar, A., 2024. Toxic heavy metal ions contamination in water and their sustainable reduction by eco-friendly methods: Isotherms, thermodynamics and kinetics study. *Sci. Rep*. 14, 1–13. https://doi.org/10.1038/s41598-024-58061-3

Singla, R., Alex, T.C., Kumar, R., 2020. On mechanical activation of glauconite: Physicochemical changes, alterations in cation exchange capacity and mechanisms. *Powder Technol*. 360, 337–351. https://doi.org/10.1016/j.powtec.2019.10.035

Sirianuntapiboon, S., Sansak, J., 2008. Treatability studies with granular activated carbon (GAC) and sequencing batch reactor (SBR) system for textile wastewater containing direct dyes. *J. Hazard. Mater*. 159, 404–411. https://doi.org/10.1016/j.jhazmat.2008.02.031

Siu, P.W., Brown, Z.J., Farha, O.K., Hupp, J.T., Scheidt, K.A., 2013. A mixed dicarboxylate strut approach to enhancing catalytic activity of a de novo urea derivative of metal–organic framework UiO-67. *Chem. Commun.* 49, 10920–10922. https://doi.org/10.1039/c3cc47177b

Soltani, R., Marjani, A., Shirazian, S., 2020. A hierarchical LDH/MOF nanocomposite: Single, simultaneous and consecutive adsorption of a reactive dye and Cr(vi). *Dalt. Trans.* 49, 5323–5335. https://doi.org/10.1039/d0dt00680g

Subash, A., Naebe, M., Wang, X., Kandasubramanian, B., 2023. Biopolymer – A sustainable and efficacious material system for effluent removal. *J. Hazard. Mater.* 443, 130168. https://doi.org/10.1016/j.jhazmat.2022.130168

Suiyi, Z., Yanong, R., Yuxin, Z., Minglin, Z., Weilu, Y., Xinfeng, X., Yang, Y., Jiancong, L., Zhan, Q., Jialin, L., Yu, C., 2024. A novel clinoatacamite route to effectively separate Cu for recycling Ca/Zn/Mn from hazardous smelting waterwork sludge. *J. Environ. Chem. Eng.* 12, 112024. https://doi.org/10.1016/j.jece.2024.112024

Tan, X., Liu, Y., Zeng, G., Wang, X., Hu, X., Gu, Y., Yang, Z., 2015. Application of biochar for the removal of pollutants from aqueous solutions. *Chemosphere* 125, 70–85. https://doi.org/10.1016/j.chemosphere.2014.12.058

Tchernev, D.I., 2001. Natural zeolites in solar energy heating, cooling, and energy storage. *Rev. Mineral. Geochemistry* 45, 588–617. https://doi.org/10.2138/rmg.2001.45.17

Than-ardna, B., Hiranphinyophat, S., Matsumoto, M., Meeyoo, V., Kitiyanan, B., 2025. Synthesized structure of Mg-MOF-74 decorated on ZIF-8 as solid adsorbent for CO_2 capture. *Resour. Chem. Mater.* 4. https://doi.org/10.1016/j.recm.2025.100116

Tsai, W.T., Chang, C.Y., Lee, S.L., 1998. A low cost adsorbent from agricultural waste corn cob by zinc chloride activation. *Bioresour. Technol.* 64, 211–217. https://doi.org/10.1016/S0960-8524(97)00168-5

Ukanwa, K.S., Patchigolla, K., Sakrabani, R., Anthony, E., Mandavgane, S., 2019. A review of chemicals to produce activated carbon from agricultural waste biomass. *Sustain.* 11, 1–35. https://doi.org/10.3390/su11226204

Upadhyay, U., Sreedhar, I., Singh, S.A., Patel, C.M., Anitha, K.L., 2021. Recent advances in heavy metal removal by chitosan based adsorbents. *Carbohydr. Polym.* 251, 117000. https://doi.org/10.1016/j.carbpol.2020.117000

Vardhan, K.H., Kumar, P.S., Panda, R.C., 2019. A review on heavy metal pollution, toxicity and remedial measures: Current trends and future perspectives. *J. Mol. Liq.* https://doi.org/10.1016/j.molliq.2019.111197

Vaz-Ramos, J., Le Calvé, S., Begin, S., 2025. Polycyclic aromatic hydrocarbons in water environments: Impact, legislation, depollution processes and challenges, and magnetic iron oxide/graphene-based nanocomposites as promising adsorbent solutions. *J. Hazard. Mater.* 490. https://doi.org/10.1016/j.jhazmat.2025.137726

Velempini, T., Ahamed, M.E.H., Pillay, K., 2023. Heavy-metal spent adsorbents reuse in catalytic, energy and forensic applications- A new approach in reducing secondary pollution associated with adsorption. *Results Chem.* 5, 100901. https://doi.org/10.1016/j.rechem.2023.100901

Walker, G.M., Weatherley, L.R., 2000. Textile wastewater treatment using granular activated carbon adsorption in fixed beds. *Sep. Sci. Technol.* 35, 1329–1341. https://doi.org/10.1081/SS-100100227

Wang, B., Lan, J., Bo, C., Gong, B., Ou, J., 2023. Adsorption of heavy metal onto biomass-derived activated carbon: Review. *RSC Adv.* 13, 4275–4302. https://doi.org/10.1039/d2ra07911a

Wang, F., Xiang, L., Sze-Yin Leung, K., Elsner, M., Zhang, Y., Guo, Y., Pan, B., Sun, H., An, T., Ying, G., Brooks, B.W., Hou, D., Helbling, D.E., Sun, J., Qiu, H., Vogel, T.M., Zhang, Wei, Gao, Yanzheng, Simpson, M.J., Luo, Yi, Chang, S.X., Su, G., Wong, B.M., Fu, T.M., Zhu, D., Jobst, K.J., Ge, C., Coulon, F., Harindintwali, J.D., Zeng, X., Wang, H., Fu, Y., Wei, Z., Lohmann, R., Chen, C., Song, Y., Sanchez-Cid, C., Wang, Y., El-Naggar,

A., Yao, Y., Huang, Y., Cheuk-Fung Law, J., Gu, Chenggang, Shen, H., Gao, Yanpeng, Qin, C., Li, Hao, Zhang, Tong, Corcoll, N., Liu, M., Alessi, D.S., Li, Hui, Brandt, K.K., Pico, Y., Gu, Cheng, Guo, J., Su, J., Corvini, P., Ye, M., Rocha-Santos, T., He, H., Yang, Y., Tong, M., Zhang, Weina, Suanon, F., Brahushi, F., Wang, Z., Hashsham, S.A., Virta, M., Yuan, Q., Jiang, G., Tremblay, L.A., Bu, Q., Wu, J., Peijnenburg, W., Topp, E., Cao, X., Jiang, X., Zheng, M., Zhang, Taolin, Luo, Yongming, Zhu, L., Li, X., Barceló, D., Chen, J., Xing, B., Amelung, W., Cai, Z., Naidu, R., Shen, Q., Pawliszyn, J., Zhu, Y. G., Schaeffer, A., Rillig, M.C., Wu, F., Yu, G., Tiedje, J.M., 2024. Emerging contaminants: A one health perspective. *Innovation* 5, 100612. https://doi.org/10.1016/j.xinn.2024.100612

Wang, J., Guo, X., 2020. Adsorption kinetic models: Physical meanings, applications, and solving methods. *J. Hazard. Mater.* 390, 122156. https://doi.org/10.1016/j.jhazmat.2020.122156

Wang, J., Wang, Z., Vieira, C.L.Z., Wolfson, J.M., Pingtian, G., Huang, S., 2019. Review on the treatment of organic pollutants in water by ultrasonic technology. *Ultrason. Sonochem.* 55, 273–278. https://doi.org/10.1016/j.ultsonch.2019.01.017

Wang, J., Xu, B., 2023. Removal of radionuclide 99Tc from aqueous solution by various adsorbents: A review. *J. Environ. Radioact.* 270, 107267. https://doi.org/10.1016/j.jenvrad.2023.107267

Wang, S., Peng, Y., 2010. Natural zeolites as effective adsorbents in water and wastewater treatment. *Chem. Eng. J.* 156, 11–24. https://doi.org/10.1016/j.cej.2009.10.029

Wang, Z., Fernández-Blanco, C., Chen, J., Veiga, M.C., Kennes, C., 2024. Effect of electron acceptors on product selectivity and carbon flux in carbon chain elongation with Megasphaera hexanoica. *Sci. Total Environ.* 912. https://doi.org/10.1016/j.scitotenv.2023.169509

Wibowo, Y.G., Anwar, D., Safitri, H., Surya, I., Sudibyo, S., Yuliansyah, A.T., Murti Petrus, H.T.B., 2025. Functionalized magnetite-biochar with live and dead bacteria for adsorption-biosorption of highly toxic metals: Cd, Hg, and Pb. *Next Mater.* 6, 100487. https://doi.org/10.1016/j.nxmate.2025.100487

Wu, C.H., Tsai, S.C., 2022. Adsorption of sulfonamides from aqueous solution by carbon nanotubes: Determination of kinetics, equilibrium and thermodynamic parameters. *Desalin. Water Treat.* 249, 74–86. https://doi.org/10.5004/dwt.2022.28019

Wu, J., Qiu, X., Chen, S., 2022. Preparation and characterization of an amine-modified graphene aerogel for enhanced carbon dioxide adsorption. *J. Mater. Sci.* 57, 1727–1737. https://doi.org/10.1007/s10853-021-06768-3

Xie, S., Huang, L., Su, C., Yan, J., Chen, Z., Li, M., Du, M., Zhang, H., 2024. Application of clay minerals as adsorbents for removing heavy metals from the environment. *Green Smart Min. Eng.* 1, 249–261. https://doi.org/10.1016/j.gsme.2024.07.002

Xu, J., Du, P., Bi, W., Yao, G., Li, S., Liu, H., 2020. Graphene oxide aerogels co-functionalized with polydopamine and polyethylenimine for the adsorption of anionic dyes and organic solvents. *Chem. Eng. Res. Des.* 154, 192–202. https://doi.org/10.1016/j.cherd.2019.12.014

Yaashikaa, P.R., Senthil Kumar, P., Karishma, S., 2022. Review on biopolymers and composites – Evolving material as adsorbents in removal of environmental pollutants. *Environ. Res.* 212. https://doi.org/10.1016/j.envres.2022.113114

Yahia, E.H., Serafin, J., Román-Martínez, M.C., Sreńscek-Nazzal, J., Dziejarski, B., Saidi, M., Ouzzine, M., 2025. Preparation of activated carbon from Moroccan argan press cake using KOH activation and its application for CO_2 adsorption. *Fuel* 393. https://doi.org/10.1016/j.fuel.2025.134922

Yan, X., Ge, H., 2023. Preparation of metal organic frameworks modified chitosan composite with high capacity for Hg(II) adsorption. *Int. J. Biol. Macromol.* 232, 123329. https://doi.org/10.1016/j.ijbiomac.2023.123329

Yang, I., Jung, M., Kim, M.S., Choi, D., Jung, J.C., 2021. Physical and chemical activation mechanisms of carbon materials based on the microdomain model. *J. Mater. Chem. A* 9, 9815–9825. https://doi.org/10.1039/d1ta00765c

Yang, X., Zhou, Y., Hu, J., Zheng, Q., Zhao, Y., Lv, G., Liao, L., 2024. Clay minerals and clay-based materials for heavy metals pollution control. *Sci. Total Environ.* 954, 176193. https://doi.org/10.1016/j.scitotenv.2024.176193

Yi, H., Nakabayashi, K., Yoon, S.H., Miyawaki, J., 2021. Pressurized physical activation: A simple production method for activated carbon with a highly developed pore structure. *Carbon* 183, 735–742. https://doi.org/10.1016/j.carbon.2021.07.061

Yue, B., Liu, S., Chai, Y., Wu, G., Guan, N., Li, L., 2022. Zeolites for separation: Fundamental and application. *J. Energy Chem.* 71, 288–303. https://doi.org/10.1016/j.jechem.2022.03.035

Zango, Z.U., Ramli, A., Jumbri, K., Abdurrahman, M., 2021. UiO-66 and ZIF-8 metal-organic frameworks for acenaphthene adsorption. *Springer Proc. Complex.* 239–253. https://doi.org/10.1007/978-981-16-4513-6_21

Zare, E.N., Fallah, Z., Le, V.T., Doan, V.D., Mudhoo, A., Joo, S.W., Vasseghian, Y., Tajbakhsh, M., Moradi, O., Sillanpää, M., Varma, R.S., 2022. Remediation of pharmaceuticals from contaminated water by molecularly imprinted polymers: A review, *Environ. Chem. Lett.* https://doi.org/10.1007/s10311-022-01439-4

Zaynab, M., Al-Yahyai, R., Ameen, A., Sharif, Y., Ali, L., Fatima, M., Khan, K.A., Li, S., 2022. Health and environmental effects of heavy metals. *J. King Saud Univ. - Sci.* 34, 101653. https://doi.org/10.1016/j.jksus.2021.101653

Zhai, L., Zheng, X., Liu, M., Wang, X., Li, W., Zhu, X., Yuan, A., Xu, Y., Song, P., 2023. Tuning surface functionalizations of UiO-66 towards high adsorption capacity and selectivity eliminations for heavy metal ions. *Inorg. Chem. Commun.* 154, 110937. https://doi.org/10.1016/j.inoche.2023.110937

Zhang, G., Lin, Y., Wang, M., 2011. Remediation of copper polluted red soils with clay materials. *J. Environ. Sci.* 23, 461–467. https://doi.org/10.1016/S1001-0742(10)60431-7

Zhang, J., Guo, H., Yang, F., Wang, M., Zhang, H., Zhang, T., Sun, L., Yang, M., Yang, W., 2021. Walnut shell-derived porous carbon integrated with Ni-MOF/SPANI composites for high-performance supercapacitor. *Colloids Surfaces A Physicochem. Eng. Asp.* 630, 127584. https://doi.org/10.1016/j.colsurfa.2021.127584

Zhang, Q.H., Yang, W.N., Ngo, H.H., Guo, W.S., Jin, P.K., Dzakpasu, M., Yang, S.J., Wang, Q., Wang, X.C., Ao, D., 2016. Current status of urban wastewater treatment plants in China. *Environ. Int.* 92–93, 11–22. https://doi.org/10.1016/j.envint.2016.03.024

Zhao, G., Li, J., Ren, X., Chen, C., Wang, X., 2011. Few-layered graphene oxide nanosheets as superior sorbents for heavy metal ion pollution management. *Environ. Sci. Technol.* 45, 10454–10462. https://doi.org/10.1021/es203439v

Zheng, M., Xu, L., Chen, C., Labiadh, L., Yuan, B., Fu, M.L., 2022. MOFs and GO-based composites as deliberated materials for the adsorption of various water contaminants. *Sep. Purif. Technol.* 294, 121187. https://doi.org/10.1016/j.seppur.2022.121187

Zhou, N., Chen, H., Xi, J., Yao, D., Zhou, Z., Tian, Y., Lu, X., 2017. Biochars with excellent Pb(II) adsorption property produced from fresh and dehydrated banana peels via hydrothermal carbonization. *Bioresour. Technol.* 232, 204–210. https://doi.org/10.1016/j.biortech.2017.01.074

5 Treatment of Wastewater Using Biochars and Hybrid Composites

Mohssine Ghazoui, Aboubacar Sidigh Sylla, Siham Dabali, Aziz Hasib, Khaoula Esslimani, Belkacem Benadda, Otmane Boudouch, and Reda Elkacmi

5.1 INTRODUCTION

Sustainable wastewater management has become a global priority in the face of increasing water pollution caused by human activities (Kishor et al., 2019). Demographic expansion, increasing urbanisation, and industrialisation have led to a significant deterioration in the quality of water resources. Domestic, agricultural, hospital, and industrial effluents contain a wide range of chemical compounds, some of which are barely or not at all eliminated by conventional treatment (Lellis et al., 2019; Wato, 2020).

Among these substances of concern are emerging contaminants, a heterogeneous group including pharmaceutical residues, personal care products, pesticides, synthetic dyes, heavy metals, radionuclides, and various persistent organic compounds (Tripathi et al., 2022). Although present in low concentrations (of the order of ng L^{-1} to µg L^{-1}), these pollutants are likely to induce toxic, endocrine-disrupting, genotoxic, or carcinogenic effects in both aquatic organisms and humans. Their non-biodegradable nature and potential for bioaccumulation raise growing environmental and health concerns (Ghazoui et al., 2024, 2025). A schematic overview of the main sources, pollutant categories, and associated effects is presented in Figure 5.1.

Conventional treatment technologies, such as coagulation–flocculation, membrane filtration, advanced oxidation, or ion exchange, have shown limited performance in effectively eliminating emerging contaminants. These processes also have several drawbacks: high energy consumption, generation of secondary by-products (sludge, oxidised compounds), operational complexity, and high operating and maintenance costs. These limitations compromise their large-scale applicability in a context of environmental sustainability (Abhraham et al., 2020; Alsukaibi, 2022; Kishor et al., 2021).

As a result, there is a growing need for alternative solutions capable of effectively treating a wide range of pollutants, while being economically viable, environmentally responsible, and easy to implement. Research is now focusing on the development of

 DOI: 10.1201/9781003624905-5

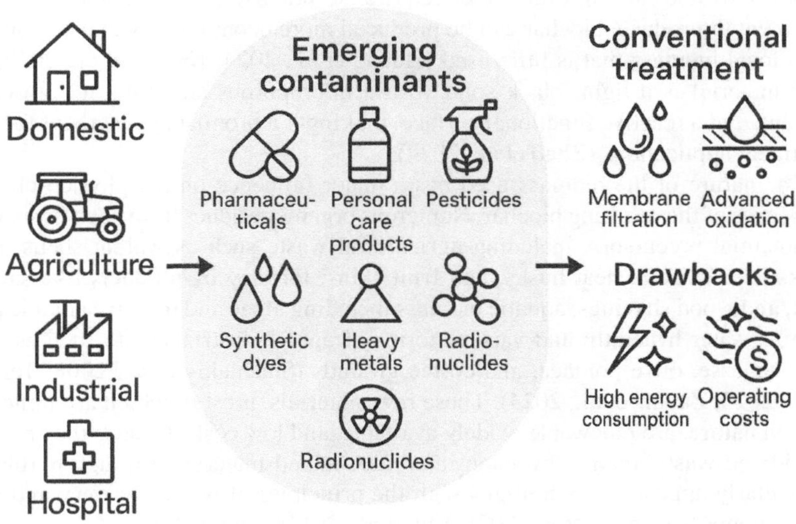

FIGURE 5.1 Schematic representation of the sources, types of emerging contaminants, and their impacts in wastewater.

new functional materials derived from renewable resources, with high adsorption properties, good chemical stability, easy regeneration, and low environmental impact (Zhao et al., 2025a).

Among the various treatment strategies under development, adsorption has emerged as one of the most promising due to its simplicity, efficiency, and versatility (Bernal et al., 2021). This physico-chemical process involves the accumulation of pollutants at the surface of a solid material, called an adsorbent, through mechanisms such as van der Waals interactions, hydrogen bonding, ion exchange, or surface complexation (Wang et al., 2024). Adsorption processes are particularly well suited to removing low-concentration contaminants from aqueous environments, with the advantage of operating under mild conditions and generating little to no harmful by-products (Hama Aziz et al., 2023; Pathak & Gupta, 2020).

In this context, this chapter explores an innovative approach based on the use of adsorbent materials of biosourced origin and hybrid composites for the treatment of contaminated wastewater. This strategy is based on the use of plant residues that can be converted into biochar, combined with inorganic or nanostructured materials, to optimise pollutant removal performance under realistic conditions. The following sections detail the properties, mechanisms, and prospects of these advanced adsorption systems.

5.2 BIOCHARS AS ADSORBENTS FOR WASTEWATER TREATMENT

5.2.1 Definition and Origin of Biochars

Biochar is a porous, carbon-rich material obtained by pyrolysis – a process of thermal decomposition of biomass in a controlled atmosphere, generally low in oxygen.

Unlike activated carbon, which often requires several stages of chemical and thermal treatment to produce, biochar can be produced more economically and ecologically, from local biomass that is little used (Huang et al., 2024; Kasera et al., 2022). The final material is a light, black solid with a microporous structure, a high carbon content, and a reactive functional surface, making it a promising adsorbent for water treatment applications (Zhao et al., 2024).

The nature of the biomass used has a major influence on the physico-chemical properties of the resulting biochar. Numerous organic residues have been investigated as potential precursors, including agricultural waste such as walnut shells, maize stalks, rice straw, wheat husks, and fruit skins; forestry by-products like sawdust, bark, and wood shavings; aquatic biomass including algae and invasive aquatic plants such as water hyacinth; and various forms of agro-industrial waste such as sugarcane bagasse, olive pomace, and coffee grounds (Stasinakis et al., 2008; Ton-That et al., 2023; Zubair et al., 2024). These raw materials, most of which are lignocellulosic in nature, are renewable, widely available, and low cost. Because they are often considered waste, their conversion into value-added materials through pyrolysis is particularly appealing, as it aligns with the principles of waste recovery and circular economy (Asemave et al., 2021; Liu et al., 2014). An overview of these biomass categories and representative examples is provided in Table 5.1.

The pyrolysis process typically takes place at temperatures of between 300°C and 700°C, under anaerobic conditions, or in an inert atmosphere. The production parameters – temperature, heating time, rate of temperature rise, and ambient gas – directly influence the porous structure, elemental composition, and surface reactivity of the biochar obtained (Chen et al., 2019; Dinc & Yel, 2018).

In addition to their technical potential, biochars offer numerous environmental advantages. Their production reduces the volume of organic waste, contributes to carbon sequestration, and enables the development of low-cost treatment solutions that are particularly suited to rural areas and developing countries. As part of a circular economy approach, biochars are a sustainable way of developing effective adsorbents for wastewater treatment (Abid et al., 2022; Wei et al., 2023).

TABLE 5.1
Classification of Biomass Sources for Biochar Production and Their Typical Applications in Wastewater Treatment

Biomass Type	Source	Examples	References
Agricultural residues	Crop by-products	Nut shells, corn stalks, rice husks, wheat straw	Condeço et al. (2021), Singh et al. (2014)
Aquatic biomass	Aquatic plants (freshwater/marine)	Algae, water hyacinth, invasive aquatic plants	Häder et al. (2020), Khan et al. (2022)
	By-products of the wood industry	Sawdust, bark, wood chips	Bosch et al. (2022), Karić et al. (2022)
Agro-industrial waste	Food/agro-processing residues	Bagasse, olive pomace, coffee grounds	Aslan and Şirazi (2020), Wekoye et al. (2020)

5.2.2 Physico-Chemical Properties of Biochars

The effectiveness of biochars in adsorption processes depends largely on their physico-chemical properties. The most important parameters include specific surface area, total pore volume, pore size distribution (micropores, mesopores), and the presence of surface-active functional groups such as hydroxyls (-OH), carboxyls (-COOH), and carbonyls (C=O) (Ahmed et al., 2021).

Pyrolysis conditions have a direct influence on these characteristics. In general, higher temperatures (>500°C) favour greater aromaticity, higher fixed carbon content, and a more stable structure, while lower temperatures allow more oxygenated functional groups to be retained, which are essential for interaction with polar contaminants. The heating rate, thermal retention time, and pyrolysis atmosphere (generally inert) are also key factors modulating the quality of the biochar produced (Yahya et al., 2015).

Other influential parameters include surface charge, pH at zero charge point, and cation exchange capacity, which determine pollutant retention mechanisms. The hierarchical porous structure of biochars (combining micro-, meso-, and macropores) favours the diffusion of solutes and enables interactions of the monomolecular or multilayer adsorption type, particularly in the case of heavy metals, dyes, or organic contaminants. The study on the efficient removal of 2,4-dinitrophenol on coconut shell biomass-derived activated carbon has been conducted (Mafo et al., 2024).

5.2.3 Techniques for Preparing and Modifying Biochars

The production of biochars is based on the controlled pyrolysis of lignocellulosic biomasses at temperatures generally between 300°C and 700°C, in an inert atmosphere (nitrogen, argon) or in a vacuum. The choice of operating conditions – temperature, heating rate, and residence time – has a major influence on the textural and chemical properties of the material obtained (Hassaan et al., 2023).

In order to improve their adsorption performance, biochars can be subjected to various modification techniques, which fall into two broad categories: physical and chemical modifications. Physical modifications mainly involve high-temperature (>700°C) thermal activation to increase the specific surface area and develop a more extensive pore network. Other approaches include microwave treatment or multi-stage carbonisation. Chemical modifications involve specific agents designed to functionalise the biochar surface or enrich it with active elements. The main strategies include activation with alkaline hydroxides (such as KOH or NaOH) or strong acids (such as H_3PO_4, HCl, or H_2SO_4). The aim is to generate additional porosity and increase the density of functional groups present on the surface. Oxidation with agents such as hydrogen peroxide (H_2O_2) or potassium permanganate ($KMnO_4$), meanwhile, allows oxygenated groups to be introduced, thus improving the material's affinity for polar compounds (Agbale et al., 2025; Elmouwahidi et al., 2012; Ghezzehei et al., 2014).

Another widely used approach involves metal impregnation, particularly with Fe, Mn, Zn, or Mg ions, in order to specifically target certain pollutants such as heavy metals or anions. More advanced techniques include magnetising biochars with iron particles (Fe_3O_4 or Fe_0), facilitating their separation after treatment, or grafting organic ligands onto the biochar surface, which improves selectivity towards complex organic contaminants such as antibiotics, dyes, or pesticides. The effectiveness of these modifications depends on a number of factors, including the method used, the type of initial biomass, the operating conditions, and the chemical compatibility between the carbon substrate and the modifying agents (Nikić et al., 2024).

5.2.4 Examples of Applications for Different Contaminants

Biochars have proven highly effective in eliminating a wide range of pollutants present in wastewater, in particular heavy metals, synthetic dyes, antibiotics, and other organic micropollutants. Their performance is highly dependent on the surface characteristics of the material, the type of contaminant targeted, and the treatment operating conditions (pH, temperature, initial concentration, etc.) (Tong et al., 2019).

In the case of heavy metals, such as lead (Pb^{2+}), cadmium (Cd^{2+}), copper (Cu^{2+}), or hexavalent chromium (Cr^{6+}), adsorption is mainly based on surface complexation, ion exchange, or precipitation mechanisms. Biochars functionalised with acid groups (carboxyls, hydroxyls) or enriched with metals (Fe, Mn) show an increased affinity for these toxic cations (Charii et al., 2024; Singh et al., 2010).

For dyes, particularly cationic dyes such as methylene blue or malachite green, adsorption is favoured by the porous structure of the biochar, the negatively charged surface, and electrostatic and π–π interactions between the carbon skeleton and the aromatic rings of the dyes (Han et al., 2020).

Antibiotics and drugs (such as tetracycline, ciprofloxacin, or sulfamethoxazole) are removed via a combination of hydrophobic interactions, hydrogen bonding, and adsorption by functional groups present on the biochar. Biochars doped or modified with specific materials (metal oxides, amine ligands) show enhanced selectivity towards these complex molecules (Han et al., 2019).

Comparable results have been obtained for other classes of contaminants, such as nutrients (ammonium, phosphate, nitrate), organochlorine pesticides, and emerging pharmaceutical compounds. Several studies have reported adsorption capacities in excess of 150–200 mg g^{-1}, notably for biochar-based hybrid materials (Huang et al., 2019). A detailed overview of the adsorption capacities, kinetics, and isotherm models for different contaminants is provided in Table 5.2. This compilation highlights the versatility of biochar in targeting both cationic and anionic pollutants under various operating conditions. Such comprehensive datasets are essential to guide the rational design and selection of biochars for wastewater treatment applications (Salleh et al., 2011).

5.3 HYBRID COMPOSITES BASED ON BIOCHAR

5.3.1 Types of Biochar Hybrid Composites

Biochar-based hybrid composites are being developed to overcome the limitations of crude biochars, particularly in terms of adsorption capacity, structural stability, and selectivity towards contaminants. The most common approaches include

TABLE 5.2
Extended Adsorption Models and Parameters for Various Contaminants using Biochar-Based Adsorbents

Contaminant	Biochar Source	Kinetic Model	Isotherm Model	q_{max} (mg g^{-1})	References
Lead (Pb^{2+})	Walnut shell	Pseudo-second-order	Langmuir	9.9	Çelebi and Gök (2017)
Cadmium (Cd^{2+})	Rice husk	Pseudo-second-order	Langmuir	11.6	Montalvo-Andía et al. (2022)
Methylene blue	Coffee grounds	Pseudo-second-order	Langmuir	142.8	Cuccarese et al. (2023)
Congo red	AMBC	-	Langmuir	89.3	Faheem et al. (2019)
Methyl red	Fennel seed	Pseudo-second-order	Freundlich	135	Paluch et al. (2023)
Crystal violet	Date palm petiole	Pseudo-second-order	Langmuir	209	Chahinez et al. (2020)
Tetracycline	Rice straw	Pseudo-first-order	Langmuir	50.72	Fan et al. (2018)
Phosphate (PO_4^{3-})	Oak sawdust	Pseudo-first-order	Langmuir	142.7	Wang et al. (2015)
Chromium (Cr^{6+})	Corn stalk biochar	Pseudo-second-order	Freundlich	17.07	Guo et al. (2020)
Ciprofloxacin	Corn cob	Pseudo-second-order	Freundlich	13	Peñafiel et al. (2019)
Malachite green	Sheep manure	Pseudo-second-order	Langmuir	208.33	Dilekoğlu (2022)
Nitrate $\left(NO_3^-\right)$	Date palm	-	Freundlich	9.21	Alsewaileh et al. (2019)
Arsenic (As^{5+})	Biosolid	-	Langmuir	67.28	Rahman et al. (2022)
Uranium (U(VI))	Rice straw	Pseudo-second-order	Langmuir	110.56	Ahmed et al. (2021)

the integration of materials such as Metal–Organic Frameworks (MOFs), metal nanoparticles, and functional polymers (Huang et al., 2019). Biochar@MOFs composites combine the large specific surface area and ordered porosity of MOFs with the thermal and mechanical stability of biochar, making it possible to broaden the range of pollutants targeted, in particular metal ions and certain organic micropollutants. Biochars supporting metal nanoparticles, such as zero-valent iron, magnetic iron oxide, or manganese dioxide, are used for their catalytic or redox properties, which are useful in the treatment of heavy metals such as lead or chromium (Agbale et al., 2025). In addition, combining biochar with functionalised polymers such as chitosan or polyaniline can enhance chemical interactions with specific contaminants by introducing additional functional groups to the surface of the material. These different composite structures represent a promising avenue for the development of more effective multifunctional adsorbents, adapted to the complexity of real effluents (Adam et al., 2022; Al-Ananzeh, 2021).

5.3.2 Synergy Mechanisms

The enhanced performance of biochar-based hybrid composites is the result of synergy mechanisms between the various constituents, which act in a complementary manner to enhance adsorption efficiency (Huang et al., 2019). The porous structure of biochar provides a stable and conductive support, ensuring the homogeneous dispersion of active materials such as MOFs, metal nanoparticles, or polymers. The integration of MOFs, with their extremely high specific surface area and fine chemical functionalisation, increases the selectivity of certain ions or molecules, while facilitating their diffusion within the pores. Metallic nanoparticles, for their part, introduce redox or catalytic properties that favour interaction with reactive species or the degradation of organic pollutants (Dou et al., 2021). Polymers enrich the surface with a variety of functional groups, favouring specific interactions such as hydrogen bonds, electrostatic forces, or π–π interactions. This combination leads to improved accessibility of active sites, increased surface charge density, and improved chemical and mechanical stability of the composite material. The synergy effects thus make it possible to optimise the adsorption capacity, the pollutant retention rate, and the regenerability of the system under complex conditions of use (Bacelo et al., 2020; Khan et al., 2022; Mbarki et al., 2022).

5.3.3 Case Studies on Specific Contaminant Adsorption

The efficiency of hybrid biochar-based composites has been widely demonstrated through numerous experimental studies addressing a variety of environmental contaminants, including heavy metals, synthetic dyes, and emerging organic pollutants. For instance, lead (Pb^{2+}), one of the most toxic and persistent heavy metals found in industrial effluents, has been successfully removed using biochar@Fe_3O_4 composites, with adsorption capacities exceeding 200 mg g^{-1} (Du et al., 2020). These materials offer the added benefit of easy magnetic recovery and good structural stability over multiple regeneration cycles. Similarly, polyaniline-coated biochars have shown strong affinity for cadmium (Cd^{2+}) and copper (Cu^{2+}), owing to the presence of amine groups that facilitate complexation mechanisms (Ho & McKay, 2002).

In the context of radioactive contaminants such as uranium (U(VI)), composites based on biochar@MOFs, particularly those incorporating ZIF-8 or UiO-66, have demonstrated high adsorption efficiency even at low concentrations and in the presence of competing ions, especially under saline conditions. The ordered microporous structure of MOFs combined with the mechanical stability of biochar promotes enhanced diffusion of metal ions toward active sites. Arsenic (As^{5+}) and chromium (Cr^{6+}) have also been effectively removed using biochar enriched with metal oxides such as MnO_2 or ZnO, synthesised from precursors like walnut shells or rice husks (Wang et al., 2025).

Synthetic dyes commonly found in textile effluents, such as methylene blue, Congo red, and malachite green, have been effectively eliminated by chitosan- or polydopamine-functionalised biochar composites. These systems benefit from a combination of physical adsorption, electrostatic interactions, and hydrogen bonding, achieving removal rates above 95% under optimal conditions. Composites with surface amine groups and negatively charged frameworks enhance affinity for these aromatic pollutants and show good reusability (Jawad et al., 2022).

Emerging organic contaminants, especially antibiotics such as tetracycline, ciprofloxacin, and sulfamethoxazole, pose significant challenges for conventional water treatment systems. Hybrid biochars modified with organic ligands or doped with transition metals have exhibited strong adsorption capacities for these compounds due to multiple mechanisms, including π–π interactions, hydrogen bonding, and surface complexation (Sabzehmeidani et al., 2021). Some studies have also evaluated these hybrid materials under simulated or real wastewater conditions containing mixed pollutants such as metals, nutrients (phosphate, ammonium), and organics. The results confirmed the robustness of biochar-based composites in complex matrices, demonstrating fast kinetics, high adsorption efficiency, and tolerance to variable pH and ionic strength (Carmalin Sophia et al., 2016). Representative examples of these studies, including the composite type, biomass origin, adsorption capacities, and experimental conditions, are summarised in Table 5.3.

5.4 PERFORMANCE UNDER REALISTIC AND COMPLEX CONDITIONS

5.4.1 Effect of pH, Temperature, and Competing Ions

The physico-chemical conditions of the environment have a decisive influence on the adsorption performance of biochars and their composites in wastewater treatment. Among these parameters, pH is one of the most decisive factors, as it affects both the ionisation state of the functional groups present on the surface of the adsorbent material and the chemical form of the pollutants in solution. At acid pH, the functional groups such as carboxyls or hydroxyls present on the biochar surface are mainly protonated, which limits the adsorption of metal cations by electrostatic repulsion (Selvaraj et al., 2021). On the other hand, at higher pH, the surface becomes progressively negative, favouring the complexation of metal ions by coordination bonding.

TABLE 5.3
Representative Examples of Hybrid Biochar Composites for the Adsorption of Selected Pollutants

Target Contaminant	Composite Type	Biomass Precursor	Adsorption Capacity ($mg\ g^{-1}$)	Experimental Conditions	References
Crystal violet	Biochar@Fe_3O_4 nanocomposite	-	349.4	pH 6.0, 40 °C, $C_0 = 400\,mg\ L^{-1}$	Sun et al. (2015)
Cd^{2+}	Biochar@ $MnFe_2O_4$@ ZIF-8 nanocomposite	Potato peel	80.52	pH 6, 45 °C, 100 minutes	Foroutan et al. (2023)
Cr, Cu, Se, Pb (multi-metal)	Biochar–chitosan	Sludge-based	0.050, 0.123, 0.014, 0.010	Real oil sands process water, natural pH, 0.5 g L^{-1} dosage	Song et al. (2021)
As(III)	Biochar@MnO_2	Rice husk	1.88	pH 7, 10 mg L^{-1}, 25 °C, 2 g L^{-1} dose, 120 minutes	Cuong et al. (2021)
Ni^{2+}, Co^{2+}	Biochar@S–Fe	Aloe vera skin	153.14 163.67	pH 10	Yu et al. (2025)

For example, studies have shown that adsorption of lead (Pb^{2+}) or cadmium (Cd^{2+}) is greatly enhanced at pH 5–6, while it is almost nil below pH 3. However, this trend is reversed for anionic contaminants such as Cr(VI), arsenate, or nitrates, which are better adsorbed under acidic conditions, where the biochar surface is positively charged (Kouotou et al., 2021). Temperature is another influential parameter that modifies thermodynamic interactions and mass transfer dynamics. A moderate rise in temperature (e.g., from 25°C to 45°C) generally increases the mobility of ions and accelerates the diffusion of pollutants through the porous biochar matrix. In addition, in the case of endothermic mechanisms, such as cation exchange adsorption or the formation of specific bonds with functional groups, a higher temperature can improve adsorption capacity. On the other hand, for certain sensitive composites, notably those containing organic polymers or thermolabile MOFs, temperatures above 50°C can cause degradation of the active structures or disorganisation of the pores, reducing their effectiveness (Foroutan et al., 2023).

The presence of competing ions in wastewater is a major constraint in real systems. These ions, such as Ca^{2+}, Mg^{2+}, K^+, Na^+, Cl^-, or SO_4^{2-}, interact with the same active sites as the targeted pollutants and can compete for adsorption. This competition is particularly problematic in high-salinity environments, such as industrial or mining wastewater, where the majority of ions can saturate the surface of the adsorbent material. For example, the presence of calcium or magnesium can strongly inhibit the adsorption of heavy metals such as Zn^{2+} or Cu^{2+}. Similarly, in municipal effluents rich in chlorides and sulphates, the adsorption of arsenate or chromate is often limited by the excess of anions. Nevertheless, certain functionalised biochars, notably those doped with metal nanoparticles or cationic polymers, have demonstrated improved selectivity and relative tolerance to these interferences, reinforcing their potential for applications in complex conditions (Zhao et al., 2024).

An in-depth study of these effects is essential for designing effective materials in real environments. It makes it possible to anticipate performance losses, select optimum operating conditions, and guide the choice of chemical modifications to be made to biochars depending on the type of effluent being treated (Sabzehmeidani et al., 2021).

5.4.2 Experiments under Simulated versus Real Conditions

The majority of studies on biochars and hybrid composites are initially carried out under controlled conditions, known as simulations, using aqueous solutions prepared in the laboratory containing a single target contaminant, at a defined concentration and without the presence of other interfering species. These tests enable the intrinsic performance of the materials to be assessed, adsorption isotherms to be established, and retention mechanisms to be modelled. However, the results obtained under these ideal conditions often tend to overestimate the real effectiveness of adsorbents when they are applied to complex effluents (Das et al., 2017).

In an applied context, industrial or municipal wastewater is highly variable in terms of chemical composition, pH, temperature, presence of dissolved organic matter, major ions, and multiple contaminants (Alsewaileh et al., 2019). This complexity makes the behaviour of adsorbents more difficult to predict. For example, a biochar that has demonstrated a high adsorption capacity for lead in a synthetic solution may see its effectiveness halved when used on mining water that also contains iron, zinc, sulphates, and colloidal matter. Similarly, the natural organic matter present in domestic effluents can cover the active sites by non-specific adsorption, blocking access to the target pollutants (Bacelo et al., 2020). Although fewer in number, studies under real conditions provide more representative results and enable us to better assess the operational robustness of biochars and their derivatives. Some studies have shown that biochar-MOFs or biochar-magnetic composites, although effective in the laboratory, are highly sensitive to the turbidity or unstable pH of industrial wastewater. Other studies, on the other hand, have shown that biochars enriched with iron or chitosan have good resistance in municipal effluents, with stable performance over several cycles of use. It is therefore essential to validate materials experimentally in complex matrices before extrapolating them to pilot or industrial-scale treatment processes (Okoniewska, 2023; Song et al., 2021; Zubair et al., 2024).

Setting up comparative evaluation protocols, combining tests in synthetic solution and tests in real wastewater, is now an essential methodological requirement for assessing the environmental and economic relevance of a modified biochar. This makes it possible not only to detect any limitations of the material, but also to identify the limit conditions of use and any need for pre-treatment of the effluent (Abid et al., 2022).

5.4.3 Observed Limitations and Optimisation Strategies

Despite the promising performance observed in the laboratory, biochars and their hybrid composites present certain limitations when used in complex environmental conditions. One of the first constraints identified concerns the limited selectivity of many crude biochars, which adsorb certain ions or molecules effectively in simple solution, but lose effectiveness in the presence of ionic competitors, natural organic matter, or other interferences common in real wastewater (Weidner et al., 2022). This lack of specificity can lead to rapid saturation of the active sites, non-targeted adsorption, and a significant drop in yield over the long term. Another major limitation lies in the regenerability of the materials. While some biochars can be regenerated simply by washing with water or solvent, others require more aggressive treatments (acid washing, calcination, thermal reactivation) that can degrade the porous structure or lead to the loss of active material (Zhao et al., 2025b). What's more, some composites, such as biochar@MOFs or biochar-polymers, are thermally or chemically sensitive, which restricts their reuse beyond a few cycles. Added to this is the issue of the mechanical stability of materials under conditions of turbulence, pressure, or variable flow, typical of continuous treatment systems. To overcome these obstacles, various optimisation strategies have been proposed. In terms of materials, targeted doping with metals, immobilisation of specific ligands, and encapsulation in robust polymer matrices can all be used to improve selectivity, chemical stability, and resistance to interference (Serhan et al., 2019). Controlling pyrolysis conditions, by adjusting the temperature, heating rate, and atmosphere used, also makes it possible to adapt the porosity and surface functionalisation of the biochar to a specific type of pollutant. More recent approaches combine biochar with multifunctional nanoparticles, capable not only of adsorbing but also of degrading or precipitating pollutants, thereby broadening the prospects for adsorption-reaction coupling (Dutta et al., 2021).

At the procedural level, the integration of biochars into hybrid systems makes it possible to compensate for certain individual limitations by capitalising on synergy effects. The use of modelling and statistical optimisation tools, such as response surface methodology or artificial intelligence algorithms, also helps to define optimal operating windows and anticipate performance drifts. Finally, pilot-scale tests in real environments are essential to validate the technical efficiency and economic feasibility of optimised materials (Boulahbal et al., 2022).

5.5 COUPLING OF BIOCHAR ADSORPTION WITH OTHER PROCESSES

5.5.1 Coupling with Advanced Oxidation Processes

Coupling biochar adsorption with advanced oxidation processes is a particularly effective treatment route for targeting persistent organic micropollutants that are resistant to conventional clean-up techniques. Advanced oxidation processes are based on the generation of powerful radical species, in particular hydroxyl radicals (-OH), which have a high oxidation potential and low selectivity, enabling them to attack a wide variety of chemical bonds. Biochar (Talfana et al., 2026), when modified or doped with transition metals such as iron, copper, or manganese, can be used in these processes either as an active catalyst or as a support for stabilising and dispersing catalytic phases. A well-documented example is the heterogeneous Fenton process, in which iron present on the biochar surface catalyses the decomposition of hydrogen peroxide (H_2O_2) to generate hydroxyl radicals in situ (Rahman et al., 2022). This system then combines rapid adsorption of contaminants onto the carbonaceous material with direct chemical degradation by radical oxidation. This mechanism is conceptually illustrated in Figure 5.2, showing UV-induced activation of the biochar-supported Fenton process. This approach has been successfully applied to the elimination of pharmaceutical compounds such as carbamazepine, pesticides such as atrazine, and highly stable industrial dyes such as Congo red. Degradation rates of over 90% are frequently achieved in just a few hours, with significant reductions in total organic carbon (Shafqat et al., 2020; Wekoye et al., 2020). Incorporating UV or sunlight into the system (photo-Fenton) regenerates the catalytic iron and enhances free-radical production, thereby increasing the degradation kinetics. Recent studies have also highlighted the role of biochar in stabilising active species, which limits the leaching of metals into the treated medium and extends the life of the material. In addition, some doped biochars can activate other oxidants such as persulphate $\left(S_2O_8^{2-}\right)$ or percarbonate, offering an alternative to hydrogen peroxide, particularly for treatments in alkaline or weakly buffered media (Li et al., 2025).

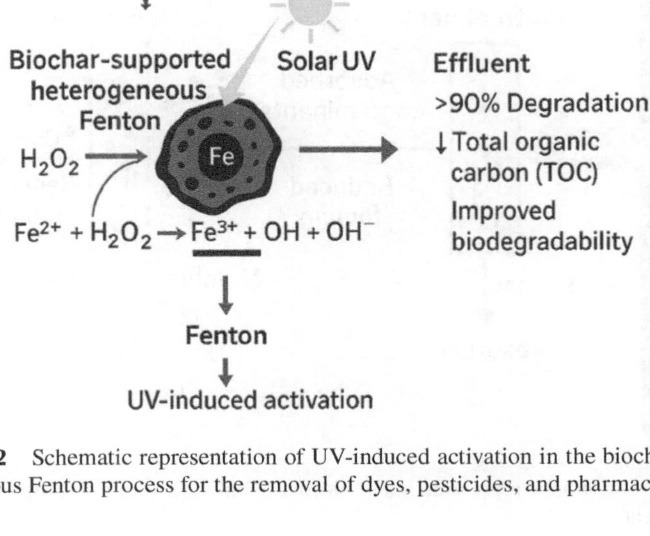

FIGURE 5.2 Schematic representation of UV-induced activation in the biochar-supported heterogeneous Fenton process for the removal of dyes, pesticides, and pharmaceuticals from wastewater.

This type of coupling also offers significant operational advantages: it reduces the amount of secondary sludge produced, allows partial regeneration of the biochar, and improves the biodegradability of effluents by transforming refractory compounds into simpler intermediates. However, control of the operating parameters remains essential to avoid degradation of the carbon support, non-selective oxidation of the compounds of interest, or loss of catalyst stability. The development of low-cost, stable, and recyclable biochars that can be activated is therefore a priority for the industrial deployment of these coupled systems (Li et al., 2024).

5.5.2 Coupling with Membrane Filtration

Integrating biochar into membrane filtration processes is a promising strategy for improving the quality of wastewater treatment while extending the life of membranes. Membrane technologies such as microfiltration, ultrafiltration, nanofiltration, and reverse osmosis are widely used today for the physical separation of dissolved or colloidal contaminants (Adam et al., 2022). However, these systems often face major limitations, including membrane pore clogging, biofilm formation, and a gradual decrease in permeation flux, leading to increased operating costs and chemical cleaning requirements (Huffman et al., 2024).

In this context, biochar acts as an effective pre-treatment, adsorbing some of the dissolved organic matter, colloidal compounds, or micropollutants before they pass through the membrane. This coupling strategy is conceptually illustrated in Figure 5.3, which highlights the role of biochar in reducing membrane fouling and improving overall system performance. This step significantly reduces the pollutant load on the membrane module and limits physical or biological fouling. Studies have shown that introducing a layer of biochar upstream of the membrane can reduce chemical oxygen demand by more than 50%, while reducing the frequency of cleaning cycles (Elayadi et al., 2020).

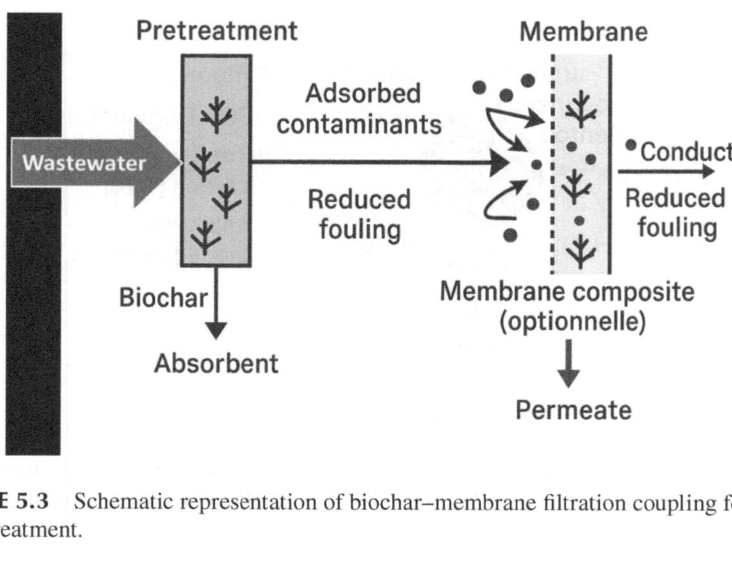

FIGURE 5.3 Schematic representation of biochar–membrane filtration coupling for wastewater treatment.

In addition to its role in pre-filtration, biochar can also be incorporated directly into membrane materials, in the form of active fillers incorporated into the polymer matrix. Biochar-polymer composite membranes, made from polymers such as polysulphone, polyvinylidene fluoride, or chitosan, show a significant improvement in surface properties, with an increase in hydrophilicity, controlled roughness, and improved resistance to biological fouling (Faheem et al., 2019). In addition, modified or doped biochars (for example, with Fe_3O_4 or metal oxides) give the membrane additional antibacterial or catalytic properties, promoting the in situ degradation of certain pollutants (Abu-Dief et al., 2025). Biochar–membrane coupling also makes it possible to envisage hybrid configurations, such as submerged membrane reactors with a fixed bed of biochar, where adsorption, biodegradation, and filtration phenomena are combined in a single compact system. This type of solution is particularly well suited to urban, hospital, or industrial wastewater with a high load, and can help to reduce the overall energy footprint of the treatment process (Ghaffar et al., 2018).

However, a number of challenges remain, particularly with regard to the durability of biochar-enriched membranes, the even distribution of the material in the polymer matrix, and the risk of releasing fine particles into the permeate. Rigorous optimisation of formulations, manufacturing processes, and operating conditions is therefore necessary to guarantee the long-term stability and performance of coupled systems (Bazan-Wozniak & Pietrzak, 2020; Maher et al., 2019).

5.5.3 Biochar-Assisted Bioremediation and Cases on a Semi-Industrial Scale

The integration of biochar into bioremediation systems represents a synergistic approach that combines the advantages of physical adsorption with biological degradation processes. Thanks to its porous structure, chemical stability, and high specific surface area, biochar is an excellent support for microbial colonisation, enabling bacterial or fungal populations active in the degradation of complex organic pollutants to thrive (Li et al., 2021). Unlike traditional inert supports, biochar encourages the establishment of stable biofilms, improves nutrient transfer, and reduces the local toxic effects of contaminants by partially adsorbing them in the immediate vicinity of the micro-organisms (Bhatia et al., 2017). This concept is illustrated in Figure 5.4, which shows the dual role of biochar as a microbial carrier and adsorbent within an integrated bioremediation system.

This role of biocompatible matrices is particularly relevant in the treatment of poorly biodegradable compounds, such as certain pesticides, polycyclic aromatic hydrocarbons, or pharmaceutical residues. For example, systems combining biochar and specific bacterial strains have improved the elimination of tetracycline or naphthalene by initial adsorption followed by enzymatic degradation. Biochar can also stabilise local conditions (pH, oxygenation, humidity) in heterogeneous environments such as planted filters, fixed-bed reactors, or lagooning zones, thereby boosting microbial metabolic activity (Lellis et al., 2019; Singla & Sit, 2021).

A number of experiments on a pilot or semi-industrial scale have been carried out in recent years to validate these concepts in real environments. For example, the use of biochar produced from agricultural residues, in a fixed bed in an urban wastewater

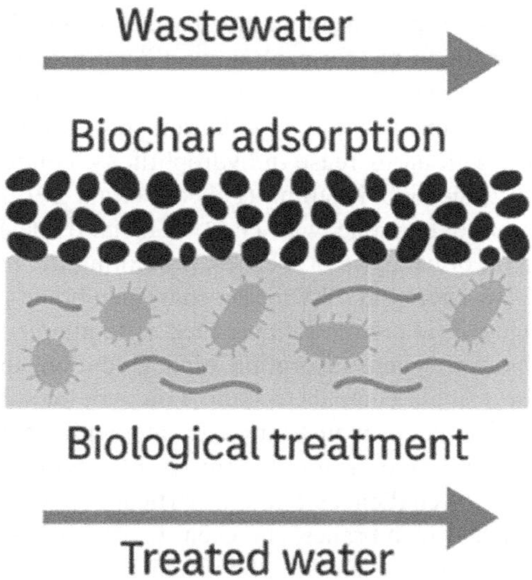

FIGURE 5.4 Conceptual illustration of biochar-assisted bioremediation in wastewater treatment.

treatment plant, significantly reduced concentrations of BOD_5, pharmaceutical compounds, and nutrients, while maintaining stable microbial activity over several weeks of operation (Briffa et al., 2020). Other projects have used biochar-enriched anaerobic bioreactors for the degradation of nitrates and trace metals in agro-industrial effluents, with performance superior to that observed with conventional substrates (Karadag et al., 2015).

These results highlight the value of biochar not only as a passive adsorbent but also as an active component of functional microbial ecosystems, capable of being integrated into sustainable ecotechnological systems. The combination of physical and biological functions offers interesting prospects for the development of more robust treatment solutions, particularly in rural areas, decentralised contexts, or low-maintenance installations.

5.6 CONCLUSION

At the end of this chapter, it is clear that biochars and their hybrid composites constitute a class of adsorbent materials with great potential for wastewater treatment, particularly for emerging pollutants, heavy metals, and persistent organic compounds. The highest performances have been observed with activated or modified biochars, particularly those doped with metal oxides (Fe, Mn, Zn), enriched with oxygenated functional groups or grafted with polymers and MOFs. These materials offer increased adsorption capacity, improved selectivity, and enhanced stability in complex aqueous environments, which puts them in a favourable position compared with conventional adsorbents such as commercial activated carbons.

From an environmental point of view, the use of abundant lignocellulosic residues from agriculture, forestry, and the agri-food industries gives biochars a highly attractive sustainability profile. Their production by pyrolysis not only enables organic waste to be recycled but also enables carbon to be sequestered in a stable form, thereby helping to reduce greenhouse gas emissions. In economic terms, the simplicity of the manufacturing processes, the partial regenerability of the materials, and their low cost make them a viable solution for contexts with limited resources, particularly in developing countries or rural areas.

The prospects for the development of these materials are numerous. Surface engineering, the design of multifunctional biochar@MOFs or biochar@nanoparticles composites, and integration into hybrid treatment devices (filters, columns, reactors) all point to the possibility of gradual industrialisation of these systems. In addition, their compatibility with the principles of the circular economy – by transforming waste into functional resources – opens the way to local production and application channels with a view to enhancing sustainability.

However, a number of scientific and technological hurdles remain. The precise mechanisms of interaction between pollutants and functional surfaces remain to be elucidated for certain complex compounds. Long-term stability, regenerability without loss of effectiveness, and risk assessment linked to the potential release of elements need to be better mastered. Future research will also need to focus on evaluation under real conditions, modelling at the process scale, and analysis of the full life cycle of materials, in order to guarantee their effectiveness, safety, and industrial viability. Finally, particular attention will need to be paid to the standardisation of tests and the reproducibility of results, which are essential conditions for transfer to large-scale operational applications.

REFERENCES

Abhraham, N., Babu, A., Gopi, A., Rajan, S., & Paul, Ti. (2020). Adsorptive treatment of dairy waste water using rubber seed shell activated carbon. *IOSR Journal of Mechanical and Civil Engineering (IOSR-JMCE)*, 17(4), 49–57. https://doi.org/10.9790/1684-1701003

Abid, N., Masmoudi, M. A., Megdiche, M., Barakat, A., Ellouze, M., Chamkha, M., Ksibi, M., & Sayadi, S. (2022). Biochar from olive mill solid waste as an eco-friendly adsorbent for the removal of polyphenols from olive mill wastewater. *Chemical Engineering Research and Design*, 181, 384–398. https://doi.org/10.1016/j.cherd.2022.02.029

Abu-Dief, A. M., Alsaedi, W. H., & Zikry, M. M. (2025). A collective study on the fabrication of nano-materials for water treatment. *Journal of Umm Al-Qura University for Applied Sciences*. https://doi.org/10.1007/s43994-025-00227-4

Adam, M. R., Othman, M. H. D., Kurniawan, T. A., Puteh, M. H., Ismail, A. F., Khongnakorn, W., Rahman, M. A., & Jaafar, J. (2022). Advances in adsorptive membrane technology for water treatment and resource recovery applications: A critical review. *Journal of Environmental Chemical Engineering*, 10(3). https://doi.org/10.1016/j.jece.2022.107633

Agbale, N. R., Ajieh, M. U., Ogunleye, O. O., Agarry, S. E., Arinkoola, A. O., & Owebor. (2025). Efficient simultaneous adsorptive removal of heavy metals and organics in produced water by a novel multi-functional alginate encapsulated kaolin, -plantain pseudo-stem and snail shell composite biochar. *NIPES-Journal of Science and Technology Research*, 7(2), 2734–2352. https://doi.org/10.37933/nipes/7.2.2025.30

Ahmed, W., Núñez-Delgado, A., Mehmood, S., Ali, S., Qaswar, M., Shakoor, A., & Chen, D. Y. (2021). Highly efficient uranium (VI) capture from aqueous solution by means of a hydroxyapatite-biochar nanocomposite: Adsorption behavior and mechanism. *Environmental Research*, *201*(May). https://doi.org/10.1016/j.envres.2021.111518

Al-Ananzeh, N. M. (2021). Treatment of wastewater from a dairy plant by adsorption using synthesized copper oxide nanoparticles: Kinetics and isotherms modeling optimization. *Water Science and Technology*, 83(7), 1591–1604. https://doi.org/10.2166/wst.2021.089

Alsewaileh, A. S., Usman, A. R., & Al-Wabel, M. I. (2019). Effects of pyrolysis temperature on nitrate-nitrogen $\left(NO_3^- \text{-} N\right)$ and bromate $\left(BrO_3^-\right)$ adsorption onto date palm biochar. *Journal of Environmental Management*, 237(February), 289–296. https://doi.org/10.1016/j.jenvman.2019.02.045

Alsukaibi, A. K. D. (2022). Various approaches for the detoxification of toxic dyes in wastewater. *Processes*, *10*(10). https://doi.org/10.3390/pr10101968

Asemave, K., Thaddeus, L., & Tarhemba, P. T. (2021). Lignocellulosic-based sorbents: A review. *Sustainable Chemistry*, 2(2), 271–285. https://doi.org/10.3390/suschem2020016

Aslan, S., & Şirazi, M. (2020). Adsorption of sulfonamide antibiotic onto activated carbon prepared from an agro-industrial by-product as low-cost adsorbent: Equilibrium, thermodynamic, and kinetic studies. *Water, Air, and Soil Pollution*, *231*(5). https://doi.org/10.1007/s11270-020-04576-0

Bacelo, H., Pintor, A. M. A., Santos, S. C. R., Boaventura, R. A. R., & Botelho, C. M. S. (2020). Performance and prospects of different adsorbents for phosphorus uptake and recovery from water. *Chemical Engineering Journal*, *381*(August 2019), 122566. https://doi.org/10.1016/j.cej.2019.122566

Bazan-Wozniak, A., & Pietrzak, R. (2020). Adsorption of organic and inorganic pollutants on activated bio-carbons prepared by chemical activation of residues of supercritical extraction of raw plants. *Chemical Engineering Journal*, 393. https://doi.org/10.1016/j.cej.2020.124785

Bernal, V., Giraldo, L., & Moreno-Piraján, J. C. (2021). Understanding the solid-liquid equilibria between paracetamol and activated carbon: Thermodynamic approach of the interactions adsorbent-adsorbate using equilibrium, kinetic and calorimetry data. *Journal of Hazardous Materials*, 419(March), 1–13. https://doi.org/10.1016/j.jhazmat.2021.126432

Bhatia, D., Sharma, N. R., Singh, J., & Kanwar, R. S. (2017). Biological methods for textile dye removal from wastewater: A review. *Critical Reviews in Environmental Science and Technology*, 47(19), 1836–1876. https://doi.org/10.1080/10643389.2017.1393263

Bosch, D., Back, J. O., Gurtner, D., Giberti, S., Hofmann, A., & Bockreis, A. (2022). Forestry residue biomass and waste wood. *Carbon Resources Conversion*, 5(4), 299–309. https://doi.org/10.1016/j.crcon.2022.09.001

Boulahbal, M., Malouki, M. A., Canle, M., Redouane-Salah, Z., Devanesan, S., AlSalhi, M. S., & Berkani, M. (2022). Removal of the industrial azo dye crystal violet using a natural clay: Characterization, kinetic modeling, and RSM optimization. *Chemosphere*, 306(June), 135516. https://doi.org/10.1016/j.chemosphere.2022.135516

Briffa, J., Sinagra, E., & Blundell, R. (2020). Heavy metal pollution in the environment and their toxicological effects on humans. *Heliyon*, 6(9), e04691. https://doi.org/10.1016/j.heliyon.2020.e04691

Carmalin Sophia, A., Lima, E. C., Allaudeen, N., & Rajan, S. (2016). Application of graphene based materials for adsorption of pharmaceutical traces from water and wastewater- A review. *Desalination and Water Treatment*, 57(57), 27573–27586. https://doi.org/10.1080/19443994.2016.1172989

Çelebi, H., & Gök, O. (2017). Evaluation of lead adsorption kinetics and isotherms from aqueous solution using natural walnut shell. *International Journal of Environmental Research*, 11(1), 83–90. https://doi.org/10.1007/s41742-017-0009-3

Chahinez, H. O., Abdelkader, O., Leila, Y., & Tran, H. N. (2020). One-stage preparation of palm petiole-derived biochar: Characterization and application for adsorption of crystal violet dye in water. *Environmental Technology and Innovation*, 19, 100872. https://doi.org/10.1016/j.eti.2020.100872

Charii, H., Boussetta, A., Benali, K., Essifi, K., Mennani, M., Benhamou, A. A., El Zakhem, H., Sehaqui, H., El Achaby, M., Grimi, N., Boutoial, K., Ablouh, E. H., & Moubarik, A. (2024). Phosphorylated chitin from shrimp shell waste: A robust solution for cadmium remediation. *International Journal of Biological Macromolecules*, 268. https://doi.org/10.1016/j.ijbiomac.2024.131855

Chen, X., Che, Q., Li, S., Liu, Z., Yang, H., Chen, Y., Wang, X., Shao, J., & Chen, H. (2019). Recent developments in lignocellulosic biomass catalytic fast pyrolysis: Strategies for the optimization of bio-oil quality and yield. *Fuel Processing Technology*, 196. 106180. https://doi.org/10.1016/j.fuproc.2019.106180

Condeço, J. A. D., Hariharakrishnan, S., Ofili, O. M., Mateus, M. M., Bordado, J. M., & Correia, M. J. N. (2021). Energetic valorisation of agricultural residues by solvent-based liquefaction. *Biomass and Bioenergy*, *147*. https://doi.org/10.1016/j.biombioe.2021.106003

Cuccarese, M., Brutti, S., De Bonis, A., Teghil, R., Di Capua, F., Mancini, I. M., Masi, S., & Caniani, D. (2023). Sustainable adsorbent material prepared by soft alkaline activation of spent coffee grounds: Characterisation and adsorption mechanism of methylene blue from aqueous solutions. *Sustainability*, *15*(3). https://doi.org/10.3390/su15032454

Cuong, D. V., Wu, P. C., Chen, L. I., & Hou, C. H. (2021). Active MnO_2/biochar composite for efficient As(III) removal: Insight into the mechanisms of redox transformation and adsorption. *Water Research*, 188, 116495. https://doi.org/10.1016/j.watres.2020.116495

Das, O., Kim, N. K., Sarmah, A. K., & Bhattacharyya, D. (2017). Development of waste based biochar/wool hybrid biocomposites: Flammability characteristics and mechanical properties. *Journal of Cleaner Production*, 144, 79–89. https://doi.org/10.1016/j.jclepro.2016.12.155

Dilekoğlu, M. F. (2022). Malachite green adsorption from aqueous solutions onto biochar derived from sheep manure: Adsorption kinetics, isotherm, thermodynamic, and mechanism. *International Journal of Phytoremediation*, 24(4), 436–446. https://doi.org/10.1080/15226514.2021.1951656

Dinc, G., & Yel, E. (2018). Self-catalyzing pyrolysis of olive pomace. *Journal of Analytical and Applied Pyrolysis*, 134, 641–646. https://doi.org/10.1016/j.jaap.2018.08.018

Dou, H., Xu, M., Wang, B., Zhang, Z., Wen, G., Zheng, Y., Luo, D., Zhao, L., Yu, A., Zhang, L., Jiang, Z., & Chen, Z. (2021). Microporous framework membranes for precise molecule/ion separations. *Chemical Society Reviews*, 50(2), 986–1029. https://doi.org/10.1039/d0cs00552e

Du, C., Song, Y., Shi, S., Jiang, B., Yang, J., & Xiao, S. (2020). Preparation and characterization of a novel Fe3O4-graphene-biochar composite for crystal violet adsorption. *Science of the Total Environment*, 711, 134662. https://doi.org/10.1016/j.scitotenv.2019.134662

Dutta, S., Gupta, B., Srivastava, S. K., & Gupta, A. K. (2021). Recent advances on the removal of dyes from wastewater using various adsorbents: A critical review. *Materials Advances*, 2(14), 4497–4531. https://doi.org/10.1039/d1ma00354b

Elayadi, F., Achak, M., Beniich, N., Belaqziz, M., & El Adlouni, C. (2020). Factorial design for optimizing and modeling the removal of organic pollutants from olive mill wastewater using a novel low-cost bioadsorbent. *Water, Air, and Soil Pollution*, *231*(7). https://doi.org/10.1007/s11270-020-04695-8

Elmouwahidi, A., Zapata-Benabithe, Z., Carrasco-Marín, F., & Moreno-Castilla, C. (2012). Activated carbons from KOH-activation of argan (Argania spinosa) seed shells as supercapacitor electrodes. *Bioresource Technology*, 111, 185–190. https://doi.org/10.1016/j.biortech.2012.02.010

Faheem, Du, J., Bao, J., Hassan, M. A., Irshad, S., & Talib, M. A. (2019). Multi-functional biochar novel surface chemistry for efficient capture of anionic Congo red dye: Behavior and mechanism. *Arabian Journal for Science and Engineering*, 44(12), 10127–10139. https://doi.org/10.1007/s13369-019-04194-x

Fan, S., Wang, Y., Li, Y., Wang, Z., Xie, Z., & Tang, J. (2018). Removal of tetracycline from aqueous solution by biochar derived from rice straw. *Environmental Science and Pollution Research*, 25(29), 29529–29540. https://doi.org/10.1007/s11356-018-2976-0

Foroutan, R., Peighambardoust, S. J., Ghojavand, S., Farjadfard, S., & Ramavandi, B. (2023). Cadmium elimination from wastewater using potato peel biochar modified by ZIF-8 and magnetic nanoparticle. *Colloids and Interface Science Communications*, 55(May), 100723. https://doi.org/10.1016/j.colcom.2023.100723

Ghaffar, A., Zhu, X., & Chen, B. (2018). Biochar composite membrane for high performance pollutant management: Fabrication, structural characteristics and synergistic mechanisms. *Environmental Pollution*, 233, 1013–1023. https://doi.org/10.1016/j.envpol.2017.09.099

Ghazoui, M., Boudouch, O., Miyah, Y., Benjelloun, M., Sylla, A. S., Moulakhnif, K., Fikri-Benbrahim, K., Touzani, I., & Elkacmi, R. (2025). Assessment of Rhus Pentaphylla-sulfuric acid activated carbon performance for cadmium ions adsorption: mechanism, response surface methodology optimisation, and cost estimation. *Indian Chemical Engineer*, 1–22. https://doi.org/10.1080/00194506.2025.2462953

Ghazoui, M., Elkacmi, R., Sylla, A. S., Moulakhnif, K., Touzani, I., & Boudouch, O. (2024). Efficient removal of methylene blue and methyl red dyes using a novel adsorbent derived from Saponaria officinalis Root via H_3PO_4, H_2SO_4, and KOH-activation: Optimization, kinetics, and isotherm study. *Desalination and Water Treatment, 318*. https://doi.org/10.1016/j.dwt.2024.100378

Ghezzehei, T. A., Sarkhot, D. V., & Berhe, A. A. (2014). Biochar can be used to capture essential nutrients from dairy wastewater and improve soil physico-chemical properties. *Solid Earth*, 5(2), 953–962. https://doi.org/10.5194/se-5-953-2014

Guo, X., Liu, A., Lu, J., Niu, X., Jiang, M., Ma, Y., Liu, X., & Li, M. (2020). Adsorption mechanism of hexavalent chromium on biochar: Kinetic, thermodynamic, and characterization studies. *ACS Omega*, 5(42), 27323–27331. https://doi.org/10.1021/acsomega.0c03652

Häder, D. P., Banaszak, A. T., Villafañe, V. E., Narvarte, M. A., González, R. A., & Helbling, E. W. (2020). Anthropogenic pollution of aquatic ecosystems: Emerging problems with global implications. *Science of the Total Environment*, 713, 136586. https://doi.org/10.1016/j.scitotenv.2020.136586

Hama Aziz, K. H., Mustafa, F. S., Omer, K. M., Hama, S., Hamarawf, R. F., & Rahman, K. O. (2023). Heavy metal pollution in the aquatic environment: Efficient and low-cost removal approaches to eliminate their toxicity: A review. *RSC Advances*, 13(26), 17595–17610. https://doi.org/10.1039/d3ra00723e

Han, H., Rafiq, M. K., Zhou, T., Xu, R., Mašek, O., & Li, X. (2019). A critical review of clay-based composites with enhanced adsorption performance for metal and organic pollutants. *Journal of Hazardous Materials*, 369, 780–796. https://doi.org/10.1016/j.jhazmat.2019.02.003

Han, Q., Wang, J., Goodman, B. A., Xie, J., & Liu, Z. (2020). High adsorption of methylene blue by activated carbon prepared from phosphoric acid treated eucalyptus residue. *Powder Technology*, 366, 239–248. https://doi.org/10.1016/j.powtec.2020.02.013

Hassaan, M. A., Yılmaz, M., Helal, M., El-Nemr, M. A., Ragab, S., & El Nemr, A. (2023). Isotherm and kinetic investigations of sawdust-based biochar modified by ammonia to remove methylene blue from water. *Scientific Reports*, 13(1), 1–20. https://doi.org/10.1038/s41598-023-39971-0

Ho, Y. S., & McKay, G. (2002). Application of kinetic models to the sorption of copper (II) on to peat. *Adsorption Science and Technology*, 20(8), 797–815. https://doi.org/10.1260/026361702321104282

Huang, F., Dong, F., Chen, L., Zeng, Y., Zhou, L., Sun, S., Wang, Z., Lai, J., & Fang, L. (2024). Biochar-mediated remediation of uranium-contaminated soils: Evidence, mechanisms, and perspectives. *Biochar*, *6*(1). https://doi.org/10.1007/s42773-024-00308-3

Huang, Q., Song, S., Chen, Z., Hu, B., Chen, J., & Wang, X. (2019). Biochar-based materials and their applications in removal of organic contaminants from wastewater: state-of-the-art review. *Biochar*, 1(1), 45–73. https://doi.org/10.1007/s42773-019-00006-5

Huffman, D. P., Pitell, S., Moncure, P., Stout, J., Millstone, J. E., Haig, S. J., & Gilbertson, L. M. (2024). Moving beyond silver in point-of-use drinking water pathogen control. *Environmental Science: Water Research and Technology*, 10(5), 1009–1018. https://doi.org/10.1039/d3ew00564j

Jawad, A. H., Sahu, U. K., Jani, N. A., ALOthman, Z. A., & Wilson, L. D. (2022). Magnetic crosslinked chitosan-tripolyphosphate/MgO/Fe_3O_4 nanocomposite for reactive blue 19 dye removal: Optimization using desirability function approach. *Surfaces and Interfaces*, *28*. https://doi.org/10.1016/j.surfin.2021.101698

Karadag, D., Köroılu, O. E., Ozkaya, B., & Cakmakci, M. (2015). A review on anaerobic biofilm reactors for the treatment of dairy industry wastewater. *Process Biochemistry*, 50(2), 262–271. https://doi.org/10.1016/j.procbio.2014.11.005

Karić, N., Maia, A. S., Teodorović, A., Atanasova, N., Langergraber, G., Crini, G., Ribeiro, A. R. L., & Đolić, M. (2022). Bio-waste valorisation: Agricultural wastes as biosorbents for removal of (in)organic pollutants in wastewater treatment. *Chemical Engineering Journal Advances*, 9. https://doi.org/10.1016/j.ceja.2021.100239

Kasera, N., Kolar, P., & Hall, S. G. (2022). Cter: A review. *Biochar*, *4*(1). https://doi.org/10.1007/s42773-022-00145-2

Khan, T. A., Nouman, M., Dua, D., Khan, S. A., & Alharthi, S. S. (2022). Adsorptive scavenging of cationic dyes from aquatic phase by H_3PO_4 activated Indian jujube (Ziziphus mauritiana) seeds based activated carbon: Isotherm, kinetics, and thermodynamic study. *Journal of Saudi Chemical Society*, 26(2), 101417. https://doi.org/10.1016/j.jscs.2021.101417

Kishor, R., Bharagava, R. N., & Saxena, G. (2018). Industrial wastewaters: the major sources of dye contamination in the environment, ecotoxicological effects, and bioremediation approaches. In *Recent advances in environmental management* (pp. 1–25). CRC Press.

Kishor, R., Purchase, D., Saratale, G. D., Saratale, R. G., Ferreira, L. F. R., Bilal, M., Chandra, R., & Bharagava, R. N. (2021). Ecotoxicological and health concerns of persistent coloring pollutants of textile industry wastewater and treatment approaches for environmental safety. *Journal of Environmental Chemical Engineering*, 9(2), 105012. https://doi.org/10.1016/j.jece.2020.105012

Kouotou, D., Ghalit, M., Ndi, J. N., Martinez, L. M. P., Ouahabi, M. El, Ketcha, J. M., & Gharibi, E. K. (2021). Removal of metallic trace elements (Pb^{2+}, Cd^{2+}, Cu^{2+}, and Ni^{2+}) from aqueous solution by adsorption onto cerium oxide modified activated carbon. *Environmental Monitoring and Assessment*, *193*(8). https://doi.org/10.1007/s10661-021-09267-9

Lellis, B., Fávaro-Polonio, C. Z., Pamphile, J. A., & Polonio, J. C. (2019). Effects of textile dyes on health and the environment and bioremediation potential of living organisms. *Biotechnology Research and Innovation*, 3(2), 275–290. https://doi.org/10.1016/j.biori.2019.09.001

Li, G., Zhang, M., Zheng, Y., Gao, J., Yue, T., Zhao, S., & Liu, C. (2024). Research progress on the synergistic treatment of multi pollutants at low temperatures using carbon-based materials. *Sustainability*, *16*(24). https://doi.org/10.3390/su162410965

Li, J., Jing, B., Tian, Y., Li, G., & Li, X. (2025). Electrochemical activation of peroxymonosulfate with commercial electrodes in the mixed reaction chamber: A cost-effective way for sulfamethazine degradation. *Desalination and Water Treatment*, 323(May), 101279. https://doi.org/10.1016/j.dwt.2025.101279

Li, Y., Wu, S., Wang, S., Zhao, S., & Zhuang, X. (2021). Anaerobic degradation of xenobiotic organic contaminants (XOCs): The role of electron flow and potential enhancing strategies. *Journal of Environmental Sciences*, 101, 397–412. https://doi.org/10.1016/j.jes.2020.08.030

Liu, C., Wang, H., Karim, A. M., Sun, J., & Wang, Y. (2014). Catalytic fast pyrolysis of lignocellulosic biomass. *Chemical Society Reviews*, 43(22), 7594–7623. https://doi.org/10.1039/c3cs60414d

Mafo, S. G. M., Tchuifon, D. R. T., Djioko, F. H. K., Kouteu, P. A. N., Fotsop, C. G., Dongmo, S. D. M., Doungmo, G., & Ndifor-Angwafor, N. G. (2024). Unravelling the efficiency removal of 2,4-dinitrophenol on coconut shell biomass-derived activated carbons theoretical and experimental investigation. *Biomass Conversion and Biorefinery*. https://doi.org/10.1007/s13399-024-05663-6

Maher, S., Brayden, D. J., Casettari, L., & Illum, L. (2019). Application of permeation enhancers in oral delivery of macromolecules: An update. *Pharmaceutics*, 11(1), 1–23. https://doi.org/10.3390/pharmaceutics11010041

Mbarki, F., Selmi, T., Kesraoui, A., & Seffen, M. (2022). Low-cost activated carbon preparation from Corn stigmata fibers chemically activated using H_3PO_4, $ZnCl_2$ and KOH: Study of methylene blue adsorption, stochastic isotherm and fractal kinetic. *Industrial Crops and Products, 178*. https://doi.org/10.1016/j.indcrop.2022.114546

Montalvo-Andía, J., Reátegui-Romero, W., Peña-Contreras, A. D., Zaldivar Alvarez, W. F., King-Santos, M. E., Fernández-Guzmán, V., Guerrero-Guevara, J. L., & Puris-Naupay, J. E. (2022). Adsorption of Cd (II) using chemically modified rice husk: Characterization, equilibrium, and kinetic studies. *Adsorption Science and Technology, 2022*. https://doi.org/10.1155/2022/3688155

Nikić, J., Watson, M., Tubić, A., Šolić, M., & Agbaba, J. (2024). Recent trends in the application of magnetic nanocomposites for heavy metals removal from water: A review. *Separation Science and Technology*, 59(2), 293–331. https://doi.org/10.1080/01496395.2024.2315626

Okoniewska, E. (2023). Removal of selected dyes on activated carbons under flow-through conditions. *Desalination and Water Treatment*, 311, 111–117. https://doi.org/10.5004/dwt.2023.29970

Paluch, D., Bazan-Wozniak, A., Wolski, R., Nosal-Wiercińska, A., & Pietrzak, R. (2023). Removal of methyl red from aqueous solution using biochar derived from fennel seeds. *Molecules, 28*(23). https://doi.org/10.3390/molecules28237786

Pathak, P., & Gupta, D. K. (2020). Strontium contamination in the environment. In *The Handbook of Environmental Chemistry* (Vol. 88). https://doi.org/10.1007/978-3-030-15314-4

Peñafiel, M. E., Vanegas, E., Bermejo, D., Matesanz, J. M., & Ormad, M. P. (2019). Organic residues as adsorbent for the removal of ciprofloxacin from aqueous solution. *Hyperfine Interactions, 240*(1), 71. https://doi.org/10.1007/s10751-019-1612-9.

Rahman, M. A., Lamb, D., Rahman, M. M., Bahar, M. M., & Sanderson, P. (2022). Adsorption–desorption behavior of arsenate using single and binary iron-modified biochars: Thermodynamics and redox transformation. *ACS Omega*, 7(1), 101–117. https://doi.org/10.1021/acsomega.1c04129

Sabzehmeidani, M. M., Mahnaee, S., Ghaedi, M., Heidari, H., & Roy, V. A. L. (2021). Carbon based materials: A review of adsorbents for inorganic and organic compounds. *Materials Advances*, 2(2), 598–627. https://doi.org/10.1039/d0ma00087f

Salleh, M. A. M., Mahmoud, D. K., Karim, W. A. W. A., & Idris, A. (2011). Cationic and anionic dye adsorption by agricultural solid wastes: A comprehensive review. *Desalination*, 280(1–3), 1–13. https://doi.org/10.1016/j.desal.2011.07.019

Selvaraj, V., Swarna Karthika, T., Mansiya, C., & Alagar, M. (2021). An over review on recently developed techniques, mechanisms and intermediate involved in the advanced azo dye degradation for industrial applications. *Journal of Molecular Structure, 1224*. https://doi.org/10.1016/j.molstruc.2020.129195

Serhan, M., Sprowls, M., Jackemeyer, D., Long, M., Perez, I. D., Maret, W., Tao, N., & Forzani, E. (2019). Total iron measurement in human serum with a smartphone. In *AIChE Annual Meeting, Conference Proceedings*. https://doi.org/10.1109/jtehm.2020.3005308

Shafqat, S. R., Bhawani, S. A., Bakhtiar, S., & Ibrahim, M. N. M. (2020). Synthesis of molecularly imprinted polymer for removal of Congo red. *BMC Chemistry*, 14(1), 27. https://doi.org/10.1186/s13065-020-00680-8

Singh, J., Ali, A., & Prakash, V. (2014). Removal of lead (II) from synthetic and batteries wastewater using agricultural residues in batch/column mode. *International Journal of Environmental Science and Technology*, 11(6), 1759–1770. https://doi.org/10.1007/s13762-013-0326-9

Singh, R., Chadetrik, R., Kumar, R., Bishnoi, K., Bhatia, D., Kumar, A., Bishnoi, N. R., & Singh, N. (2010). Biosorption optimization of lead(II), cadmium(II) and copper(II) using response surface methodology and applicability in isotherms and thermodynamics modeling. *Journal of Hazardous Materials*, 174(1–3), 623–634. https://doi.org/10.1016/j.jhazmat.2009.09.097

Singla, M., & Sit, N. (2021). Application of ultrasound in combination with other technologies in food processing: A review. *Ultrasonics Sonochemistry, 73*. https://doi.org/10.1016/j.ultsonch.2021.105506

Song, J., Messele, S. A., Meng, L., Huang, Z., & Gamal El-Din, M. (2021). Adsorption of metals from oil sands process water (OSPW) under natural pH by sludge-based Biochar/Chitosan composite. *Water Research*, 194, 116930. https://doi.org/10.1016/j.watres.2021.116930

Stasinakis, A. S., Elia, I., Petalas, A. V., & Halvadakis, C. P. (2008). Removal of total phenols from olive-mill wastewater using an agricultural by-product, olive pomace. *Journal of Hazardous Materials*, 160(2–3), 408–413. https://doi.org/10.1016/j.jhazmat.2008.03.012

Sun, P., Hui, C., Khan, R. A., Du, J., Zhang, Q., & Zhao, Y. H. (2015). Efficient removal of crystal violet using Fe_3O_4-coated biochar: The role of the Fe_3O_4 nanoparticles and modeling study their adsorption behavior. *Scientific Reports*, 5(March), 1–12. https://doi.org/10.1038/srep12638

Talfana, A. A., Kerouad, S., Forsal, I., & Kotmani, W. (2026). Natural corrosion inhibitor from Cistanche Tubulosa extract for carbon steel in HCl: Gravimetric and electrochemical characterization. *Progress in Color, Colorants and Coatings*, 19, 1–34. https://doi.org/10.30509/pccc.2025.167509.1383

Tong, Y., McNamara, P. J., & Mayer, B. K. (2019). Adsorption of organic micropollutants onto biochar: A review of adsorption of organic. *Environmental Science: Water Research and Technology*, 5, 821–838. https://epublications.marquette.edu/civengin_fac/237

Ton-That, L., Huynh, T. N. L., Duong, B. N., Nguyen, D. K., Nguyen, N. A., Pham, V. H., Ho, T. H., & Dinh, V. P. (2023). Kinetic studies of the removal of methylene blue from aqueous solution by biochar derived from jackfruit peel. *Environmental Monitoring and Assessment*, 195(11), 1–22. https://doi.org/10.1007/s10661-023-11867-6

Tripathi, A. D., Mishra, P. K., Khosarvi-Darani, K., Agarwal, A., & Paul, V. (2022). Hydrothermal treatment of lignocellulose waste for the production of polyhydroxyalkanoates copolymer with potential application in food packaging. *Trends in Food Science and Technology*, 123(March), 233–250. https://doi.org/10.1016/j.tifs.2022.03.018

Wang, A., Dang, Z., Wang, Y., Fan, H., & Miao, S. (2025). Efficient inorganic stabilization materials for chromium and arsenic pollution in water and soil. *Applied Sciences*, 15(13), 7069. https://doi.org/10.3390/app15137069

Wang, Y., Ren, D., Ye, J., Li, Q., Yang, D., Wu, D., Zhao, J., & Zou, Y. (2024). Highly efficient and targeted adsorption of Congo Red in a novel cationic copper-organic framework with three-dimensional cages. *Separation and Purification Technology*, *329*(September 2023), 125149. https://doi.org/10.1016/j.seppur.2023.125149

Wang, Z., Guo, H., Shen, F., Yang, G., Zhang, Y., Zeng, Y., Wang, L., Xiao, H., & Deng, S. (2015). Biochar produced from oak sawdust by Lanthanum (La)-involved pyrolysis for adsorption of ammonium (NH_4^+), nitrate (NO_3^-), and phosphate (PO_4^{3-}). Chemosphere, 119(August), 646–653. https://doi.org/10.1016/j.chemosphere.2014.07.084

Wato, T. (2020). The agricultural water pollution and its minimization strategies – A review. *Journal of Resources Development and Management*. https://doi.org/10.7176/jrdm/64-02

Wei, F., Jin, S., Yao, C., Wang, T., Zhu, S., Ma, Y., Qiao, H., Shan, L., Wang, R., Lian, X., Tong, X., Li, Y., Zhao, Q., & Song, W. (2023). Revealing the combined affect of active sites and intra-particle diffusion on adsorption mechanism of methylene blue on activated red-pulp pomelo peel biochar. *Molecules*, *28*(11). https://doi.org/10.3390/molecules28114426

Weidner, E., Karbassiyazdi, E., Altaee, A., Jesionowski, T., & Ciesielczyk, F. (2022). Hybrid metal oxide/biochar materials for wastewater treatment technology: A review. *ACS Omega*, 7(31), 27062–27078. https://doi.org/10.1021/acsomega.2c02909

Wekoye, J. N., Wanyonyi, W. C., Wangila, P. T., & Tonui, M. K. (2020). Kinetic and equilibrium studies erof Congo red dye adsorption on cabbage waste powd. *Environmental Chemistry and Ecotoxicology*, 2, 24–31. https://doi.org/10.1016/j.enceco.2020.01.004

Yahya, M. A., C. W. Ngah, C. W. Z., Hashim, M. A., & Al-Qodah, Z. (2015). Preparation of activated carbon from desiccated coconut residue by chemical activation with NaOH. *Journal of Materials Science Research*, 5(1), 24. https://doi.org/10.5539/jmsr.v5n1p24

Yu, M., Sun, Y., Shi, W., Wang, X., Zhang, C., & Ke, Q. (2025). Synthesis of high-performance magnetic biochar for adsorption of Ni2+ and Co2+from spent lithium-ion battery effluent. *Journal of Hazardous Materials Advances*, 18(59), 100627. https://doi.org/10.1016/j.hazadv.2025.100627

Zhao, S., Ma, Y., & Zhou, J. (2024). Removal of cadmium ion from wastewater by manganese oxides-loaded sludge biochar. *Desalination and Water Treatment*, 319(April), 100563. https://doi.org/10.1016/j.dwt.2024.100563

Zhao, W., Fan, L., Guo, Y., Wu, H., Zang, S., & Wang, H. (2025a). Study on the carbon source stability of constructed wetland prepared from modified aged straw. *Journal of Environmental Management*, 373(December). https://doi.org/10.1016/j.jenvman.2024.123708

Zhao, Y., Wang, C., Han, Q., Fang, Z., Gao, Y., Chen, H., Li, J., Yang, X., Chen, J., & Wang, H. (2025b). Recent advances in biochar-based hydrogel composites: Preparation, aquatic environmental applications, and adsorption mechanisms. *Processes*, 13(3), 10–16. https://doi.org/10.3390/pr13030664

Zubair, M., Manzar, M. S., Khan, S. A., Suleiman, M. A., Nawaz, M., Khan, T. A., Bakather, O. Y., da Gama, B. M. V, Yasir, M., Melli, L., & Khan, G. (2024). Enhanced adsorption of anionic dyes onto NaOH modified green coffee waste biochar: characteristics and performance evaluation. *International Journal of Environmental Science and Technology*. https://doi.org/10.1007/s13762-024-06090-w

6 Electrochemical Methods for Liquid Waste Treatment

Innovations, Applications, and Sustainability

Mouad Mouhsin, Mohamed Amine Kasbaji, Oumayma El Wafi, Anas Hafdi, Meriem Kasbaji, Mohamed M'Barki, Mounir El Achaby, Wiem Mezlini, Saad ALjlil, and Mustapha Oubenali

6.1 INTRODUCTION TO ELECTROCHEMICAL WASTE TREATMENT

The management of liquid industrial and urban waste represents a critical challenge in modern environmental engineering. With the increasing concentration of organic pollutants, heavy metals, and pharmaceutical compounds in water bodies, conventional treatment techniques, such as biological degradation and physico-chemical processes, are frequently unable to address the complexity and resilience of these contaminants effectively. Traditional methods often suffer from issues related to high operational costs, limited removal efficiencies, the generation of secondary pollutants, and challenges in handling dynamic pollutant loads [1]. In response to these constraints, researchers and practitioners have turned to electrochemical processes as a promising alternative that offers innovative pathways for waste liquid remediation.

Electrochemical treatment methods harness electrical energy to drive oxidation and reduction reactions, thereby converting hazardous pollutants into less-harmful substances or facilitating their physical separation. One of the primary advantages of these processes is that they eliminate or minimize the need for additional chemicals, reducing both the risks of secondary pollution and the overall treatment costs [2]. The absence of chemical additives not only simplifies process logistics but also enhances safety and sustainability by preventing the introduction of extraneous contaminants into the water stream.

Among the various electrochemical approaches, EO has emerged as one of the most effective techniques for degrading recalcitrant organic compounds. This method involves the in situ generation of potent oxidizing species, such as hydroxyl radicals, at the surface of electrode materials. These radicals are capable of indiscriminately

DOI: 10.1201/9781003624905-6

attacking a wide range of organic pollutants, facilitating their mineralization or transformation into intermediates that are more amenable to biological treatment. Recent studies have demonstrated that optimization of operating parameters and electrode materials significantly improves the production efficiency of hydroxyl radicals, making EO a viable option for treating diverse effluents [3].

In addition to EO, EC has gained attention due to its unique mechanism in which sacrificial electrodes release metal ions upon anodic dissolution. The liberated ions, typically comprising aluminum or iron, undergo hydrolysis to form insoluble hydroxides that act as coagulating agents. These coagulants aggregate pollutants through adsorption and charge neutralization, thereby facilitating the removal of suspended solids, colloidal particles, and some dissolved contaminants through sedimentation or flotation processes. The integration of EC into treatment schemes has been shown to yield significant improvements in turbidity and organic load reductions, particularly in industrial waste liquid scenarios where conventional coagulation might prove inadequate [4].

Another notable electrochemical process is electroflotation, which exploits the production of gas microbubbles generated at the electrode interfaces. In electroflotation, these microbubbles adhere to suspended particles, reducing their effective density and driving them upward to the surface, where they can be easily collected. This technique is particularly useful in clarifying turbid waste liquid, as it offers a gentle yet effective means of removing fine particles without resorting to chemical flocculants. Recent advancements have focused on enhancing bubble generation efficiency and optimizing electrode configurations to further improve separation performance in various waste liquid contexts [5].

Furthermore, hybrid systems combining electrochemical treatment with biological processes have garnered significant interest in recent years. For example, microbial fuel cells represent an innovative approach where the metabolic activity of microorganisms is coupled with electrochemical reactions to achieve pollutant degradation while simultaneously generating electrical energy. These systems not only provide dual benefits of waste remediation and renewable energy production but also offer unique insights into the synergistic interactions between biological and electrochemical processes. Although still at an early stage of development compared to standalone electrochemical techniques, the integration of microbial fuel cells into waste liquid treatment schemes holds significant promise for achieving higher overall process sustainability [6].

Electrochemical methods offer several advantages that underscore their potential role in modern waste liquid management. Among these advantages are high treatment efficiencies, decreased reliance on chemical additives, modularity that enables scalability from laboratory to industrial applications, and lower production of secondary waste such as sludge. These features contribute to a reduced environmental footprint and enhanced economic viability when compared to some traditional treatment methods. Additionally, the flexibility of electrochemical processes allows them to be tailored to treat a wide range of pollutant profiles, from high-strength industrial waste liquids to the lower contaminant levels observed in urban effluents [2].

Despite these benefits, several challenges remain in the widespread adoption of electrochemical waste liquid treatment technologies. High energy consumption, particularly in processes involving intensive oxidation reactions, continues to be a concern that may limit their economic feasibility in certain contexts. Advances in electrode material science are needed to develop more durable and efficient electrodes that can withstand the corrosive operational environments encountered during treatment [4]. Moreover, scaling up these techniques from controlled laboratory experiments to full-scale industrial applications involves addressing complex engineering challenges related to reactor design, mass transfer limitations, and process integration with existing treatment infrastructures.

Future research should aim at optimizing process parameters, enhancing energy recovery, and developing composite treatment schemes that combine the strengths of electrochemical methods with biological and physical processes. Such integrated approaches could mitigate the limitations of individual techniques while delivering superior overall performance in terms of pollutant removal and resource recovery. In addition, techno-economic assessments and life-cycle analyses will be critical in benchmarking the performance and environmental impacts of electrochemical treatments against conventional technologies, thereby guiding policy makers and industry stakeholders in their decision-making [1,6].

In conclusion, electrochemical processes offer a versatile, effective, and environmentally friendly alternative for treating liquids. By leveraging redox reactions to break down contaminants without the need for excessive chemical additives, these technologies address many of the shortcomings of traditional treatment methods. While challenges related to energy costs, electrode durability, and industrial scalability persist, ongoing research and technological innovation continue to pave the way for their broader implementation in sustainable waste liquid management practices [2,7].

6.2 FUNDAMENTALS OF ELECTROCHEMICAL PROCESSES IN WASTE TREATMENT

Electrochemical processes applied to the treatment of liquid effluents rely on the use of electrical energy to induce redox reactions, enabling the degradation, separation, or transformation of pollutants present in water [8]. These methods operate on fundamental electrochemical principles, offering a versatile and often environmentally benign approach to water purification [9].

At the heart of many electrochemical treatment techniques lies electrolysis, a process where an electric current is employed to drive a non-spontaneous chemical reaction. In the context of waste liquid treatment, electrolysis typically occurs in an electrochemical cell containing at least two electrodes: an anode (the positive electrode) and a cathode (the negative electrode), immersed in an electrolyte solution. When an external voltage is applied across these electrodes, an electric field is established, facilitating the movement of ions in the solution and driving chemical reactions at the electrode surfaces [10].

A primary application of electrolysis in waste liquid treatment is to facilitate redox reactions involving the transfer of electrons, which are central to the removal or transformation of pollutants [11]. At the anode, oxidation takes place, where

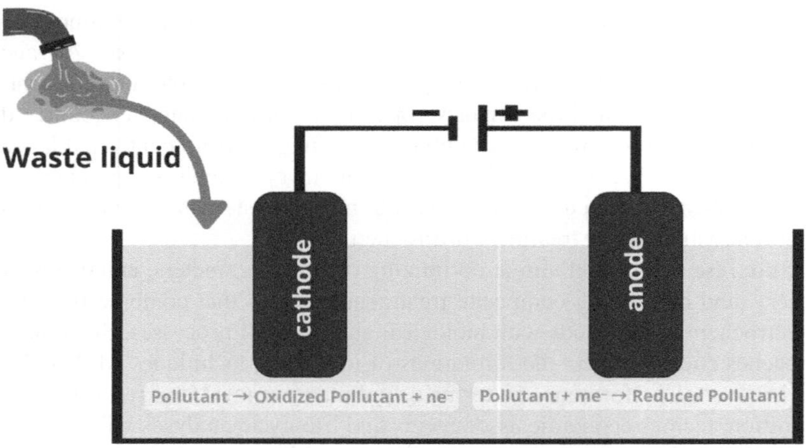

FIGURE 6.1 Schematic representation of an electrochemical cell for liquid-waste treatment.

pollutants or other species lose electrons (Figure 6.1). This can lead to the direct breakdown of organic pollutants or the generation of powerful oxidizing agents like hydroxyl radicals (•OH) from water [12].

A general representation of oxidation at the anode can be written as:

$$\text{Pollutant} \rightarrow \text{Oxidized Pollutant} + \text{ne}^-$$

For instance, in EO, particularly with efficient anodes like Boron-Doped Diamond (BDD), water can be oxidized to form hydroxyl radicals [13]:

$$H_2O \rightarrow \bullet OH + H^+ + e^-$$

Conversely, at the cathode, reduction occurs, where pollutants or other species gain electrons. This can result in the deposition of heavy metals onto the cathode surface (electrodeposition) or the reduction of pollutants like halogenated compounds or nitrates to less-harmful forms [14]. A general representation of reduction at the cathode is:

$$\text{Pollutant} + \text{me}^- \rightarrow \text{Reduced Pollutant}$$

For example, the reduction of metal ions (M^{n+}) to their metallic form ($M_{(s)}$) is represented as:

$$M^{n+} + ne^- \rightarrow M(s)$$

The feasibility and rate of these electrochemical reactions are critically governed by the electrode potentials, which represent the potential difference between the electrode and the electrolyte solution. Each electrochemical reaction has a specific standard electrode potential, indicating its tendency to occur under standard conditions

[15]. The applied voltage across the electrochemical cell must be sufficient to overcome the thermodynamic energy barriers (related to electrode potentials) and the kinetic activation energies (overpotentials) to drive the desired reactions at an acceptable rate [15]. The relationship between the applied voltage (E_{cell}) and the electrode potentials of the cathode ($E_{cathode}$) and anode (E_{anode}), considering overpotentials (η) and ohmic drop (IR_{drop}), is fundamentally described by accounting for these factors beyond the simple thermodynamic difference [16]:

$$E_{cell,Applied} \geq (E_{cathode} - E_{anode}) + |\eta_{cathode}| + |\eta_{anode}| + IR_{drop}$$

For a non-spontaneous reaction like electrolysis to occur, the applied voltage must exceed the equilibrium cell potential and account for these additional energy losses.

Many electrochemical transformations in waste liquid treatment are faradaic reactions, which are chemical or physical changes that occur due to the transfer of electrons at the electrodes [16]. Faraday's laws of electrolysis quantitatively describe the relationship between the amount of substance transformed at an electrode and the quantity of electricity passed through the electrochemical cell [17]. Faraday's first law states that the mass of a substance produced or consumed at an electrode is directly proportional to the quantity of electricity passed through the cell. Mathematically, this is expressed as:

$$m = \frac{Q.M}{n.F}$$

where:

- m is the mass of the substance produced or consumed (in grams).
- Q is the total electric charge passed through the cell (in Coulombs).
- M is the molar mass of the substance (in grams per mole).
- n is the number of moles of electrons transferred per mole of substance.
- F is the Faraday constant.

Faraday's second law relates the amounts of different substances transformed by the same charge to their equivalent weights.

An electrochemical system for treating liquid waste comprises several essential elements. First, a power supply delivers and regulates the cell potential or current to drive the desired redox reactions [18]. Second, electrodes constitute the reaction sites: anodes may be BDD, mixed metal oxides, or graphite, while cathodes often employ carbon felt, stainless steel, or platinum meshes; choice of material and surface morphology strongly influence reaction kinetics, selectivity, and long-term stability [19]. Third, the electrochemical reactor (batch, flow-through, filter-press, etc.) must be designed to optimize inter-electrode spacing, fluid dynamics, and mass transport; compact geometries and controlled turbulence improve contaminant removal rates [16]. Fourth, when selective ion transport or compartmentalization is required, ion-exchange membranes or porous separators are incorporated to prevent short-circuiting and enable paired oxidation–reduction processes. Fifth, the addition of a supporting electrolyte adjusts solution conductivity, ensuring uniform current distribution and minimizing side reactions. Sixth, fluid-handling components, pumps, piping, and mixers control influent and effluent flows and facilitate scale-up to continuous operation. Seventh, a suite of sensors and instrumentation (pH, oxidation–reduction potential, conductivity, temperature, flow rate) provides real-time monitoring, while an integrated control and data-acquisition system automates voltage/current profiles and logs performance metrics. Eighth, gas-management equipment (venting or collection systems) safely handles hydrogen and oxygen generated during electrolysis. Finally, a post-treatment stage filtration, adsorption, precipitation, or ion-exchange polishing removes residual contaminants and can recover by-products or metals from the treated stream [18].

6.3 TYPES OF ELECTROCHEMICAL METHODS

Electrochemical remediation of liquid wastes encompasses three principal categories (Figure 6.2): conversion, separation, and combined methods, each distinguished by its dominant removal mechanism. Conversion methods mineralize or chemically transform pollutants via direct redox reactions at electrode surfaces. Separation methods employ electric fields or electrogenerated bubbles to detach, migrate, or concentrate contaminants without altering their chemistry. Combined methods integrate both destruction and separation processes in a single reactor, often achieving synergistic performance. In practice, choosing and tailoring these techniques depends on pollutant type, waste liquid matrix, energy/economic constraints, and desired by-product recovery [16].

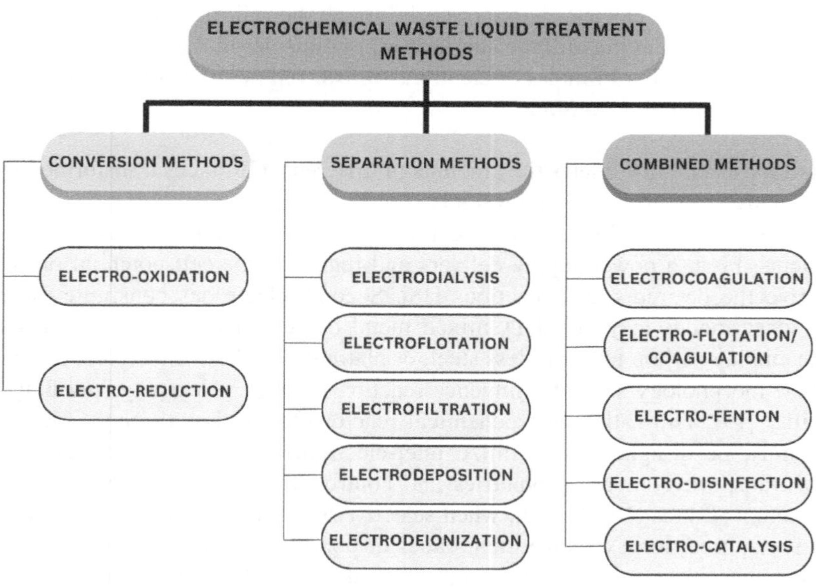

FIGURE 6.2 Classification of electrochemical liquid waste treatment methods.

In conversion methods, the core principle is the destruction or chemical modification of organic, inorganic, or metallic pollutants through either direct electron transfer at electrodes (oxidation or reduction) or indirect pathways involving in situ generated oxidants or radicals [20]. Electrochemical oxidation on high overpotential anode materials such as BDD and titanium suboxide proceeds via anodic water discharge to generate highly reactive hydroxyl radicals, which initiate pollutant mineralization through hydrogen abstraction and unsaturated bond cleavage. In chloride-rich media, simultaneous chlorine evolution on dimensionally stable anodes yields molecular chlorine, hypochlorous acid, and hypochlorite, which selectively attack electron-rich organic moieties and complement hydroxyl radical pathways. The production rates and selectivity of these oxidants depend critically on current density, solution pH, and chloride concentration, all of which must be carefully optimized to maximize mineralization efficiency while limiting parasitic reactions [21]. By contrast, cathodic electrochemical reduction enables the removal or recovery of metal ions such as copper, nickel, and lead through their precipitation as insoluble hydroxides, oxides, or native metals on the cathode surface. Nitrate reduction follows a sequential electron transfer pathway from nitrate to nitrite, nitrous oxide, and ultimately dinitrogen gas, with engineered transition metal catalysts achieving faradaic efficiencies in excess of 80%. The choice of cathode catalyst, such as copper–nitrogen–carbon composites and palladium copper alloys, alongside reactor configuration in batch or continuous flow operation, profoundly influences process selectivity, energy demand, and operational stability. Integrating anodic oxidation and cathodic reduction within a single electrochemical cell offers synergistic treatment of mixed organic and inorganic pollutants, reduces system footprint, and facilitates resource recovery [22]. Major advantages of these conversion methods include the ability to treat complex molecular structures and partially or completely destroy toxic substances. Additionally, external chemical reagent use is often reduced, limiting secondary waste formation. However, performance can vary based on effluent conductivity, required overpotential, and electrode stability [23].

Separation methods, conversely, focus on extracting or concentrating pollutants into a separate phase without necessarily altering their chemical structures [16]. These include electrodialysis, which employs stacked cation and anion exchange membranes positioned between electrodes to selectively transport ions under an applied electric potential, achieving effective desalination and removal of charged contaminants; recent advances in membrane materials have enhanced perm selectivity and fouling resistance, thereby improving energy efficiency and operational lifespan in diverse liquid wastes [16]. Capacitive deionization utilizes porous carbonaceous electrodes to adsorb ions electrostatically when a low direct current voltage is applied and to release them upon voltage reversal, enabling energy-efficient desalination as well as targeted removal of heavy metals and nutrients; the incorporation of Prussian blue analogues and other functional coatings has significantly increased ion adsorption capacity and cycle stability [24]. Electroflotation generates fine hydrogen and oxygen microbubbles directly at the electrode surfaces, which adhere to suspended particles or oil droplets and lift them to the liquid surface for skimming; optimized reactor geometries and hydrodynamic conditions now yield smaller, more uniformly distributed bubbles and faster separation rates [25].

Electrofiltration combines an applied electric field with membrane separation to repel charged foulants and sustain high permeate flux, mitigating membrane fouling through electrophoretic transport and localized pH modulation at the membrane–electrolyte interface [26]. Electrodeposition enables the direct cathodic reduction of metal ions to solid metal on suitable substrates, facilitating high-purity recovery of copper, nickel, and lead from complex waste liquids; recent developments in bio-electrochemical cell configurations and pulsed potential strategies have markedly improved current efficiency and metal selectivity in industrial applications [27]. The appeal of separation techniques lies primarily in their selectivity, reduced sludge production, and potential valorization of recovered species (metals, salts). However, high-performance membrane/electrode materials and careful fouling management remain critical considerations [27].

Combined approaches encompass a range of methods that simultaneously harness different electrochemical actions or integrate electrochemistry with complementary treatment steps. EC, often seen as a hybrid method, involves the controlled dissolution of sacrificial electrodes (typically aluminum or iron) to form in situ coagulants (e.g., metal hydroxides), while simultaneously benefiting from bubble formation (electroflotation). This dual mechanism accelerates clarification and minimizes chemical consumption [28]. Similarly, the electro-Fenton (EF) process combines cathodic production of hydrogen peroxide with iron catalysis to generate hydroxyl radicals (•OH), effectively merging advanced oxidation with electrochemical regulation of iron and oxygen species [29]. Other integrated systems couple coagulation with anodic oxidation for substantial reduction in organic load, or integrate biological treatments (bioelectrochemical systems), exploiting synergy between microbial degradation and electrochemical activation. The strengths of combined methods include versatility against diverse pollutants, synergistic effects enhancing removal efficiency, faster pollutant mineralization, and selective resource recovery with minimal environmental footprint. Nonetheless, technical configuration can become complex, and operational control often requires careful regulation of pH, current density, conductivity, and management of interactions such as membrane fouling or biomass interference [30].

Overall, this classification (conversion, separation, and hybrid methods) highlights the extensive array of available electrochemical tools, each presenting specific advantages, limitations, and optimal application domains. Practically, selecting a particular process or process combination depends heavily on waste liquid characteristics, regulatory discharge or reuse objectives, energy availability, and installation and maintenance costs.

6.4 MATERIALS AND ELECTRODE DESIGN

The choice of materials for the anode and cathode is crucial in electrochemical waste liquid treatment processes, as it directly affects pollutant degradation kinetics, electrode durability, and operational costs [31]. On the anodic side, electrode materials fall into two classes: active and non-active. Active anodes composed of titanium substrates coated with ruthenium dioxide or iridium dioxide combine excellent electrical conductivity with outstanding corrosion resistance, making them well suited for continuous operation in corrosive waste liquid environments. Their relatively low overpotential for oxygen evolution, however, limits the surface coverage of adsorbed hydroxyl radicals, which in turn constrains the oxidation mechanisms to more selective pathways that favor easily oxidizable compounds over recalcitrant species [32]. As a result, active anodes deliver lower energy consumption and longer service life but may require supplementary treatment steps when dealing with complex organic mixtures.

Non-active anodes, such as antimony-doped tin oxide, lead dioxide, and Magnéli-phase titanium suboxide (Ti_4O_7), operate at higher overpotentials for oxygen evolution and, therefore, promote the formation of weakly adsorbed hydroxyl radicals over the electrode surface. This enhanced radical coverage increases the probability of bulk oxidation reactions, enabling more comprehensive degradation of persistent organic pollutants in a single pass. The trade-off for this deeper mineralization capability is the need for higher applied voltages, which elevates energy requirements, and the potential for accelerated electrode wear under harsh operating conditions [33,34]. Consequently, the selection between active and non-active coatings must balance the treatment goals, whether prioritizing energy efficiency and electrode longevity or aiming for maximal pollutant breakdown while accounting for the specific characteristics of the liquid waste, such as ionic strength, organic load, and allowable maintenance intervals.

Common cathode materials for electrochemical treatment of liquid wastes include metallic substrates such as stainless steel, nickel, and copper, as well as carbon-based electrodes like graphite felt, carbon cloth, and porous carbon matrixes [35]. Stainless steel and nickel offer excellent electrical conductivity and mechanical strength, but can undergo surface passivation and hydrogen embrittlement under strongly reducing conditions, which may impair long-term performance. Copper exhibits superior catalytic activity for nitrate reduction and heavy metal recovery, though its susceptibility to corrosion necessitates protective coatings or alloying strategies. Carbon materials provide high specific surface area, chemical inertness, and low cost, while surface modifications such as nitrogen doping or incorporation of metal nanoparticles significantly enhance electron transfer kinetics and selectivity for target species [36]. More recently, composite cathodes combining conductive carbon frameworks with dispersed catalysts (for example, copper–nitrogen–carbon catalysts or palladium nanoparticle-decorated carbon) have demonstrated markedly improved faradaic efficiencies and reduced energy consumption. Selecting an optimal cathode material, therefore, involves balancing electrical conductivity, catalytic activity, chemical durability, and economic considerations in relation to the specific pollutants and operating conditions of the treatment system.

Electrode design and architecture are just as important as material composition in determining the overall treatment performance. Key parameters include surface area, porosity, geometry, electrode spacing, and current distribution. For instance, 3D electrodes, such as those incorporating granular activated carbon or magnetically assembled 2.5D structures, significantly increase the available electroactive surface area, improving both reaction kinetics and pollutant accessibility [37]. Microstructured or nanostructured anodes, including TiO_2 nanotubes combined with g-C_3N_4 or aerogel composites, enable high pollutant adsorption and enhanced catalytic activity. Meanwhile, microfluidic reactor systems and flow-through electrochemical cells ensure better current efficiency and uniform treatment in dynamic waste liquid environments [38].

6.5 APPLICATIONS IN VARIOUS INDUSTRIES

Electrochemical technologies have found growing applications across multiple industrial sectors due to their efficiency, environmental compatibility, and scalability for treating diverse liquid waste streams. These methods, such as EC, EF, EO, and photoelectrochemical processes, are particularly valuable in handling toxic, recalcitrant, or non-biodegradable pollutants that challenge conventional treatment techniques.

6.5.1 Textile and Dyeing Industry

The textile industry generates large volumes of waste liquid containing synthetic dyes, surfactants, and heavy metals. EO and EF processes are especially effective in degrading complex dye molecules (e.g., azo and anthraquinone dyes) through the generation of hydroxyl radicals (OH) with high oxidizing power. BDD and Ti/SnO_2–Sb anodes are commonly used to achieve complete decolorization and mineralization of dye-laden effluents. Studies have demonstrated over 90% removal efficiency of dyes such as methylene blue and reactive black using EO and EF systems [39].

6.5.2 Pharmaceutical and Healthcare Industry

Waste liquid from pharmaceutical manufacturing contains antibiotics, hormones, analgesics, and other emerging contaminants that are difficult to degrade biologically. Electrochemical advanced oxidation processes, especially EF and anodic oxidation with Ti_4O_7 or PbO_2 anodes, have shown excellent performance in degrading pharmaceuticals like ibuprofen, diclofenac, and sulfamethoxazole. These processes not only reduce chemical oxygen demand (COD) but also neutralize the biological activity of drug residues, mitigating ecological risk [40].

6.5.3 Food and Beverage Industry

In this sector, waste liquid is characterized by high organic loads, fats, oils, and variable pH. EC is often applied for the removal of suspended solids, oil and grease, and turbidity, especially in dairy and meat processing plants. Hybrid EC systems combined with EO or electroflotation further enhance treatment efficiency. Additionally, EF processes can be used to remove persistent organic residues from sugar refining or beverage production units [41].

6.5.4 Petrochemical and Oil Refining Industry

Petrochemical effluents contain aromatic hydrocarbons, phenols, and other toxic compounds. EO with high-oxygen evolution potential anodes (like BDD or Ti_4O_7) enables the direct degradation of these non-biodegradable pollutants. Combined

EC-EO or EF systems are also used to break emulsions and oxidize residual hydrocarbons. These processes reduce toxicity, COD, and color, making them suitable for pre-treatment before biological processes or for final polishing stages [42].

6.5.5 Mining and Metallurgical Industry

Waste liquid from mining operations often contains high levels of metals (e.g., arsenic, chromium, lead), sulfates, and cyanides. EC and electroreduction methods are effective for the removal or recovery of heavy metals by reducing their ionic forms and precipitating them as stable solids. Electrochemical processes can also be tailored for the selective recovery of valuable metals or the treatment of acid mine drainage [43].

6.5.6 Pulp and Paper Industry

The pulp and paper industry generates waste liquid rich in lignin, chlorinated compounds, and high COD. EO and EC have been effectively employed to treat such effluents. EO processes, utilizing anodes like BDD, facilitate the breakdown of complex organic molecules, leading to significant reductions in COD and color. EC, on the other hand, aids in the removal of suspended solids and colloidal particles, enhancing the overall treatment efficiency. These methods offer advantages over traditional treatments by reducing sludge production and chemical usage [44].

6.5.7 Tannery Industry

Tannery waste liquid contains high concentrations of organic matter, sulfides, and chromium compounds. EC has been successfully applied to remove these pollutants. Aluminum or iron electrodes, when used in EC processes, generate coagulant species in situ, which bind with contaminants, facilitating their removal. This method not only reduces pollutant levels but also minimizes the need for added chemicals, making it an environmentally friendly option for tannery waste liquid treatment [45].

6.5.8 Electroplating and Metal Finishing Industry

Waste liquid from electroplating and metal finishing industries is laden with heavy metals like nickel, copper, and zinc. Electrochemical methods, particularly electrodeposition and EC, are employed to recover these metals and treat the effluent. Electrodeposition allows for the direct recovery of metals onto cathodes, while EC facilitates the aggregation and removal of metal hydroxides. These techniques not only purify the waste liquid but also enable resource recovery, aligning with sustainable industrial practices [46].

6.5.9 Semiconductor and Electronics Industry

The semiconductor and electronics industry generates waste liquid containing fluorides, heavy metals, and organic solvents. Electrochemical oxidation processes,

TABLE 6.1
Main Industrial Electrochemical Methods for Liquid Waste Treatment

Electrochemical Method	Industrial Applications	Advantages	Limitations	References
EC	Textiles, petroleum, agri-food industries	Reduction of chemical sludge, low reagent cost	Formation of metallic sludge, electrical consumption	[48–50]
EO	Pharmaceutical, hospital wastewater	Complete degradation of organic pollutants	High electrode cost, high energy consumption	[51,52]
Electroflotation	Grease, oils, agri-food effluents	No addition of chemical reagents	Mainly effective for particulate pollutants	[5,52–54]
Electrodialysis	Desalination, leachate treatment	Selective ion separation, resource recovery	High membrane cost, fouling issues	[55,56]
Capacitive Deionization	Low salinity water, potable water	Low energy consumption, continuous operation	Mainly effective at low salinity	[57,58]
EF	Pharmaceutical, dye-containing wastewater	Powerful oxidation, complete mineralization	Optimal pH required (~3), residual iron management	[59–61]

especially those employing BDD anodes, have been effective in degrading organic contaminants. Additionally, electrodialysis has been used for the selective removal of ionic species, aiding in water reuse and resource recovery. These electrochemical treatments help in meeting stringent discharge regulations and promote sustainable manufacturing practices [47].

The main electrochemical methods applied at an industrial scale for liquid waste treatment are summarized in Table 6.1.

6.6 CASE STUDIES AND PRACTICAL IMPLEMENTATIONS

In this section, four real-world case studies from different industries are examined to illustrate the practical implementation of various electrochemical methods for liquid waste treatment. Each case highlights a distinct electrochemical technique (EC, electrodialysis, EO, and EF) applied at pilot or industrial scale within the last five years. For each case, we describe the industrial context and waste liquid characteristics, the electrochemical process and operating conditions, treatment performance (removal efficiencies and other outcomes), environmental benefits, and any noted limitations or challenges. These examples demonstrate the versatility of electrochemical treatments across sectors, while also identifying practical considerations for their use in the field. A concise overview of operating conditions and treatment outcomes for each case study is provided in Table 6.2.

TABLE 6.2
Summary of Four Pilot- and Lab-Scale Electrochemical Liquid-Waste Treatment Case Studies, Showing Industry Context, Technique, Scale, Operating Conditions, and Performance Outcomes

Technique	Key Operating Conditions	Performance Outcomes
EC	Fe electrodes, 653 A m^{-2}; pH 8–9; 30–60 min retention	99.6% color removal; 93.7% COD removal
EDR	EDR stack with polarity reversal; feed 50–100 mg L^{-1} F^-	80% fluoride removal; 70% water recovery
EO	Ti/PbO_2 anode; 90 mA cm^{-2}; 0.08% electrolyte; 3 h	78% COD; 64% NH_3-N; 87% P removal
EF + BDD	BDD anode + EF cathode; 4.2 mA cm^{-2}; pH 3; 0.2 mM Fe^{2-}; 4 h	97.1% total organic carbon removal; BOD_5/COD from 0.14 to 0.68

6.6.1 EC in the Textile Industry

Textile manufacturing generates waste liquid with intense color, high COD, dissolved solids (from dyeing salts), and often toxic or non-biodegradable organic dyes. Conventional treatments struggle with color removal, making EC an attractive option. In EC, soluble metal anodes (typically iron or aluminum) dissolve under an applied current to release metal cations that hydrolyze into coagulants in situ. These coagulants (e.g., $Fe(OH)_3$ or $Al(OH)_3$) adsorb and destabilize dyes and colloids, forming flocs that can be separated, thereby removing color and organic pollutants [62]. A recent pilot-scale study treated textile dyeing waste liquid using an integrated EC and flotation system (EC–clarifier flotation) followed by biological and adsorption steps [63]. The EC unit employed iron electrodes at an average current density of ~653 A m^{-2}, with a power consumption of 0.65 kWh m^{-3}. This electrochemical pre-treatment, combined with downstream processes, achieved 93.7% COD removal and 99.6% color removal, producing an effluent suitable for reuse in textile production. These high removal efficiencies underscore EC's effectiveness in precipitating dyes and organic matter. The EC step also drastically reduced total suspended solids, aiding subsequent biological treatment by lowering sludge loads.

Environmentally, the EC-based system enabled water recycling (50 m^3 day^{-1} pilot treating textile effluent) and halved the operational cost compared to conventional treatment, due in part to reduced chemical usage and sludge handling.

6.6.2 Electrodialysis in the Semiconductor Industry

The semiconductor fabrication industry requires ultra-pure water and generates large volumes of waste liquid containing inorganic contaminants (e.g., fluoride, nitrate, ammonia, silica, copper) from processes like wafer cleaning and chemical-mechanical polishing. A key challenge is treating fluoride-rich acidic waste from local scrubbers (air pollution control units using water to scrub fumes of hydrofluoric acid). Traditional chemical neutralization (adding calcium or lime to precipitate CaF_2) is effective but produces sludge and can be sensitive to fluctuating loads. As discharge regulations tighten (e.g., fluoride limits below 5 mg L^{-1} in some regions [64]) and water reuse becomes a priority, electrodialysis has emerged as a viable solution at pilot and full scale. Electrodialysis is a membrane separation process in which an electric field drives cations and anions through ion-exchange membranes, separating ionic species from water. A notable implementation is at a semiconductor fab in Singapore, where a 150 m^3h^{-1} electrodialysis reversal (EDR) system was commissioned to reclaim scrubber waste liquid containing ~50–100 ppm fluoride and ~65 ppm silica [65]. The EDR unit (a variant of electrodialysis that periodically swaps electrode polarity to reduce fouling) was selected for its ability to handle high silica content that would foul reverse osmosis membranes.

6.6.3 EO in the Food and Beverage Industry

Waste liquid from food and beverage industries (such as dairies, slaughterhouses, breweries, etc.) is typically high in organic load (COD/biochemical oxygen demand [BOD]) and nutrients, and often contains fats, oils, and greases. Biological treatment is commonly applied, but certain waste streams or recalcitrant compounds may require additional treatment to meet discharge standards. EO has gained attention as an advanced treatment for high-COD wastes or as a post-treatment to polish biologically treated effluents. In EO, an anodic reaction directly destroys organic pollutants by electron transfer and generates oxidants (like hydroxyl radicals, ozone, or active chlorine if chloride is present) that oxidize contaminants. A recent study explored EO on anaerobically digested dairy waste liquid (a challenging effluent rich in residual organics and ammonia) to enhance its quality before discharge or reuse [64]. The EO reactor used a Ti/PbO_2 anode (lead dioxide-coated titanium, a high-oxygen overpotential anode) and a stainless-steel cathode. By optimizing parameters (current density ~90 mA cm^{-2}, 0.08% electrolyte addition for conductivity, 3 h treatment time, moderate mixing), the process achieved significant removals: COD was reduced by ~78%, ammonia (NH_3-N) by ~64%, total phosphorus by 87%, and total nitrogen by 67%. Additionally, over 92% of orthophosphate was eliminated, indicating strong oxidation of phosphorus compounds. These removals under lab-scale optimum conditions are promising, as they indicate that EO can substantially polish effluents that have already undergone anaerobic digestion.

6.6.4 EF in the Pharmaceutical Industry

Pharmaceutical industry waste liquid and related streams often contain complex organic micropollutants (active pharmaceutical ingredients, endocrine disruptors, solvents, etc.) that are poorly biodegradable and can be environmentally persistent. Advanced oxidation processes are frequently required to destroy these contaminants. The EF can mineralize a wide range of organics under acidic conditions and has been studied extensively for treating pharmaceutical compounds. A pilot-scale EF system, often combined with anodic oxidation, was applied to treat real pharmaceutical plant waste liquid characterized by high COD (~1,200 mg L^{-1}) and very low biodegradability ($BOD_5/COD \approx 0.14$) [64]. In one such study, a paired EF process with a BDD anode was used; the BDD anode provides additional heterogeneous oxidation (direct anodic oxidation and generation of radicals on the anode surface), complementing the homogeneous Fenton chemistry. Operating conditions included an acidic pH (~3), a moderate applied current density of 4.2 mA cm^{-2}, and a catalytic Fe^{2+} dosage of 0.2 mM. After 4 h of treatment, the system achieved 96.1% total organic carbon removal, essentially mineralizing nearly all the organic content in the waste liquid. This is a remarkable performance given the complex mixture of pharmaceuticals and by-products in the waste. Moreover, the treatment vastly improved the biodegradability of the effluent: the BOD_5/COD ratio rose from 0.14 to 0.68, indicating that the remaining matter was much more amenable to biological treatment or likely innocuous. In practical terms, the EF-treated water could potentially be polished in a conventional biological process or discharged with minimal environmental impact, as most persistent toxins were eliminated.

6.7 CHALLENGES AND LIMITATIONS

Electrode degradation represents one of the most pervasive challenges in electrochemical waste liquid treatment, limiting both the longevity and performance consistency of treatment units. Over time, electrodes exhibit a progressive loss of active surface area and catalytic functionality, driven by a combination of mechanical, chemical, and electrochemical phenomena. Mechanically, repeated gas evolution (e.g., hydrogen at the cathode, oxygen or chlorine at the anode) generates micro-jets and bubbles that scour the electrode surface, causing micro-abrasion and eventual surface pitting. Chemically, aggressive species in the waste stream, such as chloride, sulfate, bromide, or organic acids, can attack the electrode material, dissolving active components into the solution. Electrochemically, high current densities accelerate material dissolution via localized over-potential hotspots, leading to uneven wear and changes in the electrode microstructure [66]. As the electrode roughness increases and active facets are eroded, the effective catalytic site density falls, reducing reaction rates and driving the need for higher applied voltages to sustain target removal efficiencies.

In practical terms, such degradation gives rise to several operational symptoms: the cell voltage may drift upward under constant-current operation, energy consumption per unit pollutant removed increases, and current distribution across the electrode becomes uneven. These fluctuations not only compromise process efficiency but also complicate process control, as automatic feedback loops struggle to compensate for the dynamic loss of active area. In pilot studies, an ohmic increase of 10%–20% over 1,000 h of continuous operation has been reported for dimensionally stable anodes in chloride-rich media, translating to a proportional rise in energy costs [66].

Passivation and corrosion are interrelated phenomena that further undermine electrode performance. Passivation refers to the formation of a thin insulating film (typically an oxide or hydroxide) on the electrode surface during operation. For example, lead dioxide anodes may develop a layer of hydrated PbO_2, while tin-based anodes accumulate SnO_2-rich films. Although such layers can initially protect the underlying metal from further dissolution, they simultaneously impede electron transfer kinetics, forcing the system to operate at higher overpotentials. This sacrificial protection thus trades reaction selectivity for energy efficiency and must be managed through periodic polarity reversal or chemical cleaning [67].

Corrosion, by contrast, denotes the irreversible chemical or electrochemical loss of electrode material into the bulk solution. Under acidic or highly alkaline conditions (common in EF processes ($pH \approx 3$) or alkaline EC ($pH > 9$)), metallic substrates such as iron or aluminum may corrode to soluble Fe^{2+}/Fe^{3+} or Al^{3+} ions. Similarly, mixed-metal oxide coatings can delaminate under repeated redox cycling. Corrosion manifests as surface roughening, crack formation, and, eventually, structural failure of electrode plates or meshes. Field trials have documented anode weight losses of up to 5 g m^{-2} after 500 h of operation in chloride electrolytes, necessitating electrode replacement or refurbishment.

The co-occurrence of passivation and corrosion creates a vicious cycle. Corrosion sites serve as nucleation points for oxide film growth, while thick passive layers trap corrosive species next to the metal interface, exacerbating localized attack. Managing these competing processes drives up operational complexity and cost. In the transition from laboratory to full-scale industrial systems, the frequency of electrode maintenance emerges as a critical economic parameter. Studies estimate that maintenance accounts for 15%–25% of total operating expenditures in large electrochemical treatment plants, with electrode costs comprising another 10%–20% [68].

Beyond material degradation, several other limitations constrain the broader adoption of electrochemical waste liquid treatments:

Selectivity and Energy Efficiency: Electrode materials optimized for high overpotential offer broad oxidant generation but at the cost of increased cell voltage. Conversely, active electrodes operate at lower potential but favor partial oxidation pathways, requiring supplemental treatment steps to achieve complete mineralization [66].

Fouling and Scaling: Organic and inorganic foulants accumulate on electrode surfaces, further impeding mass transport and electron transfer. In situ gas bubble formation can dislodge some foulants, but persistent films demand periodic chemical or mechanical cleaning, interrupting continuous operation and raising safety considerations for handling cleaning agents [66].

Mass Transport Limitations: In high-flow or low-conductivity waste streams, ionic resistance and boundary layer thickness around electrodes limit attainable current densities. To overcome this, reactor designs often employ turbulent flow, rotating electrodes, or three-dimensional electrode architectures, but these modifications increase hydraulic complexity and capital cost [66].

By-product Formation: Unintended side-reactions can generate hazardous by-products, such as chlorate, perchlorate, or brominated organics when treating chloride- or bromide-rich waters. Mitigating these pathways requires careful control of current density, pH, and residence time, often necessitating real-time monitoring and feedback controls [69,70]

Economic and Environmental Footprint: While electrochemical methods minimize chemical addition, their electrical energy demand can be high. Energy consumption for EO typically ranges from 0.5 to 5 kWh m^{-3} of treated water, depending on contaminant load and reactor efficiency. Integration with renewable energy sources or waste heat recovery may improve sustainability, but adds system complexity.

Addressing these challenges calls for a multipronged approach:

Advanced Electrode Materials: Development of corrosion-resistant, high-surface-area composites such as Ti_4O_7, doped-carbon frameworks, or novel low-cost ceramic conductors aims to extend operational lifetime while maintaining high oxidant generation [67].

Smart Reactor Designs: Incorporating dynamic flow control, polarity reversal schedules, and in situ cleaning modules can mitigate fouling and passivation, sustaining performance with minimal manual intervention [67].

Monitoring and Control: Deployment of sensor arrays (pH, ORP, current density distribution) and model-based control systems can optimize operating parameters in real time, preventing conditions that accelerate degradation [67].

Hybrid Treatment Trains: Coupling electrochemical units with biological, adsorption, or membrane processes can reduce required oxidative doses and distribute treatment burdens across complementary technologies, thus lowering both energy use and electrode wear.

6.8 FUTURE DIRECTIONS AND INNOVATIONS

Electrochemical technologies for waste liquid treatment are poised for a transformative phase, driven by advancements in reactor design, materials science, and process integration with renewable energy systems. Recent studies underscore the potential of novel electrode materials such as doped diamond, MXenes, and carbon-based nanostructures to enhance reaction kinetics and reduce energy consumption [71,72]. Innovations in reactor design, including three-dimensional electrodes and scalable flow systems, promise to overcome current mass transfer limitations and electrode degradation issues. Moreover, coupling these technologies with renewable sources such as photovoltaic solar cells is emerging as a viable strategy to mitigate operational costs and decrease the carbon footprint of the treatment process [71,73].

In addition, hybrid treatment systems that combine electrochemical methods with traditional biological or advanced oxidation processes are being explored to achieve synergistic effects in pollutant removal and resource recovery. This multidisciplinary approach, which emphasizes real-time process monitoring and life-cycle economic assessments, is anticipated to pave the way for the industrial-scale adoption of electrochemical waste liquid treatment as a cornerstone of sustainable water management [10,74].

Furthermore, to process optimization, artificial intelligence (AI) plays a pivotal role in advancing the discovery and development of new electrode materials. Machine learning models and molecular simulations are used to screen vast libraries of candidate materials, predicting their electrocatalytic properties and resistance to degradation before they are physically tested [75]. This integration significantly reduces research time and costs by identifying promising materials with enhanced durability and activity. Moreover, these AI-powered tools can simulate long-term performance and degradation patterns, providing vital information for the design of more robust and sustainable electrochemical systems. The insights gained from such simulations offer guidance for the engineering of electrodes that can better withstand harsh waste liquid environments and operate efficiently over extended periods [75].

6.9 ENVIRONMENTAL AND ECONOMIC CONSIDERATIONS

Electrochemical treatment systems are increasingly recognized for their dual role in environmental remediation and resource recovery. On the environmental side, these technologies enable not only the efficient removal of diverse contaminants such as heavy metals, organic compounds, and nutrients but also facilitate the recovery of valuable resources embedded within waste liquid streams. By recovering metals and nutrients, these systems reduce the demand for virgin materials and minimize the ecological footprint associated with mining and processing activities. This holistic approach to water treatment aligns with sustainability goals by decreasing sludge generation and lowering greenhouse gas emissions through reduced reliance on conventional energy-intensive methods.

From an economic perspective, the scalability and cost-effectiveness of electrochemical technologies have generated considerable interest. Recent techno-economic evaluations indicate that, although the initial capital investment in high-performance electrode materials and optimized reactors may be higher, the overall operational costs can be significantly reduced. Lower energy consumption, enhanced system durability, and minimized chemical usage result in reduced maintenance and disposal costs, thereby improving the return on investment. Studies have shown that when operating conditions are optimized, often through the integration of renewable energy sources and advanced process control, there are substantial savings compared to traditional waste liquid treatments.

Moreover, the development of integrated, smart treatment systems that combine electrochemical processes with advanced control strategies is poised to revolutionize the field. These integrated approaches not only optimize process parameters in real time but also support circular economy principles by recovering and reusing resources, creating closed-loop systems that benefit both the environment and the economy. In this emerging paradigm, waste streams are transformed into sources of energy and recyclable materials, thereby addressing global resource scarcity while contributing to sustainable economic growth. Such innovations are central to designing resilient waste liquid treatment infrastructures that meet future environmental and market demands.

Electrochemical waste liquid treatment processes offer distinct environmental benefits compared to conventional chemical treatments by minimizing the need for added reagents and reducing the generation of hazardous by-products. However, life cycle assessments have revealed that these systems can have significant environmental footprints, particularly due to high electricity consumption and the production of electrode materials. For example, recent studies assessing sequential EF configurations report that, although the treatment effectively removes pollutants and meets stringent discharge standards, the environmental impacts are highly sensitive to energy inputs and chemical usage [76,77]. These analyses underscore the need for optimizing operating conditions and integrating renewable energy sources in order to reduce greenhouse gas emissions and overall ecological burdens.

From an economic standpoint, techno-economic assessments of electrochemical waste liquid treatment systems demonstrate promising cost advantages over traditional treatment methods. While the initial capital expenditures for high-performance electrodes and advanced reactor designs are relatively high, these costs can be offset by lower operational expenses primarily due to reduced chemical consumption, lower sludge production, and minimized maintenance requirements. Recent research highlights that through careful process integration and the application of response surface methodologies, overall operating costs can be lowered, resulting in favorable net present values and competitive minimum selling prices for treated effluents and recovered by-products [77,78]. Such evaluations are essential for guiding investment decisions and scaling up these technologies for industrial applications.

Looking ahead, there is a clear trend toward merging environmental and economic performance with technological innovation through integrated systems that leverage digital tools such as AI. AI-driven process control and predictive maintenance can further reduce energy consumption, enhance reactor efficiency, and enable adaptive strategies that adjust operating conditions in real time based on waste liquid characteristics. In parallel, advanced computational methods, including density functional theory and machine learning-based material screening, contribute to the discovery of more durable and cost-effective electrode materials. The synergy between AI and electrochemical process optimization is expected to deliver systems with lower environmental impacts and improved economic feasibility, thereby making sustainable water treatment a more attractive option for future industrial applications [76,78].

6.10 CONCLUSION

Electrochemistry applied to waste treatment represents an innovative, versatile, and promising approach to today's environmental challenges. After exploring the scientific foundations of electrochemical processes, the diversity of methods available, and the complexity of the materials involved, it becomes clear that this technology offers solutions adaptable to a wide range of industrial sectors – from textiles and microelectronics to pharmaceuticals and food processing. The case studies presented demonstrate the real effectiveness of these processes in the field, with concrete applications such as electrodialysis, adapted to specific industrial contexts. However, despite their potential, electrochemical methods still encounter certain limitations, notably in terms of cost, energy consumption, and by-product management. Future prospects are nevertheless encouraging, thanks to technological advances, improved electrode materials, and the growing integration of these systems into global sustainable treatment strategies. By taking into account both environmental and economic aspects, electrochemistry is establishing itself as a major lever in the transition to a cleaner, circular, and more responsible industry.

REFERENCES

[1] Jayaraman, J.; Kumaraswamy, J.; Rao, Y. K. S. S.; Karthick, M.; Baskar, S.; Anish, M.; Sharma, A.; Yadav, A. S.; Alam, T.; Ammarullah, M. I. Wastewater Treatment by Algae-Based Membrane Bioreactors: A Review of the Arrangement of a Membrane Reactor, Physico-Chemical Properties, Advantages and Challenges. *RSC Adv.*, **2024**, 14 (47), 34769–34790. https://doi.org/10.1039/d4ra04417g.

[2] Wu, L.; Garg, S.; Waite, T. D. Progress and Challenges in the Use of Electrochemical Oxidation and Reduction Processes for Heavy Metals Removal and Recovery from Wastewaters. *J. Hazard. Mater.*, **2024**, 479, 135581. https://doi.org/10.1016/j.jhazmat.2024.135581.

[3] Goren, A. Y.; Recepoğlu, Y. K.; Khataee, A. Chapter 4- Language of Response Surface Methodology as an Experimental Strategy for Electrochemical Wastewater Treatment Process Optimization. In *Artificial Intelligence and Data Science in Environmental Sensing*; Asadnia, M., Razmjou, A., Beheshti, A., Eds.; Cognitive Data Science in Sustainable Computing; Academic Press, 2022; pp. 57–92. https://doi.org/10.1016/B978-0-323-90508-4.00009-5.

[4] Rizvi, S.; Singh, A.; Kushwaha, A.; Gupta, S. K. Chapter 16- Recent Advances in Melanoidin Removal from Wastewater: Sources, Properties, Toxicity, and Remediation Strategies. In *Emerging Trends to Approaching Zero Waste*; Hussain, C. M., Singh, S., Goswami, L., Eds.; Elsevier, 2022; pp. 361–386. https://doi.org/10.1016/B978-0-323-85403-0.00015-3.

[5] Mohtashami, R.; Shang, J. Q. Electroflotation for Treatment of Industrial Wastewaters: A Focused Review. *Environ. Process.*, **2019**, 6 (2), 325–353. https://doi.org/10.1007/S40710-019-00348-Z.

[6] Manikandan, S.; Vickram, S.; Sirohi, R.; Subbaiya, R.; Krishnan, R. Y.; Karmegam, N.; Sumathijones, C.; Rajagopal, R.; Chang, S. W.; Ravindran, B.; et al. Critical Review of Biochemical Pathways to Transformation of Waste and Biomass into Bioenergy. *Bioresour. Technol.*, **2023**, *372*, 128679. https://doi.org/10.1016/j.biortech.2023.128679.

[7] Imwene, K. O.; Ngumba, E.; Kairigo, P. K. Emerging Technologies for Enhanced Removal of Residual Antibiotics from Source-Separated Urine and Wastewaters: A Review. *J. Environ. Manage.*, **2022**, 322, 116065. https://doi.org/10.1016/j.jenvman.2022.116065.

[8] Ogando, F. I. B.; Xastre, T.; Simões, E.; Aguiar, C. L. de. Removal of Color and Turbidity in Sugarcane Juice Treated by Electrocoagulation with Aluminum Electrodes. *Brazilian J. Food Technol.*, **2021**, 24, e2020236.

[9] Monteil, H.; Pechaud, Y.; Oturan, N.; Oturan, M. A. A Review on Efficiency and Cost Effectiveness of Electro-and Bio-Electro-Fenton Processes: Application to the Treatment of Pharmaceutical Pollutants in Water. *Chem. Eng. J.*, **2019**, 376, 119577.

[10] Fitch, A.; Balderas-Hernandez, P.; Ibanez, J. G. Electrochemical Technologies Combined with Physical, Biological, and Chemical Processes for the Treatment of Pollutants and Wastes: A Review. *J. Environ. Chem. Eng.*, **2022**, 10 (3), 107810.

[11] Martínez-Sánchez, C.; Robles, I.; Godínez, L. A. Review of Recent Developments in Electrochemical Advanced Oxidation Processes: Application to Remove Dyes, Pharmaceuticals, and Pesticides. *Int. J. Environ. Sci. Technol.*, **2022**, 19 (12), 12611–12678.

[12] Moreira, F. C.; Boaventura, R. A. R.; Brillas, E.; Vilar, V. J. P. Electrochemical Advanced Oxidation Processes: A Review on Their Application to Synthetic and Real Wastewaters. *Appl. Catal. B Environ.*, **2017**, 202, 217–261.

[13] Nidheesh, P. V; Divyapriya, G.; Oturan, N.; Trellu, C.; Oturan, M. A. Environmental Applications of Boron-doped Diamond Electrodes: 1. Applications in Water and Wastewater Treatment. *ChemElectroChem*, **2019**, 6 (8), 2124–2142.

[14] Shen, R.; Zhao, L.; Lu, J.; Watson, J.; Si, B.; Chen, X.; Meng, H.; Yao, Z.; Feng, J.; Liu, Z. Treatment of Recalcitrant Wastewater and Hydrogen Production via Microbial Electrolysis Cells. *Int. J. Agric. Biol. Eng.*, **2019**, 12 (5), 179–189.
[15] Armstrong, D. A.; Huie, R. E.; Koppenol, W. H.; Lymar, S. V; Merényi, G.; Neta, P.; Ruscic, B.; Stanbury, D. M.; Steenken, S.; Wardman, P. Standard Electrode Potentials Involving Radicals in Aqueous Solution: Inorganic Radicals (IUPAC Technical Report). *Pure Appl. Chem.*, **2015**, *87* (11–12), 1139–1150.
[16] Brillas, E.; Martínez-Huitle, C. A. Decontamination of Wastewaters Containing Synthetic Organic Dyes by Electrochemical Methods. An Updated Review. *Appl. Catal. B Environ.*, **2015**, 166–167, 603–643. https://doi.org/10.1016/j.apcatb.2014.11.016.
[17] Akarsu, C.; Kumbur, H.; Kideys, A. E. Removal of Microplastics from Wastewater through Electrocoagulation-Electroflotation and Membrane Filtration Processes. *Water Sci. Technol.*, **2021**, *84* (7), 1648–1662. https://doi.org/10.2166/WST.2021.356.
[18] Chen, G. Electrochemical Technologies in Wastewater Treatment. *Sep. Purif. Technol.*, **2004**, 38 (1), 11–41. https://doi.org/10.1016/j.seppur.2003.10.006.
[19] Panizza, M.; Cerisola, G. Direct and Mediated Anodic Oxidation of Organic Pollutants. *Chem. Rev.*, **2009**, 109 (12), 6541–6569. https://doi.org/10.1021/cr9001319.
[20] de Vargas Brião, G.; da Costa, T. B.; Antonelli, R.; Costa, J. M. Electrochemical Processes for the Treatment of Contaminant-Rich Wastewater: A Comprehensive Review. *Chemosphere*, **2024**, 355, 141884. https://doi.org/10.1016/j.chemosphere.2024.141884.
[21] Yang, K.; He, Z. Formation and Control of Oxidation Byproducts in Electrochemical Wastewater Treatment: A Review. *Chem. Eng. J.*, **2024**, 499, 156160. https://doi.org/10.1016/j.cej.2024.156160.
[22] Xue, Y.; Jia, Y.; Liu, S.; Yuan, S.; Ma, R.; Ma, Q.; Fan, J.; Zhang, W. Electrochemical Reduction of Wastewater by Non-Noble Metal Cathodes: From Terminal Purification to Upcycling Recovery. *J. Hazard. Mater.*, **2023**, 459, 132106. https://doi.org/10.1016/j.jhazmat.2023.132106.
[23] Sillanpää, M.; Shestakova, M. Chapter 2- Electrochemical Water Treatment Methods. In *Electrochemical Water Treatment Methods*; Sillanpää, M., Shestakova, M., Eds.; Butterworth-Heinemann, 2017; pp. 47–130. https://doi.org/10.1016/B978-0-12-811462-9.00002-5.
[24] Yu, F.; Yang, Y.; Zhang, X.; Ma, J. Application of Capacitive Deionization in Drinking Water Purification. *Sep. Purif. Technol.*, **2025**, *354*, 129285. https://doi.org/10.1016/j.seppur.2024.129285.
[25] Kyzas, G. Z.; Matis, K. A. Electroflotation Process: A Review. *J. Mol. Liq.*, **2016**, 220, 657–664. https://doi.org/10.1016/j.molliq.2016.04.128.
[26] Khosravanipour Mostafazadeh, A.; Zolfaghari, M.; Drogui, P. Electrofiltration Technique for Water and Wastewater Treatment and Bio-Products Management: A Review. *J. Water Process Eng.*, **2016**, 14, 28–40. https://doi.org/10.1016/j.jwpe.2016.10.003.
[27] Yang, R.; Wang, H.; Zhu, X.; Sun, Z. Review on Electrochemical Processes for the Treatment of Heavy Metal Complexes in Wastewater: Performance, Mechanism, Application and Improvement. *Int. J. Electrochem. Sci.*, **2025**, 20 (4), 100971. https://doi.org/10.1016/j.ijoes.2025.100971.
[28] Yasasve, M.; Manjusha, M.; Manojj, D.; Hariharan, N. M.; Sai Preethi, P.; Asaithambi, P.; Karmegam, N.; Saravanan, M. Unravelling the Emerging Carcinogenic Contaminants from Industrial Waste Water for Prospective Remediation by Electrocoagulation – A Review. *Chemosphere*, **2022**, *307*, 136017. https://doi.org/10.1016/j.chemosphere.2022.136017.
[29] Temur Ergan, B.; Aydin, E. S.; Gengec, E. Improving Electro-Fenton Degradation Performance Using Waste Biomass-Derived-Modified Biochar Electrodes: A Real Environment Textile Water Treatment. *J. Environ. Chem. Eng.*, **2023**, 11 (6), 111439. https://doi.org/10.1016/j.jece.2023.111439.

[30] Nidheesh, P. V; Ganiyu, S. O.; Kuppam, C.; Mousset, E.; Samsudeen, N.; Olvera-Vargas, H.; Kumar, G. Bioelectrochemical Cells as a Green Energy Source for Electrochemical Treatment of Water and Wastewater. *J. Water Process Eng.*, **2022**, 50, 103232. https://doi.org/10.1016/j.jwpe.2022.103232.

[31] Shestakova, M.; Sillanpää, M. Electrode Materials Used for Electrochemical Oxidation of Organic Compounds in Wastewater. *Rev. Environ. Sci. Bio/Technology*, **2017**, 16 (2), 223–238. https://doi.org/10.1007/s11157-017-9426-1.

[32] Santos, M. O.; de OS Santos, G.; Mattedi, S.; Griza, S.; Eguiluz, K. I. B.; Salazar-Banda, G. R. Influence of the Calcination Temperature and Ionic Liquid Used during Synthesis Procedure on the Physical and Electrochemical Properties of Ti/(RuO_2) 0.8–(Sb_2O_4) 0.2 Anodes. *J. Electroanal. Chem.*, **2018**, 829, 116–128.

[33] Wang, G.; Liu, Y.; Ye, J.; Lin, Z.; Yang, X. Electrochemical Oxidation of Methyl Orange by a Magnéli Phase Ti_4O_7 Anode. *Chemosphere*, **2020**, *241*, 125084.

[34] Tang, H.; Shang, Q.; Tang, Y.; Yi, X.; Wei, Y.; Yin, K.; Liu, M.; Liu, C. Static and Continuous Flow Photoelectrocatalytic Treatment of Antibiotic Wastewater over Mesh of TiO2 Nanotubes Implanted with G-C_3N_4 Nanosheets. *J. Hazard. Mater.*, **2020**, 384, 121248.

[35] Lee, J.; Antonini, G.; Al-Omari, A.; Muller, C.; Mathew, J.; Bell, K.; Pearce, J. M.; Santoro, D. Electrochemical Methods for Nutrient Removal in Wastewater: A Review of Advanced Electrode Materials, Processes, and Applications. *Sustain.*, **2024**, *16* (22).

[36] Jiao, Y.; Ma, L.; Tian, Y.; Zhou, M. A Flow-through Electro-Fenton Process Using Modified Activated Carbon Fiber Cathode for Orange II Removal. *Chemosphere*, **2020**, *252*, 126483.

[37] Shao, D.; Zhang, X.; Lyu, W.; Zhang, Y.; Tan, G.; Xu, H.; Yan, W. Magnetic Assembled Anode Combining PbO_2 and Sb-SnO_2 Organically as An Effective and Sustainable Electrocatalyst for Wastewater Treatment with Adjustable Attribution and Construction. *ACS Appl. Mater. Interfaces*, **2018**, 10 (51), 44385–44395. https://doi.org/10.1021/acsami.8b14592.

[38] Santos, G. de O. S.; Gonzaga, I. M. D.; Dória, A. R.; Moratalla, A.; da Silva, R. S.; Eguiluz, K. I. B.; Salazar-Banda, G. R.; Saez, C.; Rodrigo, M. A. Testing and Scaling-up of a Novel Ti/$Ru_{0.7}Ti_{0.3}O_2$ Mesh Anode in a Microfluidic Flow-through Reactor. *Chem. Eng. J.*, **2020**, 398, 125568. https://doi.org/10.1016/j.cej.2020.125568.

[39] Cuesta-Mota, D.; Serra-Clusellas, A.; Guimerà, X.; Canals Casals, L.; López-Grimau, V. Hydrogen Production from Electrochemical Treatment of Textile Dyeing Wastewaters. *J. Water Process Eng.*, **2024**, 64, 105628. https://doi.org/10.1016/j.jwpe.2024.105628.

[40] Mosur Nagarajan, A.; Subramanian, A.; Prasad Gobinathan, K.; Mohanakrishna, G.; Sivagami, K. Electrochemical-Based Approaches for the Treatment of Pharmaceuticals and Personal Care Products in Wastewater. *J. Environ. Manage.*, **2023**, 344, 118385. https://doi.org/10.1016/j.jenvman.2023.118385.

[41] Güven, G.; Perendeci, N. A.; Tanyolaç, A. Reaction Kinetics on Treatment of Food Industry Wastewaters by Electrochemical Oxidation. *Desalin. Water Treat.*, **2014**, 52 (37), 7092–7100. https://doi.org/10.1080/19443994.2013.827780.

[42] An, C.; Huang, G.; Yao, Y.; Zhao, S. Emerging Usage of Electrocoagulation Technology for Oil Removal from Wastewater: A Review. *Sci. Total Environ.*, **2017**, 579, 537–556. https://doi.org/10.1016/j.scitotenv.2016.11.062.

[43] Sureshkumar, K.; Kempegowda, R. G.; Ramakrishnappa, T. 31- Electrochemical Recovery of Metals from Industrial Wastewaters. In *Recent Advances in Industrial Wastewater Treatment*; Sillanpää, M., Khadir, A., Gurung, K. B., Eds.; Elsevier, 2023; pp. 639–654. https://doi.org/10.1016/B978-0-323-95327-6.00035-X.

[44] Nesterov, D.; Barrera-Martínez, I.; Martínez-Sánchez, C.; Sandoval-González, A.; Bustos, E. Approaching the Circular Economy: Biological, Physicochemical, and Electrochemical Methods to Valorize Agro-Industrial Residues, Wastewater, and Industrial Wastes. *J. Environ. Chem. Eng.*, **2024**, 12 (5), 113335. https://doi.org/10.1016/j.jece.2024.113335.

[45] Phu, T. K. C.; Nguyen, P. L.; Phung, T. V. B. Recent Progress in Highly Effective Electrocoagulation-Coupled Systems for Advanced Wastewater Treatment. *iScience*, **2025**, 28(3), 111965.

[46] Rajoria, S.; Vashishtha, M.; Sangal, V. K. Treatment of Electroplating Industry Wastewater: A Review on the Various Techniques. *Environ. Sci. Pollut. Res.*, **2022**, 29 (48), 72196–72246.

[47] Sadafale, H. S.; Gaikwad, R. W. Emerging Trends in Wastewater Treatment of Semiconductor Industry: A Review. *Nat. Environ. Pollut. Technol.*, **2023**, 22 (4), 1451–1462. https://doi.org/10.46488/nept.2023.v22i04.032.

[48] Sen, S.; Prajapati, A. K.; Bannatwala, A.; Pal, D. Electrocoagulation Treatment of Industrial Wastewater Including Textile Dyeing Effluent – A Review. *Desalin. Water Treat.*, **2019**, 161, 21–34. https://doi.org/10.5004/DWT.2019.24302.

[49] Syam Babu, D.; Anantha Singh, T. S.; Nidheesh, P. V.; Suresh Kumar, M. Industrial Wastewater Treatment by Electrocoagulation Process. *Sep. Sci. Technol.*, **2020**, 55 (17), 3195–3227. https://doi.org/10.1080/01496395.2019.1671866.

[50] Jing, G.; Ren, S.; Pooley, S.; Sun, W.; Kowalczuk, P. B.; Gao, Z. Electrocoagulation for Industrial Wastewater Treatment: An Updated Review. *Environ. Sci. Water Res. Technol.*, **2021**, 7 (7), 1177–1196. https://doi.org/10.1039/D1EW00158B.

[51] Nair, G.; Soni, B.; Shah, M. A Comprehensive Review on Electro-Oxidation and Its Types for Wastewater Treatment. *Groundw. Sustain. Dev.*, **2023**, *23*, 100980. https://doi.org/10.1016/J.GSD.2023.100980.

[52] Mirshafiee, A.; Nourollahi, M.; Shahriary, A. Application of Electro Oxidation Process for Treating Wastewater from Petrochemical with Mixed Metal Oxide Electrode. *Sci. Rep.*, **2024**, 14 (1), 1–13. https://doi.org/10.1038/S41598-024-52201-5.

[53] Dimoglo, A.; Sevim-Elibol, P.; Dinç; Gökmen, K.; Erdoğan, H. Electrocoagulation/Electroflotation as a Combined Process for the Laundry Wastewater Purification and Reuse. *J. Water Process Eng.*, **2019**, 31, 100877. https://doi.org/10.1016/J.JWPE.2019.100877.

[54] De Oliveira Da Mota, I.; De Castro, J. A.; De Góes Casqueira, R.; De Oliveira Junior, A. G. Study of Electroflotation Method for Treatment of Wastewater from Washing Soil Contaminated by Heavy Metals. *J. Mater. Res. Technol.*, **2015**, 4 (2), 109–113. https://doi.org/10.1016/J.JMRT.2014.11.004.

[55] Mei, Y.; Jiang, C.; Liu, Y.; Zhang, Y.; Wang, L.; Tang, C. Y. Electrodialysis Membrane Technology for Industrial Wastewater Treatment: Recent Advances. *Electrochem. Membr. Technol. Water Wastewater Treat.*, **2022**, 265–315. https://doi.org/10.1016/B978-0-12-824470-8.00005-X.

[56] Gurreri, L.; Tamburini, A.; Cipollina, A.; Micale, G. Electrodialysis Applications in Wastewater Treatment for Environmental Protection and Resources Recovery: A Systematic Review on Progress and Perspectives. *Membranes*, **2020**, *10* (7), 146. https://doi.org/10.3390/MEMBRANES10070146.

[57] He, S.; Zhu, T.; Wang, Y.; Xiong, W.; Gao, X.; Zhang, E. Application of Capacitive Deionization Technology in Water Treatment and Coupling Technology: A Review. *Environ. Sci. Water Res. Technol.*, **2024**, 10 (10), 2313–2340. https://doi.org/10.1039/D4EW00413B.

[58] Kalfa, A.; Shapira, B.; Shopin, A.; Cohen, I.; Avraham, E.; Aurbach, D. Capacitive Deionization for Wastewater Treatment: Opportunities and Challenges. *Chemosphere*, **2020**, *241*, 125003. https://doi.org/10.1016/J.CHEMOSPHERE.2019.125003.

[59] Nidheesh, P. V.; Gandhimathi, R. Trends in Electro-Fenton Process for Water and Wastewater Treatment: An Overview. *Desalination*, **2012**, 299, 1–15. https://doi.org/10.1016/J.DESAL.2012.05.011.

[60] Sobczak, M.; Bujnowicz, S.; Bilińska, L. Fenton and Electro-Fenton Treatment for Industrial Textile Wastewater Recycling. Comparison of by-Products Removal, Biodegradability, Toxicity, and Re-Dyeing. *Water Resour. Ind.*, **2024**, *31*, 100256. https://doi.org/10.1016/J.WRI.2024.100256.

[61] Omar, B. M.; Zyadah, M. A.; Ali, M. Y.; El-Sonbati, M. A. Pre-Treatment of Composite Industrial Wastewater by Fenton and Electro-Fenton Oxidation Processes. *Sci. Rep.*, **2024**, 14 (1), 1–14. https://doi.org/10.1038/S41598-024-78846-W.
[62] Jo, S.; Kadam, R.; Jang, H.; Seo, D.; Park, J. Recent Advances in Wastewater Electrocoagulation Technologies: Beyond Chemical Coagulation. *Energies*, **2024**, 17 (23), 5863. https://doi.org/10.3390/EN17235863.
[63] Ardhianto, R.; Samudro, G.; Mangkoedihardjo, S. Pilot-Scale Integrated System for Textile Wastewater Reuse: Electrocoagulation-Clarifier Flotation, Moving Bed Biofilm Reactor, and Adsorption. **2025**. https://doi.org/10.2139/SSRN.5132296.
[64] Yen, F. C.; You, S. J.; Chang, T. C. Performance of Electrodialysis Reversal and Reverse Osmosis for Reclaiming Wastewater from High-Tech Industrial Parks in Taiwan: A Pilot-Scale Study. *J. Environ. Manage.*, **2017**, 187, 393–400. https://doi.org/10.1016/J.JENVMAN.2016.11.001.
[65] Semiconductor fabrication plant reclaims local scrubber wastewater with Veolia's Electrodialysis Reversal (EDR) | Veolia WTS https://www.watertechnologies.com/case-study/semiconductor-fabrication-plant-reclaims-local-scrubber-wastewater-edr (accessed Apr 18, 2025).
[66] Dube, A.; Malode, S. J.; Alshehri, M. A.; Shetti, N. P. Electrochemical Water Treatment: Review of Different Approaches. *J. Environ. Manage.*, **2025**, 373, 123911. https://doi.org/10.1016/J.JENVMAN.2024.123911.
[67] Wang, R.-S.; Shan, L.-L.; Zhu, Z.-B.; Liu, Z.-Q.; Liao, Z.-M.; Cui, Y.-H. Current Status of Electrode Corrosion Passivation and Its Mitigation Strategies in Electrocoagulation. *Chem. Eng. Process. - Process Intensif.*, **2025**, *209*, 110192. https://doi.org/10.1016/J.CEP.2025.110192.
[68] Ogundele, O. D.; Oyegoke, D. A.; Anaun, T. E. Exploring the Potential and Challenges of Electro-Chemical Processes for Sustainable Waste Water Remediation and Treatment. *Acadlore Trans. Geosci*, **2023**, 2 (2), 80–93.
[69] Manikandan, S.; Deena, S. R.; Subbaiya, R.; Vijayan, D. S.; Vickram, S.; Preethi, B.; Karmegam, N. Waves of Change: Electrochemical Innovations for Environmental Management and Resource Recovery from Water – A Review. *J. Environ. Manage.*, **2024**, *366*, 121879. https://doi.org/10.1016/j.jenvman.2024.121879.
[70] Kumar, A.; Pan, S.-Y. Opportunities and Challenges of Electrochemical Water Treatment Integrated with Renewable Energy at the Water-Energy Nexus. *Water-Energy Nexus*, **2020**, 3, 110–116. https://doi.org/10.1016/j.wen.2020.03.006.
[71] Frontistis, Z. Current and Future Trends in Environmental Electrochemistry for Wastewater Treatment. *Water*. MDPI 2022, p. 1817.
[72] Zainul, R.; Chaya, R.; Wijaya, K.; Rahmawati, F.; Laghari, I. A.; Rahmadiawan, D.; Abdullah, M. Advances in Electrochemical Technologies for Sustainable Wastewater Treatment and Chemical Synthesis: Mechanisms, Challenges, and Prospects. *J. Med. Pharm. Chem. Res.*, **2024**, 6 (2), 140–156.
[73] Chaplin, B. P. Advanced Electrochemical Technologies for Water Treatment, Resource Recovery, and Sustainable Energy. *ACS ES&T Engineering*. ACS Publications, 2025.
[74] Yakamercan, E.; Bhatt, P.; Aygun, A.; Adesope, A. W.; Simsek, H. Comprehensive Understanding of Electrochemical Treatment Systems Combined with Biological Processes for Wastewater Remediation. *Environ. Pollut.*, **2023**, 330, 121680.
[75] Xu, G.; Jiang, M.; Li, J.; Xuan, X.; Li, J.; Lu, T.; Pan, L. Machine Learning-Accelerated Discovery and Design of Electrode Materials and Electrolytes for Lithium Ion Batteries. *Energy Storage Mater.*, **2024**, 103710.
[76] Clack, K.; Rajagopal, D.; Hoek, E. M. V. Life Cycle and Techno-Economic Assessment of Bioresource Production from Wastewater. *npj Clean Water 2024* 71, **2024**, 7 (1), 1–17. https://doi.org/10.1038/s41545-024-00314-9.

[77] Salazar-Sogamoso, I.; Castaño-Sánchez, L.-M.; Salazar-López, J.-C.; Gómez-García, D.-O.; Environmental, M.-Á.; Dobrosz-Gómez, I.; Salazar-Sogamoso, L.-M.; Castaño-Sánchez, J.-C.; Salazar-López, D.-O.; Gómez-García, M.-Á.; et al. Environmental and Economic Evaluation of the Sequential Combination of Coagulation–Flocculation with Different Electro-Fenton-Based Configurations for the Treatment of Raw Textile Wastewater. *Water*, **2024**, 16 (15), 2154. https://doi.org/10.3390/W16152154.

[78] Sun, Y.; Bai, S.; Wang, X.; Ren, N.; You, S. Prospective Life Cycle Assessment for the Electrochemical Oxidation Wastewater Treatment Process: From Laboratory to Industrial Scale. *Environ. Sci. Technol.*, **2023**, 57 (3), 1456–1466. https://doi.org/10.1021/ACS.EST.2C04185.

7 Photo-Fenton and Ultrasound Processes

Applications and Efficiency Improvements for Liquid Waste Treatment

Imane El Mrabet, Hamza Bellouk, Zineb Chaouki, Karim Tanji, Yassine Darmane, Mourad Benzina, Corinne Ferronato, Hector Valdes, and Hicham Zaitan

7.1 INTRODUCTION

Water is a vital resource for life, supporting human health and socioeconomic development in addition to maintaining ecosystems. However, pollutants like pesticides, heavy metals, and complex organic compounds are discharged into natural water bodies as a result of the rapid industrialization and urbanization that have caused severe water pollution. While the natural water cycle offers some self-purification capacity, the sheer volume and complexity of modern pollutants often exceed its regenerative abilities, necessitating the adoption of advanced treatment solutions.

Persistent organic pollutants are frequently not entirely removed by conventional wastewater treatment techniques like coagulation, filtration, and biological degradation, which can result in secondary contamination and risks for the environment. As a result, AOPs have become an excellent alternative, using extremely reactive hydroxyl radicals ($^{\bullet}OH$) to convert complex pollutants into innocuous byproducts like carbon dioxide and water (Peralta-Hernández et al. 2018). The Fenton and Photo-Fenton processes are two of these AOPs that have drawn a lot of interest because of their high efficiency in degrading hazardous substances found in industrial wastewater, including that from the textile, petrochemical, and pharmaceutical sectors (Shestakova and Sillanpää 2020; Mukherjee et al. 2023).

Recently, the combination of photo-irradiation and ultrasound (US) techniques has been applied to improve the AOPs' performance by increasing overall efficiency and reaction kinetics (Peralta-Hernández et al. 2018; Bellouk et al. 2025; Bellouk et al. 2022). Indeed, recent studies have investigated the use of US on the lab scale to improve the efficiency of AOPs based on sulfate radicals $\left(SO_4^{\bullet-}\right)$ and hydroxyl

 DOI: 10.1201/9781003624905-7

radicals ($^{\bullet}OH$). The AOP using persulfate $\left(S_2O_8^{2-}\right)$ and H_2O_2 activated by US/UV was efficiently applied as secondary treatment for various wastewaters: tannery wastewater pretreated by air stripping in Fez, Morocco (Bahammou, Tazi, et al. 2025), young landfill leachate pretreated by aeration (Bellouk et al. 2025), and stabilized landfill leachate pretreated by coagulation-flocculation (Bellouk et al. 2022).

Similarly, the sono-photo-Fenton was successfully applied to tannery effluent, in one case pretreated by coagulation-flocculation (Bahammou, El Mrabet, Nawdali, et al. 2025) and in another by air stripping (Bahammou, El Mrabet, Chaouki, et al. 2025). These hybrid approaches reached cumulative removal values ranging from 85% to 99% in terms of COD, color number, and biological oxygen demand (BOD_5). Hence, these results indicate that UV and US-based processes are promising for large-scale wastewater treatment, particularly in addressing highly contaminated effluents such as landfill leachates and real industrial wastewater.

This chapter explores the fundamental principles and mechanisms of Fenton-based oxidation, its improvement by US irradiation, and its practical application in treating landfill leachate, one of the most complicated liquid effluents. Through case studies from Fez City, Morocco, we examine the performance of single and combined AOPs in achieving significant pollutant removal, highlighting key operational parameters, challenges, and potential for real-world implementation.

7.2 AOPs: MECHANISMS AND PRINCIPLES

7.2.1 Fenton and Photo-Fenton Processes: Chemical Mechanisms

Fenton chemistry involves the oxidation of organic pollutants using ferrous ions (Fe^{2+}) and hydrogen peroxide (H_2O_2), generating highly reactive hydroxyl radicals ($^{\bullet}OH$). The process, first described by Henry J. Fenton in 1894, occurs optimally in acidic conditions and follows a series of redox reactions where Fe^{2+} is oxidized to Fe^{3+}, which can subsequently be reduced back to Fe^{2+}, ensuring continuous radical production (Eqs. 7.1 and 7.2) (Peralta-Hernández et al. 2018).

$$Fe^{2+} + H_2O_2 \rightarrow Fe^{3+} + HO^- + HO^{\bullet} \tag{7.1}$$

$$Fe^{3+} + H_2O_2 \rightarrow Fe^{2+} + H^+ + HO_2^{\bullet} \tag{7.2}$$

Fenton-based processes exist in two forms: homogeneous and heterogeneous. Homogeneous Fenton oxidation is simple, requiring only dissolved iron salts and H_2O_2, and operates efficiently under ambient conditions. However, it has drawbacks, such as increased sludge production and sensitivity to pH variations. In contrast, heterogeneous Fenton uses solid-phase catalysts, reducing iron leaching, broadening the effective pH range, and enhancing radical generation through surface interactions, leading to improved degradation efficiency (Moradi et al. 2020).

Ferric ions can be reduced by reaction with hydrogen peroxide to reform the ferrous ion and more radicals (Eq. 7.3). This reaction is called the FL reaction. It allows the regeneration of Fe^{2+}, but it occurs at a slower rate than the Fenton reaction.

$$Fe^{3+} + H_2O_2 \rightarrow Fe^{2+} + H^+ + HO_2^{\bullet} \quad (7.3)$$

The photo-Fenton process is the combination of the Fenton process with UV irradiation, which mainly plays two roles: the direct photolysis of organic compounds and the production of OH radicals through the photoreduction of dissolved Fe^{3+} to Fe^{2+} according to the reactions (Eqs. 7.4 and 7.5) and/or the photolysis of H_2O_2 (Eq. 7.6). These reactions significantly increase the degradation rate of organic pollutants (Pouran, Abdul Aziz, and Wan Daud 2015; Ameta et al. 2018). Furthermore, the photodegradation is particularly effective at pH ~3, where Fe^{3+} and Fe(III)-hydroxy complexes (i.e., $Fe(OH)^{2+}$) absorb UV light, further boosting $^{\bullet}OH$ radical production (Eqs. 7.4 and 7.5) (Chauhan et al. 2021).

$$Fe(OH)^{2+} + h\nu \rightarrow Fe^{2+} + HO^{\bullet}; \lambda < 580 \text{ nm} \quad (7.4)$$

$$Fe^{3+} + H_2O \xrightarrow{h\nu} Fe^{2+} + H^+ + HO^{\bullet} \quad (7.5)$$

$$H_2O_2 + h\nu \rightarrow 2HO^{\bullet}; \lambda < 310 \text{ nm} \quad (7.6)$$

Both the Fenton and Photo-Fenton processes are among the most effective AOPs for degrading persistent organic pollutants in wastewater. Their ability to mineralize complex contaminants makes them highly valuable for treating industrial effluents and landfill leachates, particularly when combined with complementary techniques like US, which further improve oxidation efficiency.

7.2.2 Role of US in Enhancing Oxidation Efficiency

US enhances oxidation efficiency through acoustic cavitation, a process where micro-sized bubbles form, grow, and collapse within an aqueous medium, releasing significant localized energy (Figure 7.1). This phenomenon creates extreme conditions (temperatures up to 10,000 K and pressures exceeding 1,000 atm), generating highly reactive hydroxyl radicals ($^{\bullet}OH$), hydrogen radicals ($H^{\bullet}$), and hydroperoxyl radicals ($HO_2^{\bullet}$) (Chauhan et al. 2021). These radicals play a crucial role in breaking down complex organic pollutants through oxidation and pyrolysis.

The sonochemical process operates in three reaction zones: (i) the bulk liquid phase, where free radicals diffuse and initiate oxidation reactions, (ii) the gas–liquid interface, where enhanced mass transfer and interfacial reactions occur, and (iii) the bubble interior, where volatile compounds undergo pyrolytic degradation (Serna-Galvis and Torres-Palma 2020). The efficiency of pollutant degradation depends on US frequency, power density, and reactor configuration. While higher frequencies (800 kHz) favor volatile compound degradation through pyrolysis, lower frequencies (20–200 kHz) optimize free radical production for breaking down non-volatile and hydrophilic pollutants (Serna-Galvis and Torres-Palma 2020).

Although US alone can degrade some contaminants, its effectiveness improves significantly when combined with AOPs such as Fenton and Photo-Fenton (Figure 7.2).

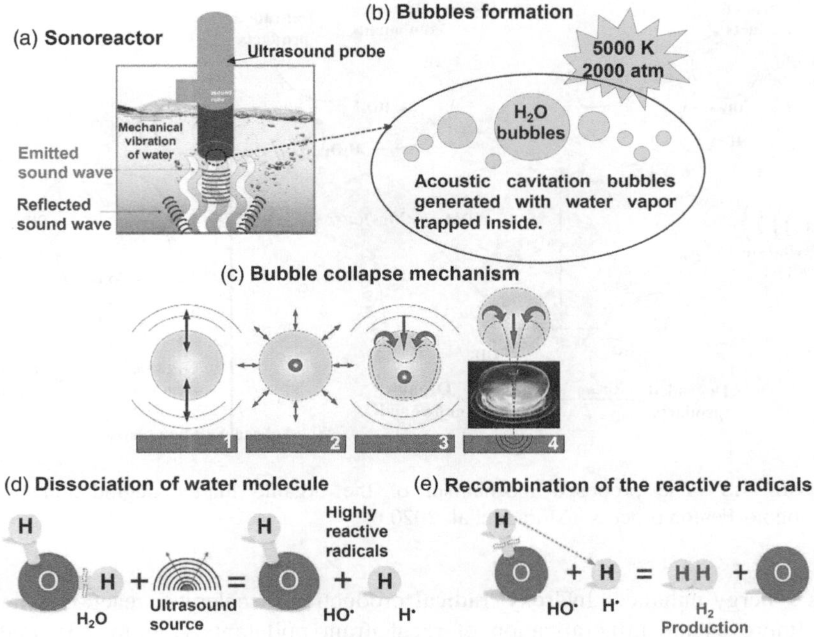

FIGURE 7.1 The mechanism of cavitation bubble formation and collapse in an aqueous medium. (Dehghani et al. 2023.)

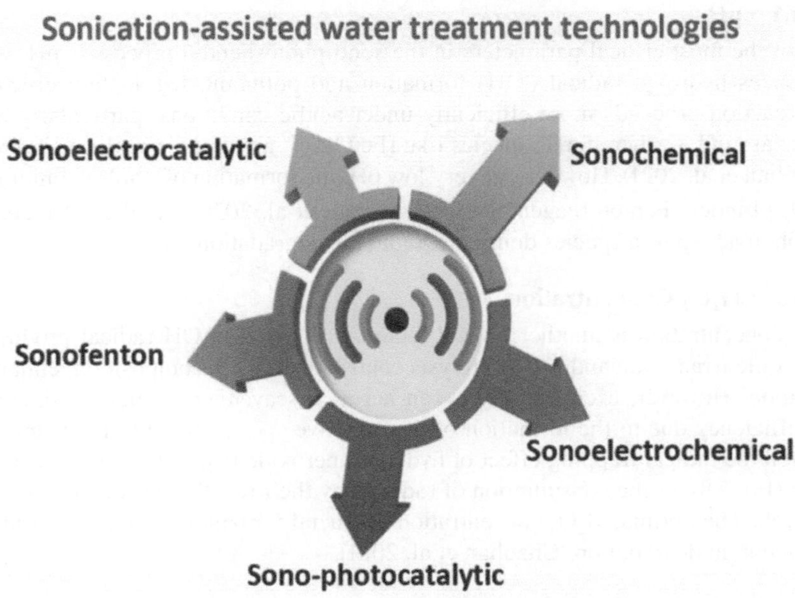

FIGURE 7.2 Various sono-assisted processes for wastewater treatment. (Elgarahy et al. 2024.)

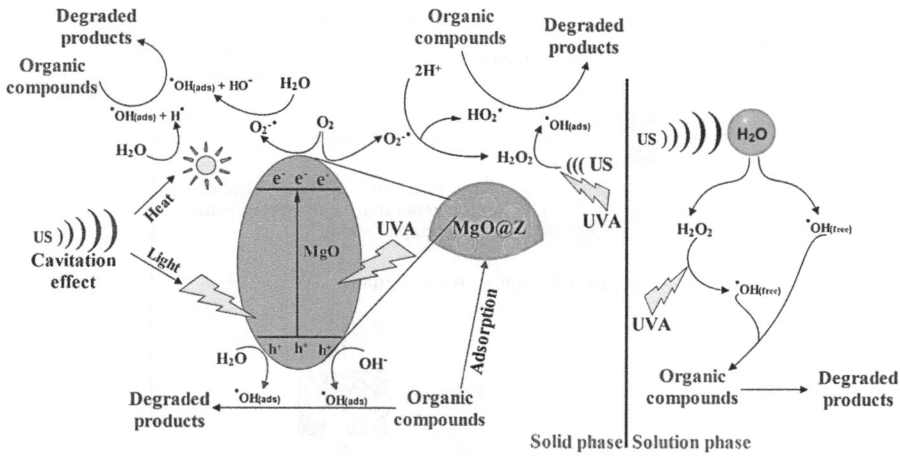

FIGURE 7.3 The proposed mechanism of the organic matter degradation by the sono-photo-Fenton process. (Moradi et al. 2020.)

This synergy enhances hydroxyl radical production, accelerates reaction kinetics, and improves the mineralization of recalcitrant pollutants (Figure 7.3), making US-assisted AOPs a promising technology for wastewater treatment.

7.2.3 Essential Factors Affecting Sono-Photo-Fenton Processes

7.2.3.1 pH

One of the most critical parameters in the sono-photo-Fenton process is pH, which influences hydroxyl radical (•OH) formation and pollutant degradation efficiency. The reaction proceeds more efficiently under acidic conditions, particularly at pH values around 3, where ferric species like $[Fe(OH)]^{2+}$ are at their peak concentration (Chauhan et al. 2021). However, at very low pH, the formation of stable oxonium ions $\left(H_3O_2^+\right)$ hinders Fenton reagent activity (Moradi et al. 2020). As the pH increases, low-photoactive iron species dominate, reducing degradation efficiency.

7.2.3.2 H_2O_2 Concentration

H_2O_2 concentration is another pivotal factor, as it governs OH radical production. Ultrasonic irradiation and UV photolysis contribute to its decomposition, enhancing oxidation. However, excessive H_2O_2 can act as a scavenger, reducing mineralization efficiency due to the formation of less-reactive species like O_2, H_2O, and $HO_2^{\bullet}$ through the radical trapping effect of hydrogen peroxide (Eq. 7.7), its self-decomposition (Eq. 7.8), or the consumption of radicals by their reaction between themselves (Eq. 7.9). The optimal H_2O_2 concentration is crucial for balancing radical formation and pollutant degradation (Chauhan et al. 2021).

$$HO^{\bullet} + H_2O_2 \rightarrow H_2O + HO_2^{\bullet} \tag{7.7}$$

$$2H_2O_2 \rightarrow 2H_2O + O_2 \quad (7.8)$$

$$HO^{\bullet} + HO^{\bullet} \rightarrow H_2O_2 \quad (7.9)$$

7.2.3.3 Initial Pollutant Concentration

The pollutant concentration also significantly impacts the efficiency of the degradation process. At optimal concentrations, active radicals effectively break down contaminants, but at higher concentrations, light penetration is hindered, limiting photoreaction efficiency, and the quantity of reactive radicals becomes insufficient for complete degradation. Additionally, pollutants may also block the catalyst's active sites, reducing process efficiency.

7.2.3.4 Light Intensity

The light intensity of the irradiation plays a vital role in increasing reaction rates, as a higher photon flux enhances reactive radicals' formation. However, at excessive intensities, radical production plateaus due to a lack of oxidants. Similarly, UV irradiation plays a key role in initiating free radical generation and ferrioxalate complex photolysis, which directly affects degradation kinetics. However, excessive irradiation can lead to photon losses due to insufficient absorbing elements (Moradi et al. 2020; Chauhan et al. 2021).

7.2.3.5 Ultrasonic Power and Frequency

Control cavitation dynamics affect free radical generation and pollutant degradation. Higher ultrasonic power enhances cavitation, increasing the number and size of bubbles, leading to greater radical formation and improved mass transfer (Shestakova and Sillanpää 2020). However, at extreme intensities, excess heating and bubble scattering reduce efficiency. The frequency of US determines cavitation bubble size and persistence, directly influencing radical production and pollutant oxidation rates (Moradi et al. 2020; Chauhan et al. 2021).

7.2.3.6 Temperature

The temperature affects redox reaction rates and iron speciation. Higher temperatures accelerate H_2O_2 degradation and ferrous ion regeneration, increasing radical formation. However, beyond an optimal threshold, $Fe(OH)_3$ precipitation and spontaneous H_2O_2 decomposition reduce degradation efficiency. Moreover, elevated temperatures enhance cavitation bubble formation, boosting pollutant breakdown until a limit where cavitation cushioning weakens radical activity (Moradi et al. 2020; Shestakova and Sillanpää 2020).

7.2.3.7 Dissolved Gases

Dissolved gases within the liquid also influence cavitation. The presence of gas nuclei reduces the pressure required to initiate cavitation, while the type of gas affects energy dissipation. Monatomic gases, due to their lower thermal conductivity, enhance the intensity of cavitation events compared to multiatomic gases, leading to greater energy release upon bubble collapse.

Ultrasonic intensity determines the effectiveness of cavitation processes. A minimum intensity threshold is required to initiate cavitation, but excessive intensity can lead to an overabundance of bubbles that interfere with uniform power distribution and reduce treatment efficiency. Optimizing ultrasonic intensity ensures efficient bubble collapse and effective radical formation (Shestakova and Sillanpää 2020).

In conclusion, the performance of the sono-photo-Fenton process is highly dependent on several operational parameters, including pH, H_2O_2 concentration, pollutant loading, light intensity, ultrasonic parameters, and system temperature. Optimizing these conditions is essential for maximizing degradation efficiency and ensuring effective wastewater treatment.

7.3 CASE STUDIES FOR LANDFILL LEACHATE TREATMENT FROM FEZ CITY, MOROCCO

7.3.1 Case Study 1: Fenton and Photo-Fenton Processes

In this study, the treatment of leachate from the municipal landfill of Fez City (Morocco) was investigated using a range of AOPs (El Mrabet et al. 2018). The evaluated methods included the conventional Fenton process (Fe^{2+}/H_2O_2), photo-Fenton (Fe^{2+}/H_2O_2/UV-A), and UV-A irradiation alone and in combination with hydrogen peroxide (UV-A/H_2O_2). The leachate sample exhibited a dark color and high turbidity (248 NTU), with elevated concentrations of COD (5,653 mg O_2L^{-1}) and BOD_5 (600 mg O_2 L^{-1}), both values significantly exceeding Morocco's discharge limits. The high COD reflects a significant load of dissolved organic matter. Electrical conductivity was also high (28 mS cm^{-1}), mainly due to elevated levels of chloride concentrations (7,100 mg L^{-1}), along with notable levels of nitrates, nitrites, and sulfates. With a BOD_5/COD ratio near 0.1 and a pH of 8, the leachate is considered stabilized and poorly biodegradable. It also contained metals such as Cr, Cu, Fe, Mn, and Zn, and had a high total suspended solid content (15.77 g L^{-1}).

The leachate treatment using UV photolysis alone proved ineffective, achieving less than 4% COD removal, likely due to low light transmittance caused by the leachate's high turbidity and dark color. Similarly, treatment with H_2O_2 alone resulted in limited efficiency, with COD reduction below 14% after 1 h (Figure 7.4a).

Fenton oxidation (Fe^{2+}/H_2O_2) at pH 3 and 60 min achieved 58% COD removal using 2,500 mg·L^{-1} H_2O_2, which increased to 70% at 10,000 mg·L^{-1}. Incorporating UV-A radiation into the Fenton system (photo-Fenton) significantly enhanced degradation, as UV-A light promotes Fe^{2+} regeneration via photolysis of Fe^{3+} complexes under acidic conditions. The photo-Fenton process achieved up to 84% COD removal and over 90% color reduction, indicating the effective degradation of complex aromatic and polymeric compounds. In contrast, UV/H_2O_2 reached only 60% color removal without sludge generation. Overall, the efficiency of AOPs for COD removal followed the order: photo-Fenton > Fenton > UV-A/H_2O_2 > UV-A, highlighting the superior performance of the photo-Fenton process due to enhanced hydroxyl radical generation.

H_2O_2 dosage is a key determinant in the efficiency of both Fenton and Photo-Fenton processes (Figure 7.4b). Theoretically, 1,000 mg·L^{-1} of H_2O_2 can oxidize 470.6 mg·L^{-1}

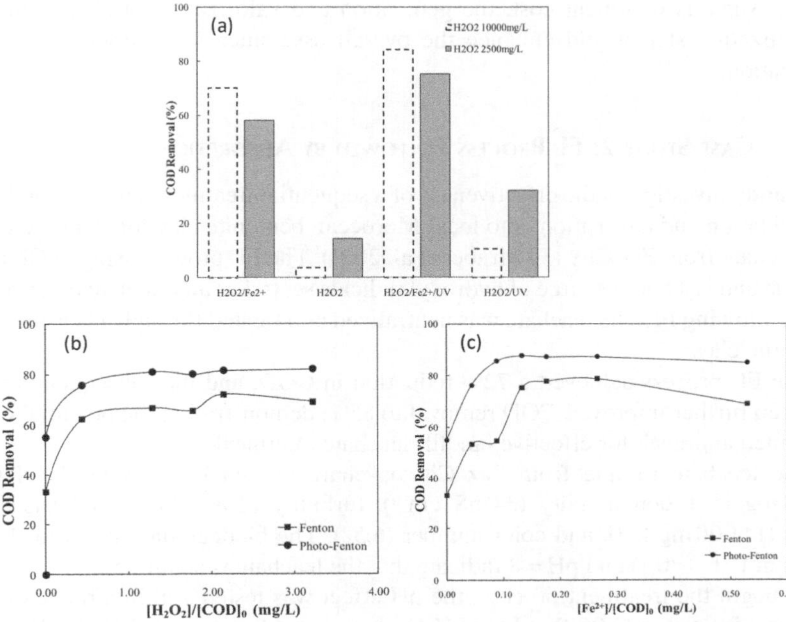

FIGURE 7.4 (a) The efficiency comparison of H_2O_2/Fe^{2+}, H_2O_2, $H_2O_2/Fe^{2+}/UV$, and H_2O_2/UV; (b) effect of $[Fe^{2+}]/COD_0$ on COD removal; (c) effect of $[H_2O_2]/COD_0$ on COD removal.

of COD. In this study, H_2O_2 concentrations from 2,500 to 19,000 mg·L^{-1} (H_2O_2/COD mass ratios of 0.44–3.22) were tested at pH 3, with a fixed Fe^{2+} concentration of 2,000 mg·L^{-1} ($H_2O_2/Fe^{2+}=0.35$). COD removal increased from 55% to 82% as H_2O_2 dosage rose to 18,214 mg·L^{-1}, attributed to the higher generation of OH radicals. However, beyond 7,285 mg·L^{-1}, additional H_2O_2 did not significantly enhance removal efficiency, likely due to $^{\bullet}OH$ radical scavenging and formation of less-reactive $HO_2^{\bullet}$ species, along with H_2O_2 decomposition and $^{\bullet}OH$ recombination. Consequently, 7,285 mg·L^{-1} was identified as the optimal H_2O_2 dose for both Fenton and Photo-Fenton.

To assess the influence of ferrous ion concentration on the Fenton and Photo-Fenton treatment of leachate (initial COD = 5,653 mg O_2 L^{-1}), experiments were conducted at a fixed H_2O_2 dose of 7,285 mg L^{-1} (H_2O_2/COD ratio = 1.29) and pH 3, while varying Fe^{2+} concentrations from 200 to 3,000 mg L^{-1} (Fe^{2+}/COD ratios = 0.04–0.53) (Figure 7.4c). Results showed that increasing the Fe^{2+}/COD ratio beyond 0.13 had a limited impact on COD removal efficiency, and no further enhancement was observed above a ratio of 0.33. Excess Fe^{2+} likely promoted side reactions that reduced overall oxidation performance. The maximum COD removal achieved was 78% for the Fenton process and 87% for the photo-Fenton process.

This case study provided valuable insights by comparing various treatment systems, including H_2O_2/Fe^{2+}, H_2O_2 alone, $H_2O_2/Fe^{2+}/UV$, and H_2O_2/UV. While the performance of these processes was thoroughly evaluated, further investigation into aspects such as treatment cost, the generation and valorization of sludge, and the neutralization step would enhance the overall assessment and support real-world application.

7.3.2 Case Study 2: FL Process Followed by Adsorption

This study investigates the effectiveness of a sequential treatment process combining FL oxidation and adsorption onto local Moroccan bentonite clay for stabilized landfill leachate from Fez City (El Mrabet et al. 2021). The FL process using $FeCl_3$ as the catalyst and H_2O_2 as a source of hydroxyl radicals served as an effective pretreatment step. Following FL, the leachate was neutralized and treated through adsorption onto bentonite clay.

The FL process achieved a 72% reduction in COD, and the subsequent adsorption step further improved COD removal to 85%, demonstrating the potential of this combined approach for effective landfill leachate treatment.

The leachate sample from Fez City is characterized by high levels of COD (5,198 mg L^{-1}), conductivity (34 mS cm^{-1}), turbidity (218 NTU), total suspended solids (14,000 mg L^{-1}), and color number (6.52). The biodegradability ratio (BOD_5/COD) of 0.09 (<0.1) and pH = 8 indicate that the leachate is stabilized.

To begin the treatment process, the pH effect was tested within a range of 2–11 (Figure 7.5a), using 11,000 mg L^{-1} of H_2O_2, based on a theoretical H_2O_2/COD ratio of 2.123, and 1,500 mg L^{-1} of Fe^{3+}, which was determined through preliminary tests. A kinetic study was then conducted under the optimal pH and the previously selected H_2O_2 and Fe^{3+} concentrations, with samples collected at intervals of 5, 15, 30, 45, 60, and 90 min.

Under experimental conditions with an initial pH of 3 and H_2O_2 concentration of 11,000 mg L^{-1}, varying Fe^{3+} dosages were tested for their impact on COD removal efficiency. As the Fe^{3+} dosage increased from 0 to 250 mg L^{-1}, COD removal efficiency improved from 34% to 79%, while the volume of sludge produced also rose from 0% to 52%. However, as the Fe^{3+} concentration was further increased to 2,000 mg L^{-1}, COD removal slightly decreased to 75%. At the highest concentration of 3,000 mg L^{-1}, the COD removal dropped significantly to 49%.

This trend suggests that while Fe^{3+} ions act as catalysts to accelerate hydroxyl radical generation, excessive Fe^{3+} leads to the production of too many radicals, reducing treatment efficiency through radical scavenging, as shown in the following reactions. Additionally, high Fe^{3+} concentrations increased the volume of sludge produced. The final pH values of the treated leachate decreased slightly with Fe^{3+} addition, likely due to the acidic nature of Fe^{3+} ions.

UV-Visible absorbance spectra, commonly used to track organic matter degradation, were recorded before and after treatment. The spectra showed a significant reduction in absorbance at wavelengths between 200 and 400 nm at an Fe^{3+} concentration of 250 mg L^{-1} (Figure 7.5b). However, at a higher concentration of 2,000 mg L^{-1} Fe^{3+}, the absorbance increased, likely due to interference from excess Fe^{3+} and residual pollutants. Visual observations confirmed that the leachate became noticeably clearer at 250 mg L^{-1} Fe^{3+} but appeared darker at higher concentrations, indicating that the optimal Fe^{3+} concentration for effective treatment is 250 mg L^{-1}. This

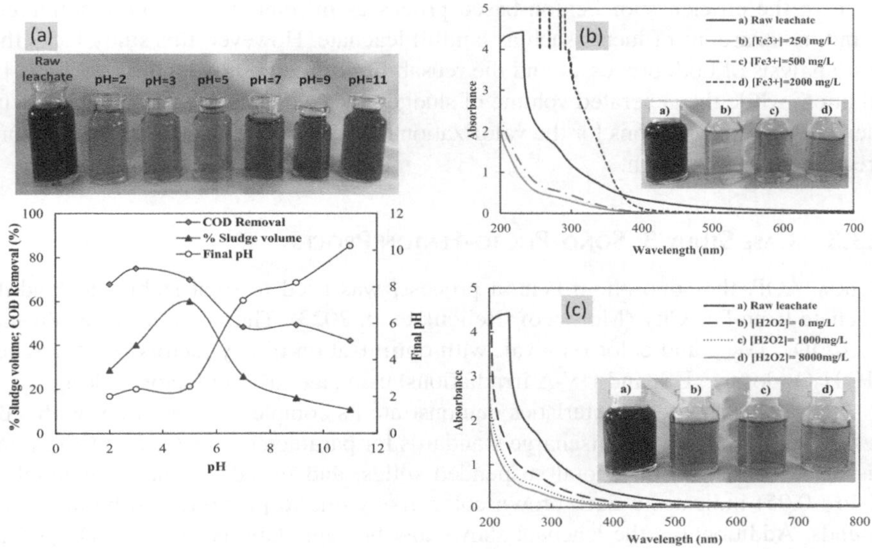

FIGURE 7.5 (a) Effect of initial pH of raw landfill leachate on COD removal and %sludge volume; (b) effect of [Fe^{3+}] on UV-visible spectra; (c) effect of [H_2O_2] on UV-visible spectra.

suggests that excessive Fe^{3+} could increase operational costs due to higher sludge production and diminished treatment efficiency.

The effect of H_2O_2 concentration on COD removal, sludge volume, and final pH was studied. Without H_2O_2, 32% of COD was removed. Adding 500 mg L^{-1} of H_2O_2 increased COD removal to 69%, and this slightly increased to 72% at 1,000 mg L^{-1}. However, further increases in H_2O_2 concentration led to a slight decrease in COD removal. This behavior is due to a balance between H_2O_2 dosage and OH production. Low H_2O_2 results in insufficient radicals, while excess H_2O_2 can self-decompose or scavenge radicals, reducing the process efficiency.

The effect of H_2O_2 on UV-visible absorbance showed that increasing [H_2O_2] significantly reduced absorbance, particularly in the 250–300 nm range, indicating improved degradation and discoloration of the leachate (Figure 7.5c).

The optimal FL process parameters were a 30-min treatment time, an initial pH of 3 (requiring neutralization post-treatment), [Fe^{3+}] = 250 mg L^{-1}, and [H_2O_2] = 1,000 mg L^{-1}, with a [Fe^{3+}]/[H_2O_2] mass ratio of 0.25. The H_2O_2 to COD ratio (0.27) was much lower than the theoretical value (2.125), reflecting the diverse nature of the effluent and the AOP system.

Variations in optimal ratios are expected depending on effluent composition, waste type, and environmental conditions. The determination of these optimal ratios remains challenging and requires further investigation for specific systems and effluents.

On the other hand, the sequential combination of FL treatment (achieving 72% COD removal) with adsorption onto Moroccan bentonite clay to reach a total efficiency of 85% highlighted the importance of combining two or more processes to improve the efficiency of Fenton-based processes in removing complex pollutants found in hazardous effluents such as landfill leachate. However, this study lacks the cost analysis of both processes and the reusability of the treated landfill leachate. In addition, while the generated volume of sludge (v/v %) during FL treatment is documented, exploring options for the valorization of these sludges would be a promising area for further research.

7.3.3 Case Study 3: Sono-Photo-Fenton Process

A new AOP, the sono-photo-Fenton process, was used to treat stabilized landfill leachate from Fez City (Morocco) (Bellouk et al. 2023). The process was evaluated for COD, UV_{254}, and color removal, with optimization of key factors (Fe^{2+} dosage, pH, H_2O_2 dosage, US, and UV-A irradiations) using a two-level factorial design.

The raw leachate characteristics demonstrate its complex composition, with values exceeding Morocco's discharge standards for parameters like COD, BOD_5, total dissolved solids, salinity, total suspended solids, and turbidity. The color number (2.16 ± 0.05) indicates a dark brown color, likely due to phenolic and humic compounds. Additionally, the leachate shows low biodegradability, with a BOD_5/COD ratio of 0.09 (<0.1) and a pH above 7.5, suggesting that it contains little biodegradable organic matter. This makes physical-chemical treatment methods more suitable than biological ones.

The test results show that US and/or UV-A irradiation alone were not effective in removing COD, UV_{254}, and color. However, a synergistic effect was observed when both were combined. Without pretreatment, the combined processes achieved over 90% COD removal, making it effective for treating stabilized leachate. The optimal conditions for maximum efficiency were identified as pH = 3, $[Fe^{2+}] = 446.19\,mg\,L^{-1}$, and $[H_2O_2] = 250\,mg\,L^{-1}$, which led to experimental removal values of COD of 93%, UV_{254} of 77%, and 98% color, confirming the high effectiveness of the sono-photo-Fenton process (Figure 7.6).

The synergistic effect between US and UV irradiation enhances disinfection and pollutant degradation by combining their distinct mechanisms. Studies have shown that US pretreatment reduces the reaction time required for UV disinfection of total coliforms, demonstrating a significant enhancement in microbial inactivation (Fetyan and Salem Attia 2020).

The synergy between US and UV/Vis light is attributed to the ultrasonic generation of H_2O_2, which undergoes photolysis under UV irradiation, boosting reactive radical formation for oxidation processes (Jyothi, Yesodharan, and Yesodharan 2014; Asaithambi, Beyene, and Alemayehu 2017). Additionally, the US prevents catalyst surface fouling, increases surface area, and enhances mass transfer of pollutants, thereby improving catalytic efficiency. The combined effect also minimizes environmental risks and addresses individual limitations of UV and US disinfection methods. When integrated with nanocomposites, US prevents nanoparticle aggregation, further increasing surface area and optimizing pollutant removal at the solid–liquid interface (Moradi et al. 2020).

Phytotoxicity tests using *Medicago sativa* seeds showed a significant reduction in phytotoxicity after treatment with the US-P-FP process, demonstrating its potential

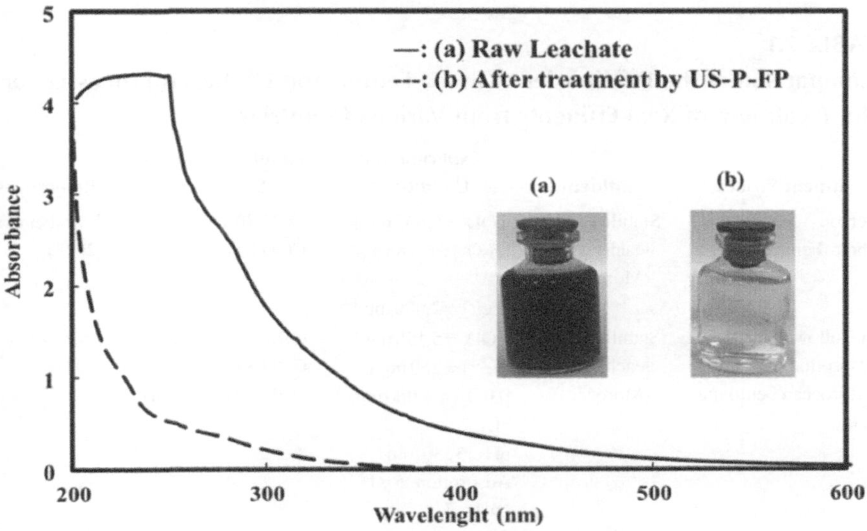

FIGURE 7.6 UV-visible spectra of raw and treated landfill leachate by sono-photo-Fenton.

for reducing the harmful effects of leachate. The US-photo-Fenton process proves to be an efficient and viable method for treating landfill leachate.

However, in this case study, the synergy index between US and UV/Fe^{2+}/H_2O_2 was estimated based on the achieved efficiencies in terms of COD, UV_{254}, and color and UV-visible spectral evolution; its calculation did not follow the adequate formula used in similar studies (Asaithambi, Beyene, and Alemayehu 2017; Poblete et al. 2020), but it was only estimated and concluded by the reached efficiencies in. Nonetheless, the results demonstrate a beneficial synergistic effect. Additionally, although this work did not evaluate the common challenges of homogeneous Fenton processes, such as sludge production and neutralization, nor include these factors in the overall process cost, it provides valuable insights into the treatment efficiency under the studied conditions.

7.3.4 Comparison of Case Studies' Findings with the Literature

The results of the three case studies are compared with similar treatment approaches reported in Table 7.1. Advanced techniques, such as sonophotocatalysis using fly ash and sono-electrochemical activation of peroxymonosulfate systems, have been applied to various wastewater types, including hospital, winery, and tannery effluents, in other countries. Reported removal efficiencies range from 23% to 99%, depending on various factors, including the nature and characteristics of the effluent (landfill leachate or industrial wastewater), the complexity of the pollutant, the oxidant combinations, the US frequency and power, the UV source, and whether the AOP was used alone or in a hybrid system. The reported case studies in this chapter demonstrated the superior performance of the Fenton-based process in removing

TABLE 7.1
Comparison of the Efficiencies of Some Fenton and US-Based Processes for the Treatment of Real Effluents from Various Countries

Treatment Process	Effluent	Experimental Conditions	Pollutant Removal Efficiency	References
Fenton photo-Fenton	Stabilized leachate (Morocco)	$COD_i = 5{,}653$ mg L^{-1} $[H_2O_2] = 7{,}285$ mg L^{-1} $[Fe^{2+}] = 2{,}000$ mg L^{-1}	COD 70% COD 84%	El Mrabet et al. (2018)
FL followed by adsorption onto local Moroccan bentonite clay	Stabilized leachate (Morocco)	$COD_i = 5{,}198$ mg L^{-1} $[Fe^{3+}] = 250$ mg L^{-1} $[H_2O_2] = 1{,}000$ mg L^{-1} pH = 3; 30 min Adsorption: 5 g L^{-1}; pH = 7	COD 72% (FL) COD 85% (FL + Adsorption)	El Mrabet et al. (2021)
US-photo-Fenton process (US-P-FP)	Stabilized leachate (Morocco)	$COD_i = 4{,}276$ mg L^{-1} $[Fe^{2+}] = 446$ mg L^{-1} $[H_2O_2] = 250$ mg L^{-1} pH = 3 UV-A (18 Watt) US (35 kHz; 168 W)	COD 93% UV_{254} 77% Color 98%	Bellouk et al. (2023)
Sono-photo-Fenton (US/UV/Fenton)	Hospital wastewater (Ethiopia)	$COD_i = 205$ mg L^{-1} $[Phosphate]_i = 3.45$ mg L^{-1} pH = 9 Reaction time = 50 min $[H_2O_2] = 20$ mg L^{-1} US (40 kHz) UV (254 nm, 32 W)	COD 94.5% Phosphate 99.15%	Endalew, Alemayehu, and Asaithambi (2025)
Sono-photo-Fenton	Landfill leachate (Coquimbo, Chile)	$COD_i = 4{,}656$ mg L^{-1} $[Fe^{2+}]=2{,}000$ mg L^{-1} $[H_2O_2] = 2{,}000$ mg L^{-1} US (864 kHz, 500 W) UV (365 nm, 10 W)	COD 80.3% TOC 97.0%	Alanís et al. (2025)
Sonophotocatalysis Fly ash catalyst	Landfill leachate (Coquimbo, Chile)	$COD_i = 6{,}750.8$ mg L^{-1} pH=3 2 g FA/L US (576 kHz, 500 W) UVc (254 nm, 40 W)	COD 40.7% Color 36.8% UV_{254} 50.8%	Poblete et al. (2024)

(Continued)

TABLE 7.1 (*Continued*)
Comparison of the Efficiencies of Some Fenton and US-Based Processes for the Treatment of Real Effluents from Various Countries

Treatment Process	Effluent	Experimental Conditions	Pollutant Removal Efficiency	References
Sonoelectrochemically activated peroxymonosulfate	Landfill leachate (Zhengzhou, China)	CODi = 1,969 mg L^{-1} Current density = 30 mA cm^{-2} US 50 W $pH_i = 2$ [PMS] = 4 Mm 120 min	COD 68.9%	Chen and He (2022)
Fenton, photo-Fenton, sono-Fenton, and sono-photo-Fenton processes	Real winery wastewater (São Roque, São Paulo, Brazil)	$COD_i = 45{,}221$ mg L^- 2 hours US (20 kHz, 400 W) UV (380 nm, 28 W)	COD 27.88% (FP) 31.51% (PFP) 23.19% (SFP) 29.29% (SPFP)	Rodrigues, Machado de Campos, and Tonello (2025)
Air stripping (aerobic pretreatment) followed by the US-UV-PS-H_2O_2 process	Real industrial tannery wastewater	$COD_i = 6{,}083$ mg L^{-1} pH= 3 $[H_2O_2] = 225$ mg L^{-1} $[K_2S_2O_8] = 3{,}670$ mg L^{-1} UV-A (18 Watt) US (35 kHz; 168 W)	COD 85% Color 93%	Bahammou, Tazi, et al. (2025)

COD, UV_{254}-absorbing compounds, and color, highlighting its potential as an effective treatment option for complex liquid effluents compared to similar processes.

7.4 CHALLENGES AND FUTURE PERSPECTIVES

Despite the promising potential of Fenton and US-based processes for the treatment of complex effluents like landfill leachate, several practical and scientific challenges persist. These include cost-effectiveness, environmental security in the long term, sludge management, and real-time monitoring. These constraints must be overcome not only by technological innovations but also by an integrated approach that includes economic, environmental, and regulatory aspects. The main difficulties and future perspectives are described in this section.

Evaluating the feasibility of a treatment method for large-scale application depends much on its cost-effectiveness. In the first case study, an estimated 0.57 USD m^{-3} was the total operational cost of the H_2O_2-Fe^{2+}-US-UV-A system in case 3. However, the cost was not estimated for the other two cases. Emphasizing that different operational parameters, such as energy demand, reaction kinetics, and process efficiencies, can change greatly with scale and mode of operation, cost estimations derived from batch systems may differ greatly from those of continuous-flow operations. Moreover, the prices of reagents and energy inputs depend on regional differences and fluctuations in world market prices for bulk chemicals, which can affect the general economic feasibility of the process in several countries.

Useful tools for evaluating the efficiency of AOPs in lowering the toxicity of treated effluent to plant life are phytotoxicity tests. But since these tests mainly measure short-term effects, more investigation is required to assess long-term phytotoxic effects, especially concerning the plant growth, reproduction, and the accumulation of residual compounds. Furthermore, since treated effluents are sometimes discharged into surface water, complementary ecotoxicity tests are absolutely essential to evaluate the larger environmental effects on aquatic life and ecosystems.

While homogeneous Fenton and FL processes are effective for contaminant removal, they often generate considerable amounts of iron-rich sludge, which poses handling and disposal challenges. To address this limitation, a promising alternative lies in developing heterogeneous Fenton systems using solid-state catalysts or composite materials such as zeolites, clay minerals, or their modified forms. These materials can be engineered into beads, pellets, or thin films, which not only facilitate their recovery and reuse but also minimize secondary pollution. Additionally, integrating these catalysts with solar-driven photoreactors offers an environmentally sustainable approach, leveraging natural sunlight to activate the process and reduce energy input, making the treatment more suitable for decentralized or low-resource settings.

Comparing the final effluent quality to national and international discharge standards is a critical step in objectively assessing the treatment efficiency of any process. Such comparisons are essential not only to validate the effectiveness of the treatment but also to determine the feasibility of reusing the treated effluent in a safe and sustainable manner. Therefore, Cases 1 and 2 should have been assessed against both the Moroccan Irrigation Standards and the Moroccan Liquid Discharge Standards, while case 3 was assessed only against the Moroccan Liquid Discharge Standards.

Characterizing water quality is a cornerstone of effective wastewater treatment and environmental management, enabling the evaluation of contamination levels, regulatory compliance, and treatment performance. Core parameters such as COD, BOD, total dissolved solids, total nitrogen, ammonium, total phosphorus, pH, turbidity, and odor concentration are routinely measured to assess water quality and monitor process efficiency. Advances in sensor technologies, ranging from electrochemical sensors based on metal nanoparticles to optical devices using cellulose membranes or fluorescent dyes, and enzyme-based biosensors, have significantly improved the accuracy and responsiveness of in-situ monitoring (Weerasinghe, Jayathilaka, and Vithanage 2025). These sensor systems, when integrated into Internet of Things platforms, allow for dynamic, real-time tracking of effluent quality and immediate feedback on operational performance (Alshami et al. 2024; Wang et al. 2023).

AI, particularly through Machine Learning algorithms, further enhances the capabilities of sensor networks by enabling sophisticated analysis and prediction of water quality parameters from complex, multivariate datasets. AI models can process spectroscopic data and sensor inputs to forecast concentrations of pollutants such as nutrients and heavy metals, minimizing reliance on time-consuming laboratory methods (Pérez-Beltrán et al. 2024). These tools support early anomaly detection, adaptive control, and predictive maintenance strategies. Studies report a notable increase in AI-focused wastewater treatment research, reflecting the growing adoption of digital tools to optimize treatment processes (Baarimah et al. 2024). Coupled with technologies like cloud computing, digital twins, and big data analytics, AI facilitates the development of smart wastewater treatment plants capable of autonomous operation, energy optimization, and resource recovery, all crucial for achieving environmental and economic sustainability (Cairone et al. 2024). Nevertheless, challenges such as limited datasets, high implementation costs, and the need for standardized frameworks persist. Addressing these issues through open data sharing, collaborative model development, and supportive regulatory changes is essential to unlocking the full potential of AI-driven water quality management systems.

7.5 CONCLUSION

This chapter has covered a critical review of the treatment of landfill leachate using AOPs based on Fenton, photo-Fenton, and sono-photo-Fenton technologies, with case studies from Fez City, Morocco. The findings indicate that, when properly optimized, AOPs can deliver high removal efficiencies for key quality parameters such as COD, color, and UV_{254}. Notably, the combination of UV-A irradiation and US significantly enhances process performance, particularly in complex leachate matrixes.

This chapter also elaborates on the valuable role of process monitoring and water quality characterization. It highlighted the importance of monitoring standard parameters and highlighted the contribution of advanced sensor technologies, covering electrochemical and biosensors, to enabling real-time data collection. These tools are essential for ensuring treatment effectiveness, consistency, and regulatory compliance.

In addition, the chapter addressed the growing potential of AI and Internet of Things technologies in wastewater treatment. These technologies offer transformative possibilities for real-time monitoring, process control, and predictive maintenance. However, their widespread adoption continues to be challenged by issues related to cost, data management, and scalability.

In brief, the investigated case studies show that AOPs, particularly in combination with adsorption and sono-chemical enhancement, are efficient and flexible methods for treating highly polluted landfill leachates. Successful industrial-scale operation will depend, however, on the realization of a balance between treatment efficiency and cost-effectiveness, as well as the installation of effective monitoring and control infrastructure. There should be greater research priority given to long-term pilot tests, life cycle analysis, and integration treatment plans incorporating chemical, biological, and digital technologies toward the sustainability of the leachate treatment.

REFERENCES

Alanís, Flavio, Rainer Nordenflycht, Matías Guerrero, Katherine Villalobos, Rodrigo Poblete, C. A. Rodríguez, Norma Pérez, Ernesto Cortés, and Ricardo A. Torres-Palma. 2025. "Sonoluminescence and H_2O_2 Produced in Water under Different Ultrasound Operating

Conditions Applied to the Sono-Photo-Fenton Landfill Leachate Treatment." *Chemical Engineering and Processing - Process Intensification* 211 (October 2024). doi:10.1016/j.cep.2025.110246.

Alshami, Ahmad, Eslam Ali, Moustafa Elsayed, Abdelrahman E.E. Eltoukhy, and Tarek Zayed. 2024. "IoT Innovations in Sustainable Water and Wastewater Management and Water Quality Monitoring: A Comprehensive Review of Advancements, Implications, and Future Directions." *IEEE Access* 12 (March): 58427–58453. doi:10.1109/ACCESS.2024.3392573.

Ameta, Rakshit, Anil K. Chohadia, Abhilasha Jain, and Pinki B. Punjabi. 2018. "Chapter 3-Fenton and Photo-Fenton Processes." In *Advanced Oxidation Processes for Wastewater Treatment: Emerging Green Chemical Technology*, edited by Suresh C. Ameta and Rakshit Ameta, 49–87. doi:10.1016/B978-0-12-810499-6.00003-6.

Asaithambi, P., Dejene Beyene, and Esayas Alemayehu. 2017. "Treatment of Landfill Leachate Waste Using Sono (US) and Photo (UV) Based Advanced Oxidation Processes: Studies on Various Operating Parameters." *Desalination and Water Treatment* 94: 147–155. doi:10.5004/dwt.2017.21583.

Baarimah, Abdullah O., Mahmood A. Bazel, Wesam Salah Alaloul, Motasem Y.D. Alazaiza, Tharaa M. Al-Zghoul, Basheer Almuhaya, Arsalaan Khan, and Ahmed W. Mushtaha. 2024. "Artificial Intelligence in Wastewater Treatment: Research Trends and Future Perspectives through Bibliometric Analysis." *Case Studies in Chemical and Environmental Engineering* 10 (August): 100926. doi:10.1016/j.cscee.2024.100926.

Bahammou, Oumaima, Imane El Mrabet, Zineb Chaouki, Mostafa Nawdali, Karim Tanji, Jamel Bouaziz, Fouad Khalil, et al. 2025. "Advanced Treatment of Tannery Effluent from Fez City (Morocco) Using a Sequence of Aerobic and Sono-Photo-Fenton Processes." *International Journal of Phytoremediation*: 1–14. doi:10.1080/15226514.2025.2499721.

Bahammou, Oumaima, Imane El Mrabet, Mostafa Nawdali, Karim Tanji, Fouad Khalil, Corinne Ferronato, and Hicham Zaitan. 2025. "Sequential Treatment of Tannery Effluent Bycoagulation-Flocculation and Sono-Photo-Fenton: Per Formance, Optimisation and Phytotoxicity Evaluation." *Journal of Environmental Engineering and Science*: 1–13. doi:10.1680/jenes.24.00136.

Bahammou, Oumaima, Imane Tazi, Imane El Mrabet, Mostafa Nawdali, Karim Tanji, Fouad Khalil, Corinne Ferronato, Mourad Benzina, and Hicham Zaitan. 2025. "Treatment of Real Tannery Industrial Wastewater via Sequential Biological and US/UV/Activated Persulfate-Hydrogen Peroxide Processes." *Water, Air, and Soil Pollution* 236 (5): 1–23. doi:10.1007/s11270-025-07904-4.

Bellouk, Hamza, Mohamed Danouche, Imane El Mrabet, Karim Tanji, Fouad Khalil, Mostafa Nawdali, Naima El Ghachtouli, and Hicham Zaitan. 2023. "Remediation of the Landfill Leachate of Fez City (Morocco) by Sono-Photo-Fenton Process: Cost and Phytotoxicity Assessment." *Journal of Water Process Engineering* 56 (September): 104565. doi:10.1016/j.jwpe.2023.104565.

Bellouk, Hamza, Imane El Mrabet, Karim Tanji, Mostafa Nawdali, Mourad Benzina, Mabrouk Eloussaief, and Hicham Zaitan. 2022. "Performance of Coagulation-Flocculation Followed by Ultra-Violet/Ultrasound Activated Persulfate/Hydrogen Peroxide for Landfill Leachate Treatment." *Scientific African* 17 (September): e01312. doi:10.1016/j.sciaf.2022.e01312.

Bellouk, Hamza, Ilham Zouitane, Imane El Mrabet, Naima El Ghachtouli, Karim Tanji, Fouad Khalil, and Hicham Zaitan. 2025. "Sequential Treatment of Young Landfill Leachate by Aeration and UV/US/Activated Persulfate-Hydrogen Peroxide: Performance, Optimization and Phytotoxicity Evaluation." *Environmental Science and Pollution Research*, no. 0123456789. doi:10.1007/s11356-025-36100-4.

Cairone, Stefano, Shadi W. Hasan, Kwang Ho Choo, Demetris F. Lekkas, Luca Fortunato, Antonis A. Zorpas, Gregory Korshin, Tiziano Zarra, Vincenzo Belgiorno, and Vincenzo Naddeo. 2024. "Revolutionizing Wastewater Treatment toward Circular Economy and Carbon Neutrality Goals: Pioneering Sustainable and Efficient Solutions for Automation and Advanced Process Control with Smart and Cutting-Edge Technologies." *Journal of Water Process Engineering* 63 (May): 105486. doi:10.1016/j.jwpe.2024.105486.

Chauhan, Rohit, G. Kumaravel Dinesh, Bablu Alawa, and Sankar Chakma. 2021. "A Critical Analysis of Sono-Hybrid Advanced Oxidation Process of Ferrioxalate System for Degradation of Recalcitrant Pollutants." *Chemosphere* 277: 130324. doi:10.1016/j.chemosphere.2021.130324.

Chen, Shuxun, and Zhengguang He. 2022. "Sonoelectrochemical Activation of Peroxymonosulfate: Influencing Factors and Mechanism of FA Degradation, and Application on Landfill Leachate Treatment." *Chemosphere* 296 (October 2021): 133365. doi:10.1016/j.chemosphere.2021.133365.

Dehghani, Mohammad Hadi, Rama Rao Karri, Janardhan Reddy Koduru, Sivakumar Manickam, Inderjeet Tyagi, Nabisab Mujawar Mubarak, and Suhas. 2023. "Recent Trends in the Applications of Sonochemical Reactors as an Advanced Oxidation Process for the Remediation of Microbial Hazards Associated with Water and Wastewater: A Critical Review." *Ultrasonics Sonochemistry* 94 (March): 106302. doi:10.1016/J.ULTSONCH.2023.106302.

El Mrabet, I., M. Kachabi, M. Nawdali, A. Harrach, F. Khalil, M. Ijjaali, M. Benzina, and H. Zaitan. 2018. "Treatment of Landfill Leachate from Fez City (Morocco) Using Fenton and Photo-Fenton Processes." *IOP Conference Series: Earth and Environmental Science* 161 (1): 012025. doi:10.1088/1755-1315/161/1/012025.

El Mrabet, Imane, Hamza Bellouk, Zineb Bencheqroun, Mostafa Nawdali, and Hicham Zaitan. 2021. "Treatment of Stabilized Landfill Leachate Using Coupled Fenton-like and Adsorption Process onto Moroccan Bentonite Clay." *Desalination and Water Treatment* 240 (November): 43–54. doi:10.5004/dwt.2021.27754.

Elgarahy, Ahmed M., M. G. Eloffy, A. K. Priya, V. Yogeshwaran, Khalid Z. Elwakeel, Zhen Yang, and Eduardo Alberto Lopez-Maldonado. 2024. "A Review on the Synergistic Efficacy of Sonication-Assisted Water Treatment Process with Special Attention given to Microplastics." *Chemical Engineering Research and Design* 206 (April): 524–552. doi:10.1016/j.cherd.2024.05.027.

Endalew, Meseret, Esayas Alemayehu, and Perumal Asaithambi. 2025. "Hospital Wastewater Treatment Using Integrated Sono-Photo-Fenton Process: Experimental Design through RSM." *Scientific African* 27 (August 2024): e02585. doi:10.1016/j.sciaf.2025.e02585.

Fetyan, Nashwa A.H., and Tamer Mohamed Salem Attia. 2020. "Water Purification Using Ultrasound Waves: Application and Challenges." *Arab Journal of Basic and Applied Sciences* 27 (1): 194–207. doi:10.1080/25765299.2020.1762294.

Jyothi, K. P., Suguna Yesodharan, and E. P. Yesodharan. 2014. "Ultrasound (US), Ultraviolet Light (UV) and Combination (US + UV) Assisted Semiconductor Catalysed Degradation of Organic Pollutants in Water: Oscillation in the Concentration of Hydrogen Peroxide Formed in Situ." *Ultrasonics Sonochemistry* 21 (5): 1787–1796. doi:10.1016/j.ultsonch.2014.03.019.

Moradi, Masoud, Ali Elahinia, Yasser Vasseghian, Elena Niculina Dragoi, Fariborz Omidi, and Amin Mousavi Khaneghah. 2020. "A Review on Pollutants Removal by Sono-Photo-Fenton Processes." *Journal of Environmental Chemical Engineering* 8 (5): 104330. doi:10.1016/j.jece.2020.104330.

Mukherjee, Jayanti, Bibhab Kumar Lodh, Ramesh Sharma, Nibedita Mahata, Maulin P. Shah, Subhasis Mandal, Susanta Ghanta, and Biswanath Bhunia. 2023. "Advanced Oxidation Process for the Treatment of Industrial Wastewater: A Review on Strategies, Mechanisms, Bottlenecks and Prospects." *Chemosphere* 345 (October): 140473. doi:10.1016/j.chemosphere.2023.140473.

Peralta-Hernández, Juan M., Shraddha Vijay, Oscar Rodríguez-Narváez, and Martin A. Pacheco-Álvarez. 2018. "Photo and Solar Fenton Processes for Wastewater Treatment." In *Electrochemical Water and Wastewater Treatment*, 223–237. doi:10.1016/b978-0-12-813160-2.00009-2.

Pérez-Beltrán, C. H., A. D. Robles, N. A. Rodriguez, F. Ortega-Gavilán, and A. M. Jiménez-Carvelo. 2024. "Artificial Intelligence and Water Quality: From Drinking Water to Wastewater." *TrAC - Trends in Analytical Chemistry* 172 (October 2023). doi:10.1016/j.trac.2024.117597.

Poblete, Rodrigo, Ernesto Cortes, Norma Pérez, C. A. Rodríguez, and Yolanda Luna-Galiano. 2024. "Treatment of Landfill Leachate by Combined Use of Ultrasound and Photocatalytic Process Using Fly Ash as Catalyst." *Journal of Environmental Management* 349 (November 2023): 119552. doi:10.1016/j.jenvman.2023.119552.

Poblete, Rodrigo, Ernesto Cortes, Guray Salihoglu, and Nezih Kamil Salihoglu. 2020. "Ultrasound and Heterogeneous Photocatalysis for the Treatment of Vinasse from Pisco Production." *Ultrasonics Sonochemistry* 61 (April 2019): 104825. doi:10.1016/j.ultsonch.2019.104825.

Pouran, Rahim Shima, A. R. Abdul Aziz, and Wan Mohd Ashri Wan Daud. 2015. "Review on the Main Advances in Photo-Fenton Oxidation System for Recalcitrant Wastewaters." *Journal of Industrial and Engineering Chemistry* 21: 53–69. doi:10.1016/j.jiec.2014.05.005.

Rodrigues, Ricardo Augusto, Mariana Bizari Machado de Campos, and Paulo Sergio Tonello. 2025. "Degradation of Phenolic Compounds and Organic Matter from Real Winery Wastewater by Fenton and Photo-Fenton Processes Combined with Ultrasound." *Water* 17 (5): 1–13. doi:10.3390/w17050763.

Serna-Galvis, Efraím A., and Ricardo A. Torres-Palma. 2020. "Recent Developments in Sonochemical Treatments of Contaminated Wastewaters." *Green Chemistry and Water Remediation: Research and Applications*, 299–315. doi:10.1016/B978-0-12-817742-6.00010-4.

Shestakova, Marina, and Mika Sillanpää. 2020. "Chapter 2- Ultrasound-Assisted Electrochemical Treatment of Wastewaters Containing Organic Pollutants by Using Novel Ti/Ta_2O_5–SnO_2 Electrodes." In *Advanced Water Treatment*, 79–161. Elsevier Inc. doi:10.1016/B978-0-12-819227-6.00002-4.

Wang, Bingyang, Xiaodan Li, Donghui Chen, Xiaohui Weng, and Zhiyong Chang. 2023. "Development of an Electronic Nose to Characterize Water Quality Parameters and Odor Concentration of Wastewater Emitted from Different Phases in a Wastewater Treatment Plant." *Water Research* 235 (November 2022): 119878. doi:10.1016/j.watres.2023.119878.

Weerasinghe, Manura, Keshani Jayathilaka, and Meththika Vithanage. 2025. "Sensors for Detection and Monitoring of Contaminants in Wastewater." *Current Opinion in Environmental Science and Health* 45: 100609. doi:10.1016/j.coesh.2025.100609.

8 Hybrid Adsorptive–Photocatalytic Systems

Synergistic Approaches for Liquid Effluent Treatment

Zineb Chaouki, Imane El Mrabet, Zineb Bencheqroun, Karim Tanji, Samira Bnikkou, Salah Rafqah, Mourad Benzina, Corinne Ferronato, Hector Valdes, and Hicham Zaitan

8.1 INTRODUCTION

Water resources are vital to human survival and a major pillar of economic development. Although water covers around 70% of the planet's surface, its unequal distribution makes access difficult for many populations. In some regions, water scarcity is exacerbated by population growth and climate change. In addition to scarcity issues, water quality is seriously affected by pollution caused by human activities. The agricultural, industrial, and domestic sectors discharge large quantities of pollutants into freshwater, making it unsafe for use. Consequently, the quantity and quality of freshwater resources constitute a critical issue for the international community and represent one of the most pressing challenges facing our planet, as stated in the Sustainable Development Goals (UN-Water, 2024).

In this context, treating these liquid effluents for reuse presents a promising alternative. It is therefore essential to develop effective processes capable of degrading pollutants more efficiently than conventional methods. Various techniques have been developed to reduce the toxicity of wastewater, including coagulation–flocculation (Chaouki et al., 2017), membrane filtration (Colakoglu et al., 2025), advanced oxidation (Lu et al., 2025), adsorption (Bencheqroun et al., 2018), evaporation (Wang et al., 2025), photocatalysis (Lin et al., 2025), etc. Because of the limitations of individual methods, combining processes has proved to be a practical solution (Belkodia et al., 2024). Thus, adsorption and photocatalysis are two widely used techniques for wastewater treatment worldwide. The adsorption is known for its remarkable efficacy, low cost, and ease of use (Paredi et al., 2025). Heterogeneous photocatalysis has shown exceptional effectiveness because of its ease of use and capacity to degrade complex pollutants without producing harmful byproducts (Danish et al., 2021; Pattnaik et al., 2023). Hence, the synergistic approach of both processes effectively removes contaminants from wastewater (Ma et al., 2025; Moles et al., 2021).

DOI: 10.1201/9781003624905-8

Current research is therefore focusing on the development of composite materials that can be used for this combined treatment to put these techniques into practice, like AC (Dhayal et al., 2025), clay (Karaman et al., 2016), zeolite (Liaquat et al., 2024), metal–organic frameworks, and GO (Absalan et al., 2025). They are advantageous materials to combine with metal oxide photocatalysts because of their high surface area and porous structure. They therefore enhance stability and prevent catalyst agglomeration (Ahmadi & Ganjidoust, 2021).

This chapter explores the fundamental principles and mechanisms of adsorption and heterogeneous photocatalytic processes, as well as their synergistic combination for environmental remediation. It provides a comprehensive overview of how these processes can be combined to enhance pollutant removal efficiency. Two case studies are presented: the first evaluates the performance of photocatalytic degradation of pretreated landfill leachate from the Mediouna site (Morocco) using TiO_2 as the photocatalyst (Chaouki et al., 2025). The second focuses on the removal of a pharmaceutical pollutant using a combined treatment employing nanocomposite photocatalysts based on AC/GO/TiO_2 (Prado Mendonça et al., 2025). These examples underscore the potential of hybrid systems to address complex wastewater challenges through improved performance and versatility.

8.2 ADSORPTION PROCESS

A common physico-chemical method for removing both organic and inorganic pollutants from liquid and gaseous effluents is adsorption. The process is based on mass transfer phenomena due to the various interactions between the molecules in the fluid phase and the surface of the adsorbent (Chen et al., 2025).

This technique uses different types of adsorbents with a high specific surface. The most widely studied materials for the removal of a variety of pollutants are carbon-based (Dogra et al., 2022; Moradi & Panahandeh, 2022). Zeolites, ion-exchange resins, and clays, among others, are also valued for their cost-effectiveness, exchange capacity, and regeneration potential (Abd El-Magied et al., 2025; Das & Sengupta, 2025; Hadri et al., 2022).

8.2.1 Adsorption Steps and Mechanism

The adsorption process follows several successive steps that govern the transfer and immobilization of adsorbates onto the solid surface, as illustrated in Figure 8.1.

1. Transport of the adsorbate from the bulk liquid phase to the vicinity of the adsorbent surface.
2. External diffusion: transfer of the solute through the liquid film to the surface of the solid particles.
3. Internal diffusion: transfer of the adsorbate inside the pores of the adsorbent under the effect of the concentration gradient.
4. Adsorption on active sites; once adsorbed, the molecule is considered immobile.

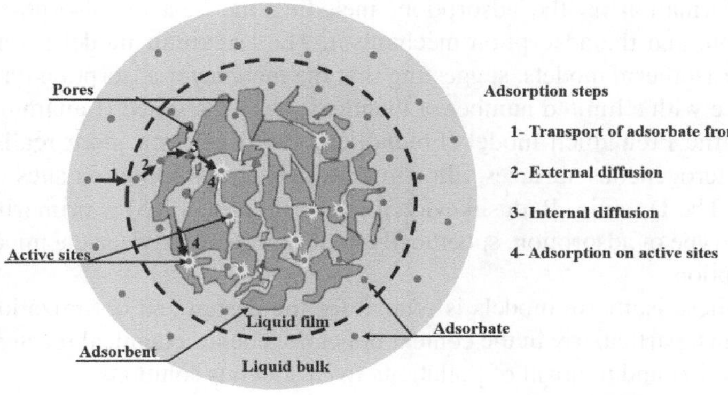

FIGURE 8.1 Adsorption mass transfer steps. (Adapted by permission from Ref. Wang and Guo, 2020.)

Adsorption is classified as physisorption or chemisorption, depending on the type of interaction between the adsorbate and the adsorbent. Physisorption involves weak attractive forces such as van der Waals, ionic forces, or electrostatic polarization interactions. It is typically a reversible phenomenon with a low adsorption enthalpy (ΔH) between 20 and 40 KJ mol^{-1} (Mohanty et al., 2005). This process is more advantageous at low temperatures and becomes less efficient as the temperature rises due to the reduction of these weak intermolecular forces (Qiu et al., 2022). However, chemisorption also involves the formation of powerful chemical bonds, like ionic or covalent ones. It is characterized by a high adsorption enthalpy between 40 and 400 KJ mol^{-1}, an irreversible phenomenon, and it can also occur at different temperatures (Calvet et al., 1980).

8.2.2 Adsorption Kinetic and Isotherm Models

Kinetic and isotherm models are commonly used to describe adsorption processes. A crucial indicator of an adsorbent's effectiveness is its adsorption kinetics. It estimates the amount of pollutant adsorbed over time, and the time required to reach equilibrium between the adsorbate and the adsorbent. The majority of kinetic curves show a common pattern: a fast initial adsorption rate, followed by a slower phase as the process progresses toward saturation of the adsorbent's active sites. Furthermore, adsorption mechanisms are frequently studied using kinetic models created from experimental data. The pseudo-first-order (Lagergren, 1898) and pseudo-second-order models (Ho et al., 2000) are commonly applied, as they allow the calculation of adsorption capacities by linearizing the data and comparing them with experimental results. Other models, such as intraparticle diffusion (Allen et al., 1989) and Elovich (Roginsky & Zeldovich, 1934), are also used depending on the system's characteristics.

On the other hand, adsorption isotherms describe the relationship between the concentration of an adsorbate in solution and the amount adsorbed onto a solid surface at a constant temperature. Analyzing isotherm data provides both quantitative and qualitative information on the adsorption, including the heat of adsorption, active site saturation, and the adsorption mechanism. The Langmuir model is one of the most popular isotherm models, suggesting that the monolayer adsorption on a homogenous surface with a limited number of identical sites is assumed (Langmuir, 1918). In contrast, the Freundlich model (Freundlich, 1906) offers a more realistic approach for heterogeneous surfaces, allowing multilayer adsorption on sites of different affinities. The Dubinin–Radushkevich model (Dubinin, 1965) is primarily used to identify the type of adsorption, specifically to differentiate between chemical and physical adsorption.

Understanding these isotherm models is crucial for the design and optimization of adsorption systems, particularly in the context of heterogeneous chemical reactors used for the purification and removal of pollutants from aqueous solutions.

8.2.3 Key Operational Parameters Affecting Adsorption Performance

The performance of an adsorption system is influenced by several key parameters. One such parameter is the **initial adsorbate concentration.** As this concentration increases, the active sites on the adsorbent surface may gradually become saturated, leading to a reduction in adsorption efficiency (Georgin, 2023). Another important parameter is **the nature of the adsorbent**. Some natural clays and zeolites with a high specific surface area and high porosity have shown excellent adsorption capacities for various contaminants present in water (Jiang et al., 2021).

The **pH** of the solution also strongly influences this process. It depends on the point of zero charge (PZC) of the adsorbent and directly affects the charge state of both the adsorbent and the adsorbate. Alkaline conditions favor cation adsorption, while acidic conditions favor anion adsorption (Pimentel et al., 2024). **Temperature** is another significant parameter controlling adsorption. In most cases, the adsorption rate decreases with increasing system temperature, possibly due to the exothermic nature of the interaction (Du et al., 2021).

The contact time between the adsorbent and the adsorbate is also an important parameter for achieving equilibrium. In the physical adsorption case, most species adsorb rapidly onto the adsorbent surface in the early stages of contact time and become slow near equilibrium (Georgin et al., 2023). However, a strong chemical bond between the adsorbate and the adsorbent requires a longer contact time to reach equilibrium.

Agitation plays a key role in promoting effective contact between the adsorbent and the solution. At low stirring speeds, adsorption mass transfer controls the process because of the thick fluid film around the particle. However, as stirring speed increases, this film decreases, and the process is controlled by intraparticle diffusion (Kyzas et al., 2020).

8.3 HETEROGENEOUS PHOTOCATALYSIS PROCESS

Recent advances in water and wastewater treatment processes have led to the development of oxidative destruction procedures for emerging pollutants. As an alternative to traditional wastewater treatment techniques, advanced oxidation processes have been used. The basis of advanced oxidation processes is the in situ production of hydroxyl radicals (OH), which efficiently degrade newly formed organic compounds in aqueous media (Wang & Zhuan, 2020).

Because of its capacity to use molecular oxygen and its ability to mineralize organic pollutants, heterogeneous photocatalysis has become a viable alternative to other advanced oxidation processes for wastewater treatment. Furthermore, its interest in environmental remediation has increased due to the possibility of using sunlight as a sustainable energy source (Bahrudin & Nawi, 2018; Bora & Mewada, 2017).

8.3.1 Heterogeneous Photocatalysis Mechanism

Typically, semiconductor catalysts are used in heterogeneous photocatalysis processes to break down organic compounds. When a semiconductor is exposed to an appropriate light source, its surface undergoes a sequence of oxidation and reduction reactions with an energy (hν) equal to or greater than its bandgap, as illustrated in Figure 8.2.

This process can be divided into five steps (Byrne et al., 2018):

1. **Adsorption** of the target pollutant onto the photocatalyst surface.
2. **Light absorption** by the photocatalyst with photon energy larger than its band gap energy, and the generation of photogenerated electron (e^-)–hole (h^+) pairs in the bulk phase (Eq. 8.1).

$$\text{Photocatalyst} + \text{hn} \rightarrow e_{BC}^{-} + h_{BV}^{+} \tag{8.1}$$

3. **Migration** of e^- and h^+ to the surface of the photocatalyst to participate in the redox reaction, and simultaneously, some recombine, releasing the absorbed energy as heat (Eq. 8.2).

$$h_{VB}^{+} + e_{CB}^{-} \rightarrow \text{energy}(\text{heat}) \tag{8.2}$$

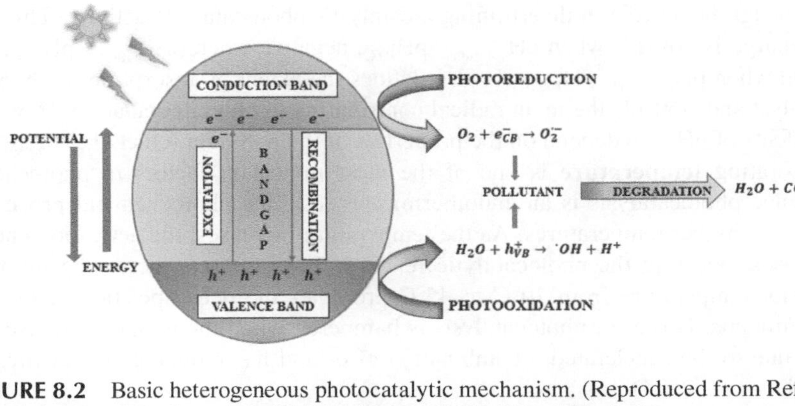

FIGURE 8.2 Basic heterogeneous photocatalytic mechanism. (Reproduced from Ref. Bora and Mewada (2017). Copyright (2017), with permission from Elsevier.)

4. **Oxidation** and **reduction** of H_2O molecules and O_2 adsorbed on the surface of the photocatalyst to hydroxyl radicals (OH) and superoxide radicals $\left(\bullet O_2^-\right)$ by h^+ in the valence band and e^- in the conduction band (Eqs. 8.3 and 8.4). These radicals further react with the pollutant and subsequently mineralize it to carbon dioxide and water, forming a number of intermediate species en route (Eqs. 8.5 and 8.6).

$$h_{BV}{}^{+} + H_2O \rightarrow {}^{\bullet}OH(\text{hydroxyl radical}) + H^{+} \quad (8.3)$$

$$e_{BC}{}^{-} + O_2 \rightarrow O_2{}^{\bullet -}(\text{superoxide radical}) \quad (8.4)$$

$${}^{\bullet}OH + \text{pollutant} \rightarrow \text{Intermediates} \rightarrow H_2O + CO_2 \quad (8.5)$$

$${}^{\bullet}O_2{}^{-} + \text{pollutant} \rightarrow \text{Intermediates} \rightarrow H_2O + CO_2 \quad (8.6)$$

5. **Desorption** of degraded products from the photocatalyst surface to the bulk solution, and the photoreaction continues.

8.3.2 Essential Factors Affecting Heterogeneous Photocatalysis Processes

The performance of heterogeneous photocatalysis depends on several parameters. One of the most critical is the **photocatalyst dose**. Degradation performance is improved as the dose rises because more active sites and hydroxyl radicals are produced (Błędzka et al., 2012). However, an excessive amount of catalyst can reduce efficiency by increasing the opacity of the suspension and decreasing photon penetration and activated surface area (Ba-Abbad et al., 2012).

Another important factor is **the initial concentration of organic compounds**. Because too many molecules are adsorbed on the catalyst, hydroxyl radical formation is reduced at high concentrations, which reduces photocatalysis efficiency (Khader et al., 2024).

Because **pH** regulates the surface charge characteristics known as the PZC, it plays a significant role in determining a catalyst's photocatalytic activity. This surface charge is positive when $pH_{solution} < pH_{PZC}$, negative when $pH_{solution} > pH_{PZC}$, and neutral when $pH_{solution} = pH_{PZC}$. pH also defines the extent of adsorption on the photocatalyst and controls the main radical contributing to photodegradation. However, the effects of pH also depend on the properties of the pollutant (Qi et al., 2023).

Operating **temperature** is one of the most important factors in photocatalysis. Since photocatalysis is an endothermic process, the photochemical process is favored at higher temperatures. As the temperature increases, the activation energy decreases, boosting the photocatalytic reaction. For instance, increasing the photoreactor temperature from 20°C to 45°C promotes the decomposition of organic contaminants. However, photocatalysis is hampered when the temperature exceeds 80°C due to the accelerated recombination of e- and h+, reducing carrier lifespan (Khader et al., 2024).

Heterogeneous photocatalysis also depends on **light intensity**, which promotes catalyst activation by exciting electrons from the valence to the conduction band. This increases the formation of electron–hole pairs and accelerates degradation rates at high intensities (Das & Ahn, 2022; Martín-Sómer et al., 2017). Conversely, at low intensities, pair recombination reduces the production of free radicals, thereby decreasing the efficiency of the process. However, if the light intensity exceeds the necessary threshold, it does not enhance electron–hole pair formation, resulting in wasted energy (Jia et al., 2010; Yang et al., 2007).

Two main parameters determine the choice of a light source: the spectral distribution of the light emitted by the source, and the power of the light source. The light sources most commonly used in photocatalysis are sunlight (natural) and artificial light created by lamps.

8.4 ADSORPTION–PHOTOCATALYSIS SYNERGY

The combined adsorption and photocatalysis greatly improve wastewater treatment efficiency, according to several studies (Dhayal et al., 2025). Through a synergy between the two methods, this integration produces high pollutant removal efficiencies: adsorption retains a variety of contaminants, which enables photocatalysis to degrade them into harmless inorganic compounds, as shown in Figure 8.3 (Khalaf et al., 2025). Overall efficiency is limited because the photocatalytic reaction rate is comparatively slow in comparison to the system's adsorption rate.

Furthermore, there are drawbacks to the adsorption method, including the potential for leaching and the requirement to discard used adsorbents. This challenge can be solved by employing photocatalysis to remove pollutants and transform them into non-harmful products (Ahmadi et al., 2025).

Moreover, the degradation of pollutants by photocatalysis is not always associated with their total removal from the system. Once a contaminant has degraded on the photocatalyst, the adsorption site is free to absorb new molecules. As a result, an on-site photocatalyst combined with a continuous adsorption phenomenon is necessary to guarantee efficient decontamination.

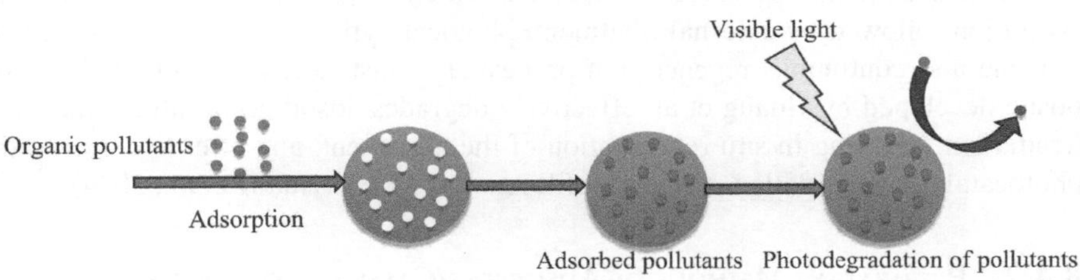

FIGURE 8.3 Schematic representation of adsorption–photocatalytic degradation of organic pollutants. (Copyright Khalaf et al. (2025), with permission from Elsevier.)

8.4.1 Importance of Adsorbent/Photocatalyst Integration for Enhanced Pollutant Removal

8.4.1.1 Narrowing of Bandgap and Enhancement of Light Absorption

Reducing the bandgap and enhancing light absorption are two important advantages of integrating adsorbents with semiconductor photocatalysts. When combined with other semiconductors, adsorbent materials may reduce the bandgap by acting as charge acceptors and photosensitizers. Covalent C–O–C bonds, which are necessary for cross-linking, are formed between the photocatalyst and oxygen-rich active sites on the adsorbent surface, such as -OH or -COOH groups (Li et al., 2013).

8.4.1.2 Improvement of Electron Transport and Charge Separation

Enhancing electron transport and charge separation, two processes essential to raising the overall effectiveness of pollutant removal, is another role of adsorbents in photocatalysis. Some adsorbents can modify the surface properties of the photocatalyst, promoting better interaction with reactants and facilitating charge separation. Additionally, certain adsorbents can create localized electron trapping sites below the photocatalyst conduction band, thereby preventing electron–hole recombination (Kusiak-Nejman et al., 2021).

8.4.1.3 Increase of Surface Area and Number of Active Sites

Photocatalysis is not the sole mechanism occurring in photoreactors; adsorption, particularly of organic compounds on the catalyst surface, plays a crucial role in enhancing photodegradation efficiency (Poblete et al., 2011; Hassan et al., 2016). Various studies have concentrated on surface modification strategies to increase the surface area and number of active sites, notably by coupling semiconductors with adsorbents to enhance photocatalytic efficiency (El Mouchtari et al., 2022).

8.4.1.4 Contaminant Accumulation and Photocatalytic Self-Regeneration

Due to their high specific surface area, adsorbents can accumulate substantial amounts of contaminants on their surfaces when integrated into photocatalytic systems. This not only increases the availability of pollutants for photocatalytic degradation but also facilitates the self-regeneration of the composite material's surface. Unlike conventional regeneration methods, which require the adsorbent to reach saturation followed by external treatment, photocatalytic systems offer a more sustainable and continuous regeneration process. For instance, the $ZnO/g\text{-}C_3N_4$ composite developed by Huang et al. effectively degrades adsorbed pyridine under UV irradiation, enabling in situ regeneration of the adsorbent, and after five cycles, the photocatalytic degradation efficiency still reaches 97.3% (Huang et al., 2024).

8.4.2 Preparation Methods of Adsorbent/Catalyst Composites

Various techniques can be used to fix semiconductor nanoparticles onto adsorbents to produce adsorbent–semiconductor hybrid materials. These attachment methods are commonly classified into three main categories (El-Sheikh et al., 2004): the sol–gel route, vapor deposition, and wet impregnation.

8.4.2.1 Sol–Gel Method

This is one of the most widespread techniques for synthesizing adsorbent–semiconductor materials (Gupta & Tripathi, 2012; Leary & Westwood, 2011). It involves using a semiconducting gel to impregnate the adsorbent particles. Gels are obtained by hydrolysis of an alkoxide either in an organic solvent (Lettmann et al., 2001; Zeng et al., 2013) or in excess water, as reported by Bosc et al. and Andronic et al., which acts as both solvent and reagent to reduce the energy costs associated with pressure conditions when using organic solvents (Andronic et al., 2014; Bosc et al., 2006).

8.4.2.2 Vapor Deposition

Initially, this technique was used in the chemical industry to deposit thin layers of compounds on support materials, starting from various precursors. Several authors have subsequently used it to deposit semiconductors on the surface of carbonaceous materials, including the adsorbent (Ma et al., 2014). The most widely used precursors are titanium alkoxides (isopropoxide, isobutoxide, ethoxide) and sometimes $TiCl_4$ (Li Puma et al., 2008). Despite the advantages noted by proponents of the technique, notably the speed of the process and the one-step operation, vapor deposition requires sophisticated equipment for regulated temperature and pressure. Furthermore, the low pressures and high temperatures applied to the system imply relatively high energy costs (Velasco et al., 2010).

8.4.2.3 Wet Impregnation

This involves the impregnation of the adsorbent with preformed semiconductor nanoparticles in aqueous (El-Sheikh et al., 2004) or alcoholic (Velasco et al., 2010) suspension. In practice, the adsorbent is introduced into the suspension, and the mixture is stirred to allow the infiltration of the solution into the porosity developed in the adsorbent. An evaporation step ensures that the semiconductor nanoparticles remain adhered to the surfaces of the adsorbent particles. A subsequent heat treatment (annealing) usually strengthens the bonds between the adsorbent and semiconductor phases.

The method has the advantage of being simple, and the photoactivity of the resulting hybrid material is virtually guaranteed (Araña et al., 2003). However, a common drawback is the relatively weak bonding forces between the adsorbent and semiconductor (Lim et al., 2011).

8.4.3 Composite Materials for Combined Adsorption and Photocatalysis

To successfully integrate photocatalysis and adsorption processes in wastewater treatment, different materials have been studied. AC is unique among these because of its high specific surface area and capacity to improve semiconductor performance. A bio-sourced composite material (AC/3D-TiO_2) was created by El Mountassir et al, by combining AC and three-dimensional TiO_2. This composite had a morphology that was favorable to photocatalytic activity, a high specific surface area, and a pure anatase phase. After 4 h of exposure to light, about 90% of the carbamazepine was degraded, indicating an efficiency that was 1.5 times higher than that of commercial TiO_2 P25 (El Mouchtari et al., 2022).

Advanced TiO_2/zeolite composites made by solid-state dispersion have also demonstrated promising findings, removing up to 93% of the malachite green dye under UV light (Imessaoudene et al., 2024).

Finally, metal–organic frameworks, like MIL-101(Fe)/TiO_2, have a high adsorption capacity and high stability under radiation, breaking down more than 95% of tetracycline and maintaining good reusability over five cycles (You et al., 2025).

8.5 CASE STUDY 1: TREATMENT OF PRETREATED MEDIOUNA LANDFILL LEACHATE USING TIO_2

This study investigates the combined treatment of landfill leachate from the Mediouna site (Morocco) through a sequential process involving coagulation–flocculation followed by heterogeneous photocatalysis with TiO_2-P25 under UV-A (300 W) irradiation. The efficiency of the photocatalytic step was evaluated by examining the effects of initial chemical oxygen demand (COD) concentration, pH, and catalyst mass on COD removal efficiency (Chaouki et al., 2025).

The analyzed leachate contains high levels of pollutants, as shown by its elevated COD (22,000 mg $O_2 \cdot L^{-1}$), BOD_5 (8,500 mg $O_2 \cdot L^{-1}$), turbidity (1,007 NTU), and total suspended solids (33,900 mg·L^{-1}). Given its pH (>7.5) and BOD_5/COD ratio (>0.1), the leachate is classified as an intermediate in age. Hence, physicochemical treatments are generally efficient in eliminating the refractory compounds (Babaei et al., 2021; Rookesh et al., 2022).

Additionally, the leachate exhibits high electrical conductivity, which is indicative of the presence of heavy metals and inorganic ions like Cl^- (5,236 mg·L^{-1}). The coexistence of metals and refractory organics raises the possibility of competitive interactions during treatment, which could affect the whole removal performance. This emphasizes the necessity of combined treatment to successfully manage leachate risks and conform to environmental discharge standards (Amigh & Mokhtarani, 2022).

Coagulation–flocculation with $FeCl_3$ was employed as a pretreatment to reduce this effluent's pollutant load. This pretreatment was able to reduce turbidity by up to 86% at an optimal dose of 12 g Fe^{3+} L^{-1}. Additionally, it removed up to 99% of $BOD_5$and reduced COD to 10,680 mg O^2L^{-1} (51.4%). Additionally, $UV_{(254)}$ absorbance, which indicates the presence of aromatic organics, dropped to 84%, showing partial removal of these compounds. The pH was adjusted from 7.6 to 5.8, closer to the optimal range for TiO_2 photocatalysis (Hassan et al., 2017).

Photodegradation experiments were performed in a reactor equipped with an Osram Ultra Vitalux UV-A lamp (300 W), a magnetic stirrer to homogenize the reaction mixture, and a double-jacketed glass reactor to maintain the solution temperature at 25°C during irradiation.

An adsorption step is conducted in the dark before the start of UV-A irradiation. This step improves the efficiency of the photocatalysis process by increasing the proximity of the pollutant to the catalyst surface. After 60 min of contact between the effluent and TiO_2P25, equilibrium is reached, according to the analysis of the adsorption kinetics, resulting in a 21% COD removal rate. These findings, which point to advantageous surface characteristics and active site availability, are consistent with

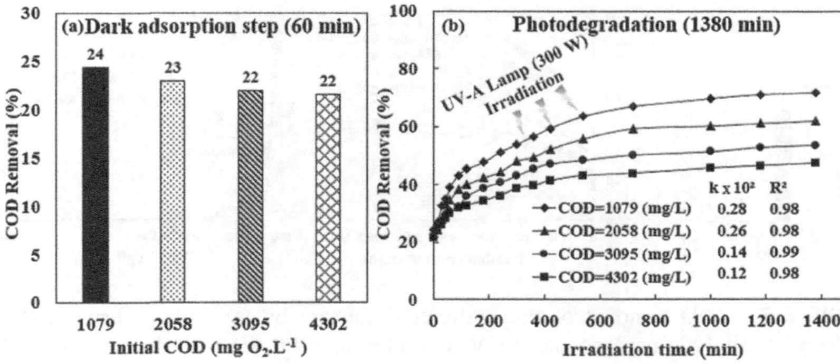

FIGURE 8.4 (a) COD removal of the pretreated leachate by (a) dark adsorption (60 min) and (b) photocatalytic process at different initial COD concentrations under UV-A irradiation (300 W). Operating conditions: m(TiO_2) = 1 g L^{-1}, pH = 5.8, T = 25°C.

a pseudo-first-order model. Improved photocatalytic performance following dark pre-adsorption was also reported by Jiménez-Relinque et al. (2022).

In the first photocatalysis test, the effect of initial COD concentration was evaluated by varying it between 1,079 and 4,300 mg O_2·L^{-1} under UV-A (1 g·L^{-1} TiO_2, pH = 5.8). As shown in Figure 8.4a, the adsorption stage (60 min) reduced COD by 21%–24%, depending on the initial concentration. Under UV-A irradiation, COD removal reached a maximum of 72% and 67% after 1,440 min for samples initially containing 1,079 and 2,058 mg O_2·L^{-1}, respectively. This corresponds to residual COD levels of 304 and 687 mg O_2·L^{-1}, close to discharge standards. However, for higher initial COD_0 levels (3,095 and 4,302 mg O_2·L^{-1}), removal efficiencies decreased to 54% and 44%, with residual levels of 1,453 and 2,264 mg O_2·L^{-1}. As the concentration of organic matter in the leachate increases, the apparent photocatalytic degradation rate constant (k) decreases from 0.0028 to 0.0012 min^{-1} for initial COD_0 values of 1,079–4,302 mg O_2·L^{-1}, respectively, resulting in a reduction in treatment efficiency. This phenomenon can be attributed to the limited availability of active sites due to adsorbed organics and light attenuation by the leachate's dark color, which reduces TiO_2 activation (Hassan et al., 2016).

The second parameter examined in this study is the effect of photocatalyst dose on the degradation of pretreated leachate with an initial COD_0 of 2,058 mg O_2·L^{-1}. The experimental data presented in Figure 8.5a and b indicate that as the TiO_2 dosage increased, COD removal efficiency improved, reaching 67% at a catalyst loading of 1 g·L^{-1} with an apparent rate constant of k = 0.0026 min^{-1} under UV-A irradiation. However, beyond this optimal dose, a decrease in performance was observed, with COD removal dropping to 47% at 2 g·L^{-1}, corresponding to a lower kinetic constant of k = 0.0012 min^{-1}. This decline can be attributed to excessive catalyst loading, which leads to light scattering, agglomeration of TiO_2 particles, and shielding of active sites, thereby limiting the penetration of photons into the suspension and reducing photocatalytic efficiency (Miao et al., 2019). Consequently, the optimal catalyst dose for the photocatalytic degradation of the pretreated leachate is 1 g·L^{-1} under the studied conditions.

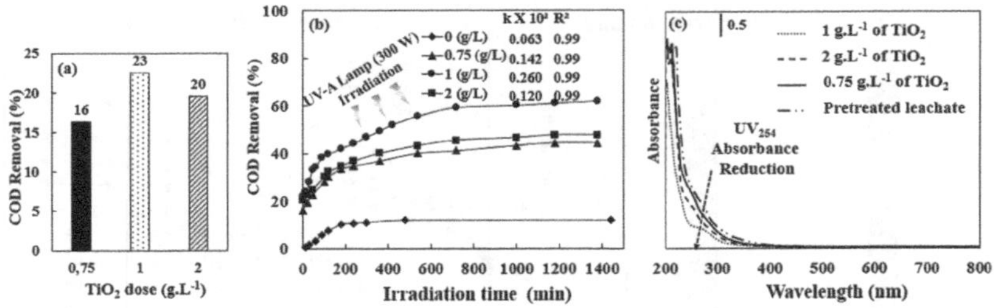

FIGURE 8.5 COD removal of the pretreated leachate by (a) dark adsorption (60 min) and (b) photocatalytic process under UV-A irradiation (300 W) with different TiO_2 doses; (c) UV-Vis spectra of $FeCl_3$-pretreated leachate before and after photocatalytic treatment. Operating conditions: $COD_0 = 2{,}058$ mg $O_2{\cdot}L^{-1}$, pH = 5.8, T = 25°C.

UV-Vis spectrophotometry further confirmed these results by analyzing the UV-Vis absorbance spectra (200–700 nm) of the pretreated leachate before and after treatment (Figure 8.5c). The results indicate a marked decrease in absorbance with increasing TiO_2 dose, with a significant reduction of UV_{254} observed at an optimal dose of 1 $g{\cdot}L^{-1}$.

The influence of pH was also investigated over a range of 2 to 10, using a TiO_2 dose of 1 $g{\cdot}L^{-1}$ and pretreated leachate with an initial COD_0 of 2,058 mg $O_2{\cdot}L^{-1}$. The results, presented in Figure 8.6a and b, show maximum degradation at pH 4, with a COD removal of 73% and an apparent kinetic constant of 0.0034 min^{-1} under UV-A. At this pH, the TiO_2 surface is positively charged, which enhances the attraction of negatively charged organic compounds. In contrast, COD removal decreased significantly under alkaline conditions, reaching only 37% at pH 10.

UV-Vis absorbance spectra depicted in Figure 8.6c further confirmed these findings, showing a greater decrease in absorbance at acidic pH, particularly at 254 nm (UV_{254}), which indicates an enhanced degradation of aromatic and refractory organics in agreement with the COD removal trends.

Despite promising results, this study presents some limitations. The treatment process consists of two separate processes: coagulation–flocculation and photocatalysis. Each step requires specific conditions and separate handling. In practical

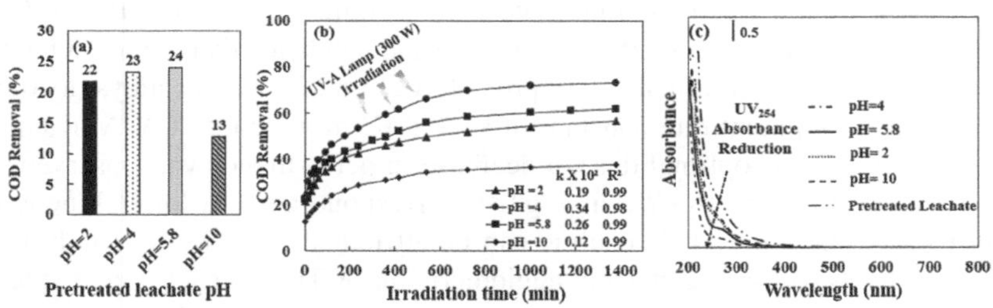

FIGURE 8.6 COD removal of the pretreated leachate by (a) dark adsorption (60 min) and (b) photocatalytic process under UV-A irradiation (300 W) at different pH values; (c) UV-Vis spectra of $FeCl_3$-pretreated leachate before and after photocatalytic treatment. Operating conditions: $m(TiO_2) = 1$ g L^{-1}, $COD_0 = 2{,}058$ mgO_2L^{-1}, T = 25°C.

applications, this setup may be complex and costly, particularly due to catalyst recovery and equipment scale-up challenges in batch mode.

To simplify and optimize the process, a promising solution is the development of hybrid materials that combine an adsorbent and a photocatalyst (e.g., AC-TiO_2). These multifunctional materials allow both adsorption and photocatalytic degradation to occur simultaneously, using a single reactor and a unified treatment system. This integrated approach reduces equipment needs, facilitates continuous operation, and can improve pollutant removal through synergistic effects. Future work should focus on these hybrid systems, along with immobilization techniques and cost-performance evaluations, to support more efficient and sustainable real-world applications.

8.6 CASE STUDY 2: DEGRADATION OF PHARMACEUTICAL POLLUTANTS USING THE AC/GO/TIO_2 PHOTOCATALYST

In this study, a hybrid material combining AC, GO, and TiO_2 was developed to remove SMT, a commonly used antibiotic that persists in aquatic environments. Three photocatalyst composites were synthesized and compared: pure TiO_2, AC/GO/TiO_2 prepared at neutral pH, and AC/GO/TiO_2 synthesized under acidic conditions to enhance interaction and dispersion. The photocatalysts were prepared through a multi-step process: AC was first functionalized with GO and then combined with TiO_2 obtained via the sol–gel method, with the final composite synthesis performed under controlled pH conditions to optimize surface and electronic properties (Prado Mendonça et al., 2025).

SMT solutions with an initial concentration of 10 mg L^{-1} were used to simulate contaminated effluents. The treatment process involved exposing the photocatalyst to UV light in a batch photoreactor under constant stirring and controlled temperature. Material characterization was conducted using scanning electron microscopy with energy-dispersive X-ray spectroscopy (SEM-EDS), X-ray diffraction (XRD), Brunauer–Emmett–Teller (BET), ultraviolet–visible diffuse reflectance spectroscopy, and X-ray photoelectron spectroscopy (XPS) to evaluate structure and composition. Photodegradation experiments were performed in triplicate, and photocatalyst stability was assessed over multiple cycles. This approach explores the synergy between adsorption and photocatalysis to develop a sustainable and efficient solution for pharmaceutical removal from water.

The acid-treated composite, AC/GO/TiO_2(acid), demonstrated superior physicochemical properties compared to the untreated version. Synthesized at pH 4, it showed better TiO_2 dispersion on the AC/GO surface, with a Ti content of 35% versus 23% for AC/GO/TiO_2. This improvement is attributed to stronger electrostatic interactions at acidic pH. XRD confirmed the exclusive presence of the anatase phase, with similar crystallite sizes (~13.7–13.9 nm), ideal for photocatalysis. Raman spectra indicated more organized D and G bands in the acid-treated composite. BET analysis showed that TiO_2, AC/GO/TiO_2, and AC/GO/TiO_2(acid) had surface areas of 110, 160, and 170 $m^2 g^{-1}$, respectively, with AC/GO/TiO_2(acid) having the highest specific surface area, supporting enhanced adsorption and reactivity. XPS analysis confirmed stronger Ti–O–C interactions and more functional groups. With a band gap of 3.46 eV, it remains active under UV light.

The photocatalytic performance of the synthesized materials was clearly demonstrated in Figure 8.7. The AC/GO/TiO_2(acid) composite showed superior SMT adsorption, removing 85% of the pollutant within the first 40 min, significantly outperforming both pure TiO_2 (no adsorption) and AC/GO/TiO_2 prepared at neutral pH (65% adsorption). This improvement is closely linked to the larger BET surface area (170 $m^2\ g^{-1}$) and enhanced TiO_2 dispersion on the AC/GO matrix under acidic conditions, which was confirmed by SEM-EDS and XPS analyses. The high surface area promotes increased pollutant adsorption, thereby making more contaminant molecules available at reactive sites for photodegradation.

As illustrated in Figure 8.7b, photocatalytic degradation over 80 min further confirmed the enhanced activity of the composites. AC/GO/TiO_2(acid) achieved 95% degradation efficiency of SMT, surpassing AC/GO/TiO_2 (78%) and pure TiO_2 (53%). The degradation kinetics (Figure 8.7c) fitted well with a pseudo-first-order model, with the rate constant (k) for AC/GO/TiO_2(acid) calculated at 0.019 min^{-1}, which was 1.6 times higher than that of AC/GO/TiO_2 and more than three times higher than pure TiO_2, highlighting the impact of acidic functionalization on photocatalytic performance.

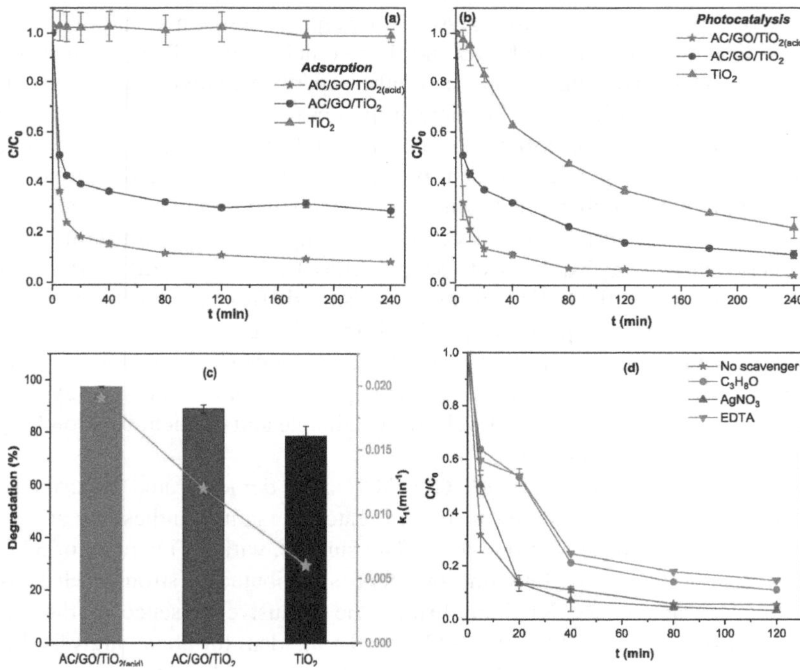

FIGURE 8.7 (a) Adsorption degradation of SMT, (b) photocatalytic degradation of SMT during UV light degradation, and (c) photocatalytic degradation (120 min), which correlates with the pseudo-first-order kinetic constant. Conditions: 100 mL of solution in [SMT] = 10 mg L^{-1}, 26°C ± 1, 100 mg of photocatalyst. (d) The performance of the AC/GO/$TiO_{2(acid)}$ sample was evaluated with various scavengers. (Copyright Prado Mendonça et al. (2025), with permission from Elsevier.)

Radical scavenger tests (Figure 8.7d) revealed that photogenerated holes (h^+) and hydroxyl radicals (OH) are the main reactive species responsible for SMT degradation. ethylenediaminetetraacetic acid, a hole scavenger, significantly suppressed degradation, confirming the key role of h^+ in the process. The addition of isopropyl alcohol, which reduces •OH radicals, also decreases degradation rates, while silver nitrate, an electron scavenger, caused a moderate inhibition, indicating the involvement of electrons in forming reactive oxygen species.

The reusability of the photocatalysts was assessed over four cycles (Figure 8.8), where the adsorption capacity saturated after two cycles due to active site occupancy, but photocatalytic degradation remained effective throughout all cycles, especially for AC/GO/TiO_2(acid). This sustained photocatalytic activity underscores the stability and robustness of the composite under repeated use.

Structural stability was confirmed by XRD patterns before and after cycling, showing that the anatase crystalline phase remained intact (Figure 8.8d). SEM-EDS analysis further demonstrated the consistent Ti content (~35%) after multiple cycles, indicating no significant leaching or degradation of the photocatalyst.

Although the surface area of AC/GO/TiO_2(acid) decreased from 170 to 80 $m^2\ g^{-1}$ after repeated adsorption cycles, likely due to pore blockage by SMT molecules, the photocatalytic efficiency was not compromised. This is attributed to the synergistic effect, where adsorption concentrates SMT near active sites and photocatalysis continuously degrades it.

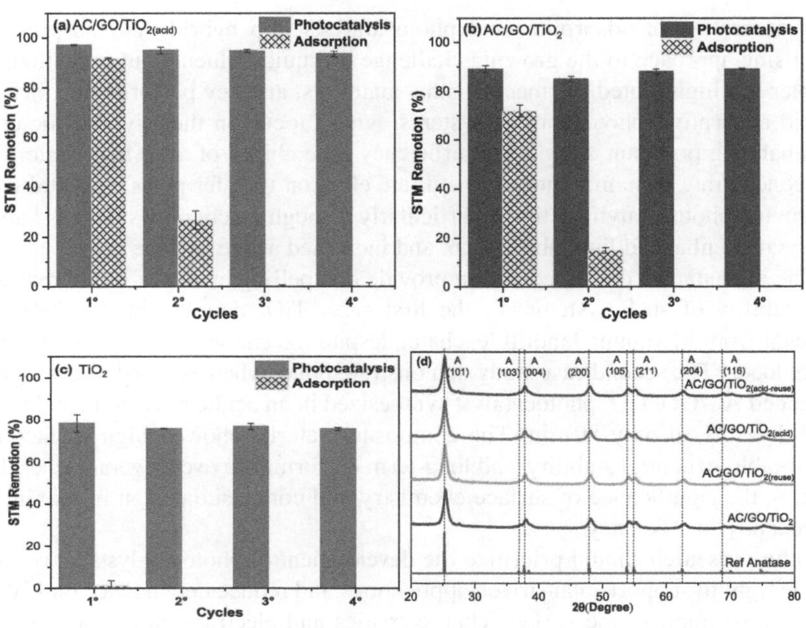

FIGURE 8.8 Photocatalytic stability of (a) AC/GO/$TiO_{2(acid)}$, (b) AC/GO/TiO_2, and (c) TiO_2. (d) XRD patterns of AC/GO/$TiO_{2(acid)}$ and AC/GO/TiO_2 photocatalysts after four cycles. (Copyright Prado Mendonça et al. (2025) with permission from Elsevier.)

Overall, these results demonstrate that the acid-treated AC/GO/TiO_2 composite is a highly effective and durable photocatalyst, combining rapid SMT removal with excellent structural and chemical stability, making it promising for sustainable water treatment applications.

This study successfully developed and characterized an AC/GO/TiO_2 photocatalyst synthesized under acidic conditions, achieving rapid and efficient degradation of SMT (95% in 80 min). The key to this performance lies in the enhanced TiO_2 dispersion, high surface area, and effective electron transfer facilitated by the GO functional groups and acidic treatment. The material's stability after multiple photocatalysis cycles, confirmed by structural and elemental analysis, underscores its durability and suitability for repeated use in real-world wastewater treatment.

Scaling up synthesis while preserving homogeneous dispersion and functionalization still presents difficulties. Furthermore, surface saturation during adsorption indicates that continuous flow systems or regeneration techniques need to be optimized in order to avoid pore blockage. To improve sustainability and energy efficiency, future research could examine the photocatalyst's performance in the presence of visible light or solar radiation. Perspectives include integrating it into hybrid wastewater treatment systems combining adsorption, photocatalysis, and biological processes, in addition to investigating regeneration techniques to restore adsorption capacity, which could further enhance reusability.

8.7 CONCLUSION AND PERSPECTIVES

The integration of adsorption and photocatalysis into hybrid systems presents a promising approach to the growing challenge of liquid effluent contamination. This chapter has highlighted the mechanisms, materials, and key performance factors of hybrid adsorptive–photocatalytic systems, with a focus on their synergistic effects in enhancing pollutant degradation efficiency. The ability of adsorbent materials to pre-concentrate contaminants and facilitate electron transfer plays a pivotal role in improving photocatalytic activity, particularly through mechanisms such as bandgap narrowing, enhanced light absorption, and increased active surface area.

The case studies discussed herein provide compelling evidence of the real-world applicability of such systems. In the first case, TiO_2 demonstrated a 73% COD removal from Mediouna landfill leachate, despite the challenges posed by high pollutant loads. The second case study demonstrated that, when exposed to UV light, an advanced AC/GO/TiO_2 photocatalyst synthesized in an acidic environment degraded SMT by 95% in only 80 min. The composite material showed high surface area, remarkable structural stability, and long-term performance over several cycles, highlighting the significance of surface chemistry and composite design in maximizing photocatalytic efficiency.

Future research should prioritize the development of photocatalysts active under visible light to support solar-driven applications and reduce dependence on UV light sources. To improve the surface characteristics and electrical properties of hybrid composites, more research should be done on doping methods, band structure engineering, and effective synthesis techniques like sol–gel and hydrothermal methods.

Scaling up hybrid systems necessitates consideration of fouling resistance, catalyst recovery, and cost-effective regeneration methods. Furthermore, to narrow the gap between laboratory performance and field deployment, long-term performance under actual wastewater matrixes, which include competing ions and variable pH, must be evaluated.

Lastly, this chapter's insights highlight the potential of hybrid adsorptive–photocatalytic systems as essential methods of a resilient and circular water economy, in addition to their potential as efficient treatment technologies.

REFERENCES

Abd El-Magied, M. O., Rezk, M. M., Youssef, M. A., Eldesouky, E. M., Abd El-Hamid, A. A. M., Morse, W. M., Hafez, W. S., & Manaa, E. S. A. (2025). Utilizing modified glycidyl methacrylate resins to remove chromium (VI) species from aqueous solutions. *Journal of Dispersion Science and Technology*. https://doi.org/10.1080/01932691.2025.2502874

Absalan, Y., Asadinamin, M., Minko, S., & Sharma, S. (2025). Key parameters in photocatalyst design for effective textile dye degradation through designing MOF. *Comments on Inorganic Chemistry*. https://doi.org/10.1080/02603594.2025.2490643

Ahmadi, S., & Ganjidoust, H. (2021). Using banana peel waste to synthesize BPAC/ZnO nanocomposite for photocatalytic degradation of Acid Blue 25: Influential parameters, mineralization, biodegradability studies. *Journal of Environmental Chemical Engineering*, 9(5), 106010. https://doi.org/10.1016/J.JECE.2021.106010

Ahmadi, S., Quimbayo, J. M., Yaah, V. B. K., de Oliveira, S. B., & Ojala, S. (2025). A critical review on combining adsorption and photocatalysis in composite materials for pharmaceutical removal: Pros and cons, scalability, TRL, and sustainability. *Energy Nexus*, 17, 100396. https://doi.org/10.1016/J.NEXUS.2025.100396

Allen, S. J., McKay, G., & Khader, K. Y. H. (1989). Intraparticle diffusion of a basic dye during adsorption onto sphagnum peat. *Environmental Pollution*, 56(1), 39–50. https://doi.org/10.1016/0269-7491(89)90120-6

Amigh, P., & Mokhtarani, N. (2022). Leachate post treatment, using Ag-TiO_2 nanoparticles immobilized on rotating vanes. *Journal of Water Process Engineering*, 47, 102842. https://doi.org/10.1016/J.JWPE.2022.102842

Andronic, L., Enesca, A., Cazan, C., & Visa, M. (2014). TiO2-active carbon composites for wastewater photocatalysis. *Journal of Sol-Gel Science and Technology*, 71(3), 396–405. https://doi.org/10.1007/S10971-014-3393-6

Araña, J., Doña-Rodríguez, J. M., Tello Rendón, E., Garriga I Cabo, C., González-Díaz, O., Herrera-Melián, J. A., Pérez-Peña, J., Colón, G., & Navío, J. A. (2003). TiO_2 activation by using activated carbon as a support: Part I. Surface characterisation and decantability study. *Applied Catalysis B: Environmental*, 44(2), 161–172. https://doi.org/10.1016/S0926-3373(03)00107-3

Ba-Abbad, M. M., Kadhum, A. A. H., Mohamad, A. B., Takriff, M. S., & Sopian, K. (2012). Synthesis and catalytic activity of TiO_2 nanoparticles for photochemical oxidation of concentrated chlorophenols under direct solar radiation. *International Journal of Electrochemical Science*, 7(6), 4871–4888. https://doi.org/10.1016/s1452-3981(23)19588-5

Babaei, S., Sabour, M. R., & Moftakhari Anasori Movahed, S. (2021). Combined landfill leachate treatment methods: An overview. *Environmental Science and Pollution Research*, 28(42), 59594–59607. https://doi.org/10.1007/S11356-021-16358-0

Bahrudin, N. N., & Nawi, M. A. (2018). Immobilized titanium dioxide/powdered activated carbon system for the photocatalytic adsorptive removal of phenol. *Korean Journal of Chemical Engineering*, 35(7), 1532–1541. https://doi.org/10.1007/S11814-018-0062-4

Belkodia, K., El mersly, L., Edaala, M. A., Achtak, H., Alaoui Tahiri, A., Briche, S., & Rafqah, S. (2024). Cheese wastewater treatment through combined coagulation-flocculation and photo-Fenton-like advanced oxidation processes for reuse in irrigation: Effect of operational parameters and phytotoxicity assessment. *Environmental Science and Pollution Research*, 31(8), 11801–11814. https://doi.org/10.1007/S11356-024-31828-X

Bencheqroun, Z., Chaouki, Z., Hadri, M., Nawdali, M., Draoui, K., Valdés, H., & Zaitan, H. (2018). Removal of textile dyes from aqueous solutions using low cost Moroccan clay. *IOP Conference Series: Earth and Environmental Science*, 161(1), 012009. https://doi.org/10.1088/1755-1315/161/1/012009

Błędzka, D., Gmurek, M., Olak-Kucharczyk, M., & Ledakowicz, S. (2012). *Photodegradation of n-butylparaben in natural water from Sulejow Reservoir*. https://www.researchgate.net/publication/235353101

Bora, L. V., & Mewada, R. K. (2017). Visible/solar light active photocatalysts for organic effluent treatment: Fundamentals, mechanisms and parametric review. *Renewable and Sustainable Energy Reviews*, 76, 1393–1421. https://doi.org/10.1016/J.RSER.2017.01.130

Bosc, F., Edwards, D., Keller, N., Keller, V., & Ayral, A. (2006). Mesoporous TiO_2-based photocatalysts for UV and visible light gas-phase toluene degradation. *Thin Solid Films*, 495(1–2), 272–279. https://doi.org/10.1016/J.TSF.2005.08.361

Byrne, C., Subramanian, G., & Pillai, S. C. (2018). Recent advances in photocatalysis for environmental applications. *Journal of Environmental Chemical Engineering*, 6(3), 3531–3555. https://doi.org/10.1016/J.JECE.2017.07.080

Calvet, R., Tercé, M., & Arvieu, J. C. (1980). Mise au point bibliographique - Adsorption des pesticides par les sols et leurs constituants. *Annales Agronomiques*, 31(4), 385–427. https://doi.org/10.34894/VQ1DJA

Chaouki, Z., El Mrabet, I., El Mouchtari, E. M., Bencheqroun, Z., Tanji, K., Nawdali, M., Rafqah, S., & Zaitan, H. (2025). Enhanced treatment of pretreated Mediouna landfill leachate (Morocco) using TiO_2 photocatalysis under UV-A and UV-C irradiation. *Water, Air, & Soil Pollution. Under Second Round of Revision*, *236*(13), 855. https://doi.org/10.1007/s11270-025-08334-y

Chaouki, Z., El Mrabet, I., Khalil, F., Ijjaali, M., Rafqah, S., Anouar, S., Nawdali, M., Valdés, H., & Zaitan, H. (2017). Use of coagulation-flocculation process for the treatment of the landfill leachates of Casablanca city (Morocco). *Journal of Materials and Environmental Science*, 8(8), 2781–2791.

Chen, F., Ma, L., Wang, R., Chai, Y., Zhan, Y., & Chen, C. (2025). Enhanced adsorption of organic matter from coal chemical membrane concentrate by applying positive potential to mesoporous activated carbon fiber. *Separation and Purification Technology*, 369, 133128. https://doi.org/10.1016/J.SEPPUR.2025.133128

Colakoglu, E. B., Uyanik, I., Elbir, H., Sahinkaya, E., & Yurtsever, A. (2025). A novel gravity-driven dynamic membrane filtration reactor for microplastic removal from plastic recycling facility wastewater. *Journal of Environmental Chemical Engineering*, 13(2), 115793. https://doi.org/10.1016/J.JECE.2025.115793

Danish, M. S. S., Estrella, L. L., Alemaida, I. M. A., Lisin, A., Moiseev, N., Ahmadi, M., Nazari, M., Wali, M., Zaheb, H., & Senjyu, T. (2021). Photocatalytic applications of metal oxides for sustainable environmental remediation. *Metals*, 11(1), 80. https://doi.org/10.3390/MET11010080

Das, D., & Sengupta, S. (2025). Modified zeolite in a sustainable, circular economy-based process for enhanced ammonium recovery from wastewater. *Cleaner Water*, 3, 100075. https://doi.org/10.1016/J.CLWAT.2025.100075

Das, S., & Ahn, Y. H. (2022). Synthesis and application of CdS nanorods for LED-based photocatalytic degradation of tetracycline antibiotic. *Chemosphere*, 291, 132870. https://doi.org/10.1016/j.chemosphere.2021.132870

Dhayal, M., Sharma, S., Jakhar, P., Sain, Y., & Sharma, H. (2025). Carbon-supported photocatalysts: A sustainable approach for enhanced environmental remediation. *Comments on Inorganic Chemistry*. https://doi.org/10.1080/02603594.2025.2494273

Dogra, I., Kumar, B. R., Etika, K. C., Chavali, M., Khalifa, A. S., Gharib, A. F., & Askary, A. El. (2022). Environmentally friendly low-cost graphene oxide-cellulose nanocomposite filter for dye removal from water. *Journal of King Saud University - Science*, 34(5), 102122. https://doi.org/10.1016/J.JKSUS.2022.102122

Du, C., Zhang, Z., Yu, G., Wu, H., Chen, H., Zhou, L., Zhang, Y., Su, Y., Tan, S., Yang, L., Song, J., & Wang, S. (2021). A review of metal organic framework (MOFs)-based materials for antibiotics removal via adsorption and photocatalysis. *Chemosphere*, 272, 129501. https://doi.org/10.1016/J.CHEMOSPHERE.2020.129501

Dubinin, M. M. (1965). Modern state of the theory of volume filling of micropore adsorbents during adsorption of gases and steams on carbon adsorbents. *Zhurnal Fizicheskoi Khimii*, 39, 1305–1317.

El Mouchtari, E. M., El Mersly, L., Jhabli, O., Anane, H., Piram, A., Briche, S., Wong-Wah-Chung, P., & Rafqah, S. (2022). Hydrothermal synthesis of 3D cauliflower anatase TiO_2 and bio sourced activated carbon: Adsorption and photocatalytic activity in real water matrices. *International Journal of Environmental Analytical Chemistry*. https://doi.org/10.1080/03067319.2022.2118586

El-Sheikh, A. H., Newman, A. P., Al-Daffaee, H., Phull, S., Cresswell, N., & York, S. (2004). Deposition of anatase on the surface of activated carbon. *Surface and Coatings Technology*, 187(2–3), 284–292. https://doi.org/10.1016/j.surfcoat.2004.03.012

Freundlich, H. (1906). Over the adsorption in solution. *The Journal of Physical Chemistry*, 57, 385–471.

Georgin, J. (2023). A review of the antibiotic ofloxacin: Current status of ecotoxicology and scientific advances in its removal from aqueous systems by adsorption technology. *Chemical Engineering Research and Design*, 193, 99–120. https://doi.org/10.1016/J.CHERD.2023.03.025

Georgin, J., Franco, D. S. P., Meili, L., Dehmani, Y., dos Reis, G. S., & Lima, E. C. (2023). Main advances and future prospects in the remediation of the antibiotic amoxicillin with a focus on adsorption technology: A critical review. *Journal of Water Process Engineering*, 56, 104407. https://doi.org/10.1016/J.JWPE.2023.104407

Gupta, S. M., & Tripathi, M. (2012). A review on the synthesis of TiO_2 nanoparticles by solution route. *Central European Journal of Chemistry*, 10(2), 279–294. https://doi.org/10.2478/S11532-011-0155-Y

Hadri, M., El Mrabet, I., Chaouki, Z., Draoui, K., Hamdaoui, M., Douhri, H., & Zaitan, H. (2022). Valorization of natural diatomite mineral: Application to removal of anionic dye from aqueous solution in a batch and fixed-bed reactor. *Journal of Central South University*, *29*(6), 2084–2098. https://doi.org/10.1007/s11771-022-5065-y

Hassan, M., Wang, X., Wang, F., Wu, D., Hussain, A., & Xie, B. (2017). Coupling ARB-based biological and photochemical (UV/TiO2 and UV/S2O82−) techniques to deal with sanitary landfill leachate. *Waste Management*, 63, 292–298. https://doi.org/10.1016/J.WASMAN.2016.09.003

Hassan, M., Zhao, Y., & Xie, B. (2016). Employing TiO_2 photocatalysis to deal with landfill leachate: Current status and development. *Chemical Engineering Journal*, 285, 264–275. https://doi.org/10.1016/J.CEJ.2015.09.093

Ho, Y. S., McKay, G., Wase, D. A. J., & Forster, C. F. (2000). Study of the sorption of divalent metal ions on to peat. *Adsorption Science and Technology*, 18(7), 639–650. https://doi.org/10.1260/0263617001493693

Huang, G., Liu, S., Tian, C., & Tao, Y. (2024). Construction of S scheme $ZnO/g\text{-}C_3N_4$ heterojunction for the removal of pyridine from coal chemical wastewater. *Optical Materials*, 150, 115288. https://doi.org/10.1016/J.OPTMAT.2024.115288

Imessaoudene, A., Mechraoui, O., Aberkane, B., Benabbas, A., Manseri, A., Moussaoui, Y., Bollinger, J. C., Amrane, A., Zoukel, A., & Mouni, L. (2024). Synthesis of a TiO_2/zeolite composite: Evaluation of adsorption-photodegradation synergy for the removal of Malachite Green. *Nano-Structures & Nano-Objects*, 38, 101191. https://doi.org/10.1016/J.NANOSO.2024.101191

Jia, G., Wang, G., Zhang, Y., & Zhang, L. (2010). Effects of light intensity and H_2O_2 on photocatalytic degradation of phenol in wastewater using TiO_2/ACF. *Proceedings -2010 International Conference on Digital Manufacturing and Automation, ICDMA 2010*, 1, 623–626. https://doi.org/10.1109/ICDMA.2010.431

Jiang, W., Cui, W. R., Liang, R. P., & Qiu, J. D. (2021). Zwitterionic surface charge regulation in ionic covalent organic nanosheets: Synergistic adsorption of fluoroquinolone antibiotics. *Chemical Engineering Journal*, 417, 128034. https://doi.org/10.1016/J.CEJ.2020.128034

Jimenez-Relinque, E., Lee, S. F., Plaza, L., & Castellote, M. (2022). Synergetic adsorption–photocatalysis process for water treatment using TiO_2 supported on waste stainless steel slag. *Environmental Science and Pollution Research*, 29(26), 39712–39722. https://doi.org/10.1007/S11356-022-18728-8

Karaman, R., Khamis, M., Abbadi, J., Amro, A., Qurie, M., Ayyad, I., Ayyash, F., Hamarsheh, O., Yaqmour, R., Nir, S., Bufo, S. A., Scrano, L., Lerman, S., Gur-Reznik, S., & Dosoretz, C. G. (2016). Paracetamol biodegradation by activated sludge and photocatalysis and its removal by a micelle–clay complex, activated charcoal, and reverse osmosis membranes. *Environmental Technology*, 37(19), 2414–2427. https://doi.org/10.1080/09593330.2016.1150355

Khader, E. H., Mohammed, T. J., Albayati, T. M., Saady, N. M. C., & Zendehboudi, S. (2024). Green nanocatalyst for the photocatalytic degradation of organic pollutants in petroleum refinery wastewater: Synthesis, characterization, and optimization. *Journal of Molecular Structure*, *1304*, 137688. https://doi.org/10.1016/J.MOLSTRUC.2024.137688

Khalaf, A. S., Shnain, Z. Y., Al-Bayati, A. D. J., Majdi, H. S., Abid, M. F., Bilal, A. A., Alsalim, A. A., Abdrahman, N. H., & Alwasiti, A. A. (2025). Advancements in multifunctional nanomaterials for synergistic photocatalytic and adsorptive water treatment processes. *Results in Chemistry*, 15, 102292. https://doi.org/10.1016/J.RECHEM.2025.102292

Kusiak-Nejman, E., Sienkiewicz, A., Wanag, A., Rokicka-Konieczna, P., & Morawski, A. W. (2021). The role of adsorption in the photocatalytic decomposition of dyes on APTES-modified TiO_2 nanomaterials. *Catalysts*, 11(2), 172. https://doi.org/10.3390/CATAL11020172

Kyzas, G. Z., Favvas, E. P., Kostoglou, M., & Mitropoulos, A. C. (2020). Effect of agitation on batch adsorption process facilitated by using nanobubbles. *Colloids and Surfaces A: Physicochemical and Engineering Aspects*, 607, 125440. https://doi.org/10.1016/J.COLSURFA.2020.125440

Lagergren, S. (1898). About the theory of so-called adsorption of soluble substances. *K. Sven Vetensk.Akad. Handl*, 24, 1–39.

Langmuir, I. (1918). The adsorption of gases on plane surfaces of glass, mica and platinum. *Journal of the American Chemical Society*, 40(9), 1361–1403. https://doi.org/10.1021/JA02242A004

Leary, R., & Westwood, A. (2011). Carbonaceous nanomaterials for the enhancement of TiO_2 photocatalysis. *Carbon*, 49(3), 741–772. https://doi.org/10.1016/J.CARBON.2010.10.010

Lettmann, C., Hildenbrand, K., Kisch, H., Macyk, W., & Maier, W. F. (2001). Visible light photodegradation of 4-chlorophenol with a coke-containing titanium dioxide photocatalyst. *Applied Catalysis B: Environmental*, 32(4), 215–227. https://doi.org/10.1016/S0926-3373(01)00141-2

Li, Y., Zhang, H., Liu, P., Wang, D., Li, Y., & Zhao, H. (2013). Cross-linked g-C_3N_4/rGO nanocomposites with tunable band structure and enhanced visible light photocatalytic activity. *Small*, 9(19), 3336–3344. https://doi.org/10.1002/SMLL.201203135

Li Puma, G., Bono, A., Krishnaiah, D., & Collin, J. G. (2008). Preparation of titanium dioxide photocatalyst loaded onto activated carbon support using chemical vapor deposition: A review paper. *Journal of Hazardous Materials*, 157(2–3), 209–219. https://doi.org/10.1016/J.JHAZMAT.2008.01.040

Liaquat, I., Munir, R., Abbasi, N. A., Sadia, B., Muneer, A., Younas, F., Sardar, M. F., Zahid, M., & Noreen, S. (2024). Exploring zeolite-based composites in adsorption and photocatalysis for toxic wastewater treatment: Preparation, mechanisms, and future perspectives. *Environmental Pollution*, 349, 123922. https://doi.org/10.1016/J.ENVPOL.2024.123922

Lim, T. T., Yap, P. S., Srinivasan, M., & Fane, A. G. (2011). TiO_2/AC composites for synergistic adsorption-photocatalysis processes: Present challenges and further developments for water treatment and reclamation. *Critical Reviews in Environmental Science and Technology*, 41(13), 1173–1230. https://doi.org/10.1080/10643380903488664

Lin, Z., Fan, L., Jiang, L., & Huang, W. (2025). Synthesis of poly(diphenylbutadiyne) polyamide for efficient photocatalytic degradation of organic dyes. *Materials Research Innovations*. https://doi.org/10.1080/14328917.2025.2495589

Lu, Y., Jiang, R., & Zhao, Y. (2025). Self-Powered advanced oxidation processes for removing contaminants from wastewater. *Chemical Engineering Journal*, 510, 161443. https://doi.org/10.1016/J.CEJ.2025.161443

Ma, L., Chen, A., Lu, J., Zhang, Z., He, H., & Li, C. (2014). In situ synthesis of CNTs/Fe–Ni/TiO_2 nanocomposite by fluidized bed chemical vapor deposition and the synergistic effect in photocatalysis. *Particuology*, 14, 24–32. https://doi.org/10.1016/J.PARTIC.2013.04.002

Ma, X., Hong, B., Hu, A., Huang, K., Haleem, A., Cao, X., & Zhong, H. (2025). Integration of adsorption and advanced oxidation catalysis: Materials, synergistic mechanisms, and applications. *Critical Reviews in Environmental Science and Technology*. https://doi.org/10.1080/10643389.2025.2496389

Martín-Sómer, M., Pablos, C., van Grieken, R., & Marugán, J. (2017). Influence of light distribution on the performance of photocatalytic reactors: LED vs mercury lamps. *Applied Catalysis B: Environmental*, 215, 1–7. https://doi.org/10.1016/j.apcatb.2017.05.048

Miao, Z., Wang, G., Li, L., Wang, C., & Zhang, X. (2019). Fabrication of black TiO_2/TiO_2 homojunction for enhanced photocatalytic degradation. *Journal of Materials Science*, 54(23), 14320–14329. https://doi.org/10.1007/S10853-019-03900-2

Mohanty, K., Das, D., & Biswas, M. N. (2005). Adsorption of phenol from aqueous solutions using activated carbons prepared from Tectona grandis sawdust by $ZnCl_2$ activation. *Chemical Engineering Journal*, 115(1–2), 121–131. https://doi.org/10.1016/J.CEJ.2005.09.016

Moles, S., Berges, J., Ormad, M. P., Nieto-Monge, M. J., Gómez, J., & Mosteo, R. (2021). Photoactivation and photoregeneration of TiO_2/PAC mixture applied in suspension in water treatments: Approach to a real application. *Environmental Science and Pollution Research*, 28(19), 24167–24179. https://doi.org/10.1007/S11356-021-12542-4

Moradi, O., & Panahandeh, S. (2022). Fabrication of different adsorbents based on zirconium oxide, graphene oxide, and dextrin for removal of green malachite dye from aqueous solutions. *Environmental Research*, 214, 114042. https://doi.org/10.1016/J.ENVRES.2022.114042

Paredi, P., Tsunoji, N., Das, S., & Bandyopadhyay, M. (2025). The copper complex-based mesoporous material as an efficient, cost-effective, environment friendly sustainable adsorbent for wastewater pollutants. *Hybrid Advances*, 10, 100441. https://doi.org/10.1016/J.HYBADV.2025.100441

Pattnaik, A., Sahu, J. N., Poonia, A. K., & Ghosh, P. (2023). Current perspective of nano-engineered metal oxide based photocatalysts in advanced oxidation processes for degradation of organic pollutants in wastewater. *Chemical Engineering Research and Design*, 190, 667–686. https://doi.org/10.1016/J.CHERD.2023.01.014

Pimentel, C. H., Castro-Agra, R., Freire, M. S., Gómez-Díaz, D., & González-Álvarez, J. (2024). Adsorption of anionic wood dyes on KOH-activated carbons from Pinus radiata sawdust. *Biomass Conversion and Biorefinery*, 15(5), 7603–7622. https://doi.org/10.1007/S13399-024-05587-1

Poblete, R., Otal, E., Vilches, L. F., Vale, J., & Fernández-Pereira, C. (2011). Photocatalytic degradation of humic acids and landfill leachate using a solid industrial by-product containing TiO_2 and Fe. *Applied Catalysis B:* Environmental, 102(1–2), 172–179. https://doi.org/10.1016/J.APCATB.2010.11.039

Prado Mendonça, T. A., Giroto, A. S., Chambi, J. T., Cuffini, S. L., Silva Vieira, N. C., & Gonçalves, M. (2025). Efficient photocatalyst based on activated carbon/graphene oxide/TiO_2 synthesized under acidic conditions for environmental remediation. *Journal of Photochemistry and Photobiology A: Chemistry*, 462, 116244. https://doi.org/10.1016/J.JPHOTOCHEM.2024.116244

Qi, K., Wang, Z., Xie, X., & Wang, Z. (2023). Photocatalytic performance of pyrochar and hydrochar in heterojunction photocatalyst for organic pollutants degradation: Activity comparison and mechanism insight. *Chemical Engineering Journal*, 467, 143424. https://doi.org/10.1016/J.CEJ.2023.143424

Qiu, B., Shao, Q., Shi, J., Yang, C., & Chu, H. (2022). Application of biochar for the adsorption of organic pollutants from wastewater: Modification strategies, mechanisms and challenges. *Separation and Purification Technology*, 300, 121925. https://doi.org/10.1016/J.SEPPUR.2022.121925

Roginsky, S., & Zeldovich, Y. B. (1934). The catalytic oxidation of carbon monoxide on manganese dioxide. *Acta Phys. Chem. USSR*, *1*(554), 2019.

Rookesh, T., Samaei, M. R., Yousefinejad, S., Hashemi, H., Derakhshan, Z., Abbasi, F., Jalili, M., Giannakis, S., & Bilal, M. (2022). Investigating the electrocoagulation treatment of landfill leachate by iron/graphite electrodes: Process parameters and efficacy assessment. *Water*, 14(2), 205. https://doi.org/10.3390/W14020205

UN-Water. (2024). *World Water Development Report*. https://www.unwater.org/publications/un-world-water-development-report-2024

Velasco, L. F., Tsyntsarski, B., Petrova, B., Budinova, T., Petrov, N., Parra, J. B., & Ania, C. O. (2010). Carbon foams as catalyst supports for phenol photodegradation. *Journal of Hazardous Materials*, 184(1–3), 843–848. https://doi.org/10.1016/J.JHAZMAT.2010.08.118

Wang, C., Tian, Y., Yu, Y., Wang, L., Qu, G., & Yang, S. (2025). Enhancing wastewater treatment efficiency with stereoscopic hydrogel evaporators and renewable energy integration for sustainability. *Chinese Journal of Chemical Engineering*, 80, 38–46. https://doi.org/10.1016/J.CJCHE.2024.12.009

Wang, J., & Guo, X. (2020). Adsorption kinetic models: Physical meanings, applications, and solving methods. *Journal of Hazardous Materials*, 390, 122156. https://doi.org/10.1016/J.JHAZMAT.2020.122156

Wang, J., & Zhuan, R. (2020). Degradation of antibiotics by advanced oxidation processes: An overview. *The Science of the Total Environment*, 701. https://doi.org/10.1016/J.SCITOTENV.2019.135023

Yang, H., Zhang, K., Shi, R., & Tang, A. (2007). Sol-gel synthesis and photocatalytic activity of CeO_2/TiO_2 nanocomposites. *Journal of the American Ceramic Society*, 90(5), 1370–1374. https://doi.org/10.1111/j.1551-2916.2007.01540.x

You, M., Tang, Y., Liang, X., Deng, T., Peng, H., Tian, J., & Du, J. (2025). Efficient photocatalytic degradation of tetracycline hydrochloride by Fe-TiO_2/MIL-101(Cr) nanocomposites under visible light. *Journal of Water Process Engineering*, 69, 106590. https://doi.org/10.1016/J.JWPE.2024.106590

Zeng, M. X., Li, Y. J., Ma, M. Y., Chen, W., & Li, L. Y. (2013). Photocatalytic activity and kinetics for acid yellow degradation over surface composites of TiO_2-coated activated carbon under different photocatalytic conditions. *Transactions of Nonferrous Metals Society of China*, 23(4), 1019–1027. https://doi.org/10.1016/S1003-6326(13)62561-3

9 Engineering Innovations in Photocatalytic Systems for Treating Liquid Effluents

A Case Study on ZnO:Co Thin Films

Karim Tanji, Ikram Daou, Mohammad Saood Manzar, Abdelghani Hsini, Abdelali El Gaidoumi, Imane El Mrabet, Zineb Chaouki, Otmane Boudouch, Abdelillah Shaim, and Hicham Zaitan

9.1 INTRODUCTION

The escalating global demand for clean water, driven by population growth, unsustainable resource extraction, and rapid industrialization, has intensified pressure on aquatic ecosystems (Naciri et al., 2022). Access to safe water is fundamental to human health and ecological stability, yet contamination from natural and anthropogenic sources threatens water security worldwide (Padmavathy & Vijayaraghavan, 2008; González-González et al., 2022). Among the most pervasive pollutants are contaminants of emerging concern, including dyes, pharmaceuticals, personal care products, pesticides, and industrial chemicals, which evade conventional treatment systems and persist in the environment (Wang et al., 2019). Industrial effluents, containing dyes, phenols, and hydrocarbons, and agricultural runoff (pesticides, fertilizers) introduce persistent, toxic compounds into water bodies. These pollutants often resist biodegradation due to complex aromatic structures, posing long-term risks to human and environmental health (Ojha, Tripathi, Vishwakarma, & Mishra, 2024). Scientists and engineers have turned to more efficient and innovative strategies and approaches to address the growing issues of water pollution, particularly the presence of persistent toxic contaminants that withstand conventional treatment processes. Lagoon systems, electrocoagulation, adsorption, and biosorption remain widely used

DOI: 10.1201/9781003624905-9

as conventional treatment processes but suffer from critical drawbacks (Behera, Patel, & Parikh, 2023). While effective for bulk pollutant removal, these methods often transfer contaminants from water to solid phases (e.g., sludge), creating secondary waste streams that require further disposal or treatment (Tatarchuk, Soltys, & Macyk, 2023). For instance, activated carbon adsorption merely concentrates pollutants rather than degrading them, while biological treatments fail to eliminate refractory organic compounds (Mohod et al., 2023). However, advanced oxidation processes (AOPs) are gaining recognition as promising alternatives for the degradation of a wide range of non-biodegradable pollutants. AOPs occur through the in-situ production of highly reactive oxygen species (ROS), principally hydroxyl radicals ($^{\bullet}OH$), offering strong, non-selective oxidation potential (El Hajam et al., 2022). These technologies offer notable benefits for the degradation and mineralization of highly hazardous effluents, making them an attractive option for modern water treatment systems.

In comparison to other AOPs, photocatalysis ensures better selectivity, recycling of catalysts, and the feasibility for adoption for solar-assisted decentralized purification systems. These properties make it an attractive option for sustainable and cost-effective water treatment, especially in rural or resource-constrained environments. The advantages of photocatalysis over conventional methods include: (i) no secondary waste: complete pollutant degradation; (ii) solar utilization: potential for low-energy, visible-light-driven systems; (iii) versatility: effective against diverse pollutants (dyes, pharmaceuticals and personal care products, pesticides).

Photocatalysis has emerged as a sustainable AOP, leveraging light energy to drive pollutant degradation using semiconductor materials (e.g., TiO_2, ZnO, g-C_3N_4). The process hinges on three key steps: (i) photoexcitation: light absorption generates electron–hole pairs (e^-/h^+). (ii) ROS formation: charge carriers react with H_2O/O_2 to produce hydroxyl radicals ($^{\bullet}OH$) and superoxide ions $\left(O_2^{\bullet-}\right)$. (iii) Pollutant mineralization: ROS oxidize organic pollutants into non-toxic end products (Fahoul, Zouheir, Tanji, & Kherbeche, 2021). However, challenges persist, such as low visible-light efficiency, photocatalyst recovery difficulties, and scalability barriers, which are addressed in this chapter through material innovations and reactor engineering.

ZnO is a typical semiconductor used in photocatalysis, with similar properties to TiO_2, which offers higher electron mobility but suffers from photocorrosion (Dimitropoulos et al., 2024). It has been investigated with various morphologies (e.g., nanorods, nanotubes, nanobelts, and nanospheres) (Xin et al., 2022; Zhu & Wang, 2025). The synthesis of ZnO films is a crucial solution to address the difficulties of powder separation from solutions after treatment. ZnO films allow easy washing, recovery, and reusability, without additional processes like centrifugation and filtration (Daher, Riachi, Chamoun, Laberty-Robert, & Hamd, 2023). This material has been subjected to various modifications and combinations to make it more active in visible or solar irradiation, such as non-metal and metal transition doping (Rahal et al., 2025; Ma et al., 2022). Other alternative solutions to address the problem of ZnO powder include its combination with magnetic photocatalysts, incorporation in membrane photocatalytic reactors, entrapment in gels and polymers, and coating on capillary photocatalytic microreactors and monolithic supports (Navidpour, Xu, Ahmed, & Zhou, 2024; Hamdan et al., 2024; Hakki & Sillanpää, 2024; Gao, Zhou, & Chen, 2024).

The transition from lab-scale photocatalysis to industrial implementation hinges on effective scale-up strategies (Monsalve-Bravo, Moscoso-Vasquez, & Alvarez, 2014). Historically, scale-up relied on empirical methods: (i) trial-and-error: iterative experiments to derive empirical correlations; (ii) rule-of-thumb: constant operational parameters (e.g., power/volume ratio); (iii) fundamental modeling: physics-based simulations for predictive scaling. While empirical approaches are resource-intensive, fundamental modeling (e.g., using Aspen Plus) enables rapid scale-up (up to 1,000×) with higher reliability (Monsalve-Bravo et al., 2014). This work adopts the latter, leveraging dynamic hierarchy techniques like parameter-change simulations to optimize pilot-scale designs.

This chapter examines recent advances in nanostructured photocatalysts and reactor engineering, with a focus on scalability. Key components include: (i) doping and composite strategies to enhance visible-light activity; (ii) process modeling using Aspen Plus v11 simulations (using the Soave–Redlich–Kwong Equation of States equation of state) to translate lab-scale results (e.g., RhB degradation kinetics) into pilot-scale designs. RhB properties were manually input into Aspen Properties Desktop V11, building on prior database-supported studies (Hachhach, Akram, Hanafi, Achak, & Chafik, 2019).

A case study on Co-doped ZnO films demonstrates this integrated approach, combining experimental validation with Aspen Plus-driven scale-up to bridge the gap between research and real-world applications. The main objective of this chapter is to explore the principles, mechanisms, and influencing parameters of photocatalysis. The innovative strategies in the photocatalytic treatment of liquid effluents were also discussed, followed by an examination of the challenges and limitations of these systems. On the other hand, the design and engineering of photocatalytic reactors with a case study using the ZnO:Co catalyst were provided. Finally, perspectives and future trends in photocatalysis were presented.

9.2 PHOTOCATALYTIC PROCESS: PRINCIPLES, MECHANISMS, AND INFLUENCING PARAMETERS

Photocatalysis is a light-driven catalytic process that leverages the unique properties of semiconductor materials to initiate redox reactions. It has gained significant attention for its potential in environmental remediation, especially in degrading organic pollutants in wastewater. The fundamental principles of photocatalysis revolve around light absorption, charge carrier dynamics, and subsequent surface reactions (El Gaidoumi et al., 2022; Fahoul et al., 2023). When a semiconductor photocatalyst is irradiated with light of energy equal to or greater than its bandgap, electrons in the valence band are excited to the conduction band, leaving behind positively charged holes in the valence band. This process results in the formation of electron–hole (e^-/h^+) pairs. These charge carriers can migrate to the surface of the photocatalyst, where they participate in redox reactions with adsorbed species. The photogenerated electrons reduce molecular oxygen (O_2) to form superoxide anion radicals $\left(^{\bullet}O_2^-\right)$, while the holes can oxidize water (H_2O) or hydroxide ions (OH–) to generate hydroxyl radicals ($^{\bullet}OH$). Both $^{\bullet}OH$ and $^{\bullet}O_2^-$ are highly ROS capable of degrading a wide range of organic contaminants into harmless end products such as CO_2 and H_2O (Chakravorty & Roy, 2024). The pollutant adsorption also plays an important role in the mechanism. The pollutants must first adsorb onto the surface of the photocatalyst for effective interaction with the ROS. Once adsorbed, the ROS react with the organic molecules, breaking them down into smaller, less-toxic compounds or completely mineralizing them. The efficiency of this process is influenced by the affinity of the pollutant for the catalyst's surface and surface area, and the interaction between the pollutant and the ROS (Khader et al., 2024). One of the challenges in photocatalysis is the recombination of the excited electron and hole before they can participate in chemical reactions. Efforts to improve photocatalytic efficiency often focus on minimizing this recombination, through the design of more efficient catalysts or the use of co-catalysts, in addition to a Lorentz force-assisted charge carrier separation enhancement strategy (Tuama et al., 2024; Gao et al., 2019).

Several parameters influence the efficiency of photocatalytic processes in treating liquid effluents. These factors range from the properties of the photocatalyst to environmental conditions, all of which play a crucial role in the performance of photocatalytic systems. The relevant photocatalyst significantly enhances the process efficiency. Crucial characteristics such as the surface, crystallinity, band gap, synthesis method, and stability of the photocatalyst determine its potential for absorbing light and catalyzing the production of ROS. The light physical properties, such as wavelength and intensity, are essential for increasing the photocatalytic efficiency. Photocatalysts generally need specific light irradiation for activation, with visible light being preferred due to its availability. The physico-chemical parameters, such as concentration of pollutants, temperature, and pH, also have a direct impact on photocatalytic efficiency. The concentration of pollutants in effluents subjected to treatment plays a critical role in the photocatalytic rate. With diluted effluents, the pollutants are more likely to be decomposed by the ROS. However, with highly concentrated effluents, the produced ROS may be inadequate to degrade the pollutants completely, leading to incomplete mineralization and slower degradation reaction rates. The pH and temperature also influenced intensively both the pollutants and the photocatalyst (Ariza-Tarazona et al., 2020). For instance, the surface protonation degree has a great influence on the adsorption of pollutants. Similarly, temperature has a direct impact on the reaction kinetics. The photodegradation efficiency can be increased when higher temperatures are used, which decreases the stability of the photocatalyst (Verma, Mezgebe, Hejase, Sahle-Demessie, & Nadagouda, 2024). In addition to the above parameters, the ability of the photocatalyst to be regenerated and reused without significant loss in efficiency is another important challenge. Over time, catalysts may undergo deactivation due to surface poisoning, aggregation, or other factors. Developing robust photocatalysts with high reusability is essential for their practical application in wastewater treatment.

9.3 PHOTOCATALYSTS: CLASSICAL APPROACHES AND EMERGING TRENDS

In the last few decades, the development of photocatalysts has evolved considerably to overcome the limitations of early systems. Initial research focused on UV-active semiconductors such as ZnO and TiO_2. These materials are the most classic studied photocatalysts due to their high oxidizing potential, non-toxicity, and chemical and thermal stability. They are efficient in the degradation of organic molecules in liquid effluents. However, they are UV light-active materials due to their wide band gaps (around 3.2 eV) and fast recombination of photogenerated electron–hole pairs (Zheng et al., 2023). Beyond these traditional semiconductors, new classes of materials are rapidly gaining attention in photocatalysis, including vanadates, oxohalides, carbon-based materials (e.g., g-C_3N_4 and GO, and rGO), metal sulfides, metal–organic frameworks (MOFs), and mixed metal oxides (e.g., perovskites and spinels), which are active in visible irradiation (Table 9.1) (Ajiboye, Oyewo, & Onwudiwe, 2021; Das & Mahalingam, 2019; Liu, Du, Du, & Jiang, 2025; Nkwachukwu & Arotiba, 2025). These materials are chosen based on various properties, such as high photocatalytic performance,

TABLE 9.1
Comparison of Common Photocatalysts and Their Advantages and Limitations

Photocatalyst	Light Range	Advantages	Limitations	References
TiO_2 (anatase)	UV	Stable, low-cost, non-toxic	Limited to UV light, fast recombination	Khan and Shah (2023)
ZnO	UV	High activity, similar to TiO_2	Limited to UV light, photocorrosion in aqueous media	Zhang et al. (2023)
$SrTiO_3$	UV	Non-toxic, thermally and chemically stable, a high transporter of electrons	Limited to UV light, fast recombination	Phoon, Lai, Juan, Show, and Chen (2019)
MoS_2	Visible	Efficient charge transport, good synergy in composites	Difficult exfoliation, limited active sites, low conductivity	Liu et al. (2020)
$ZnFe_2O_4$	Visible	Stable, magnetic properties, resistance to corrosion	Poor photoelectric conversion property and low valence band potential	Nguyen and Doong (2017)
g-C_3N_4	Visible	Visible-light responsive, metal-free	Low surface area, fast recombination	Wudil et al. (2023)
rGO	Visible	High surface area and electric conductivity, best electron–hole separation	Properties highly sensitive to the reduction degree and defects, need for combination with semiconductors	Mondal, Prabhakaran, Gupta, and Subramanian (2021)
MOF	Visible	High surface area, controllable porosity, adjustable band gap via metal or ligand	Low conductivity, sensitivity to humidity, and low thermal and chemical stability for certain MOFs, high cost	Laeim et al. (2025), Morsy, Hassan, Shaker, and Ismail (2025)

light absorption, good charge separation, stability, and overall reusability. For example, MOFs, crystalline materials composed of metal nodes and organic linkers, offer tunable porosity, high surface areas, and modular functionality. Photocatalytically active MOFs (e.g., Ti-MOFs, Zr-based MOFs) can be designed to capture light efficiently and promote selective chemical transformations (Zhou et al., 2020; Erum, Alamgir, Yu, & Gao, 2025). Another representative class of photocatalysts is graphitic carbon nitride (g-C_3N_4), a metal-free polymeric semiconductor made of carbon and nitrogen, which has a layered structure with adjustable electrical characteristics, chemical stability, and visible-light activity. Because of its unique textural and structural properties, g-C_3N_4 may be designed as an effective support for various semiconductors to use solar energy for redox processes (Huang et al., 2025; Palanivel et al., 2021).

Creating composite photocatalysts is another major avenue to enhance photocatalytic efficiency by combining complementary materials and optimizing interfacial charge transfer. A conventional heterojunction is formed by coupling two semiconductors with different band structures (e.g., TiO_2/CdS) (Qian, Xu, & Xu, 2024). The resulting built-in electric field at the interface drives effective separation of electrons and holes, reducing recombination. There are different types of heterojunctions, including type-II, Schottky junctions, p-n junctions, and Z-schemes (Zhao et al., 2022). Inspired by natural photosynthesis, Z-scheme photocatalysts consist of two photocatalysts (PCI and PCII) connected in a manner that preserves strong redox potentials while facilitating efficient charge transfer (Figure 9.1). In these systems (e.g., $ZnFe_2O_4/WO_3$), electrons with higher reducing power and holes with higher oxidizing power are retained, leading to superior photocatalytic activity (Wang et al., 2023).

The integration of plasmonic nanoparticles (e.g., Au, Ag) into semiconductor matrixes leverages localized surface plasmon resonance effects. Plasmonic photocatalysts exhibit enhanced light absorption, hot electron injection, and electromagnetic field enhancement, making them highly effective under visible and near-infrared illumination (Nazir, Huo, Wang, Weiqiang, & Wan, 2023).

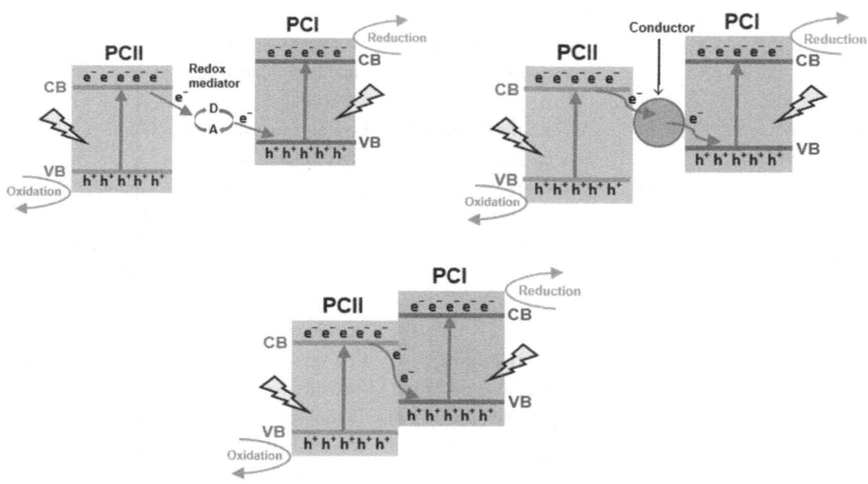

FIGURE 9.1 Photocatalytic mechanisms with the Z-scheme heterojunction.

9.4 CASE STUDY: COBALT-DOPED ZNO FOR RHB DEGRADATION

This case study focuses on simulating a pilot-scale photocatalytic reactor using cobalt-doped ZnO films. Initially, the catalyst films were tested at the laboratory scale for degrading RhB under visible light. Both pure ZnO and cobalt-doped ZnO films were synthesized using the spray pyrolysis technique (Tanji et al., 2021).

Cobalt-doped ZnO (Co:ZnO) thin films were prepared via spray pyrolysis using varying Co/Zn molar ratios (0%, 5%, and 10%) and deposited onto glass substrates at 450°C. The photocatalytic activity of the films was tested under visible light using a 300 W solar-simulated lamp, showing improved degradation performance with increased cobalt doping. Based on lab-scale data, a pilot-scale photocatalytic reactor was modeled and simulated in Aspen Plus to evaluate the process feasibility at a larger scale.

9.5 STRUCTURAL, OPTICAL, AND MORPHOLOGICAL CHARACTERIZATION

The synthesized films of ZnO:Co were characterized using diffuse reflectance spectroscopy (DRS), X-ray diffraction (XRD), and scanning electron microscopy with energy-dispersive X-ray (SEM/EDX) analysis (Figure 9.2). All observed diffraction peaks correspond to the hexagonal wurtzite structure of ZnO, in agreement with

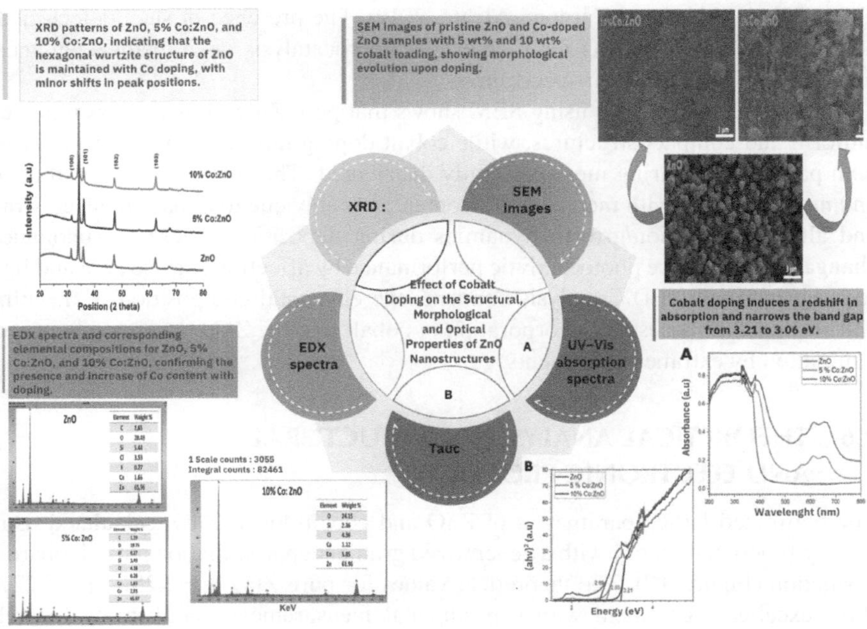

FIGURE 9.2 Characterization of ZnO:Co thin films using DRS, XRD, and SEM/EDX analysis.

JCPDS card no. 36–1451, and are indexed to the $P6_3mc$ space group with lattice constants approximately $a = 3.25$ Å and $c = 5.21$ Å. Notably, no secondary phases were detected in cobalt-doped samples, suggesting successful incorporation of Co^{2+} ions into the ZnO lattice without forming cobalt oxide impurities, consistent with literature reports (Jellal, Nouneh, Jedryka, Chaumont, & Naja, 2020). A closer examination of the XRD patterns reveals peak broadening and a systematic shift toward higher 2θ values with increasing cobalt content, indicating lattice contraction. This structural change is attributed to the substitution of Zn^{2+} ions (ionic radius: 0.74 Å) by smaller Co^{2+} ions (ionic radius: 0.65 Å). Moreover, the decreasing peak intensity and broadening with increasing doping levels imply a reduction in crystallinity or smaller crystallite domains, likely due to lattice distortion induced by Co dopant incorporation. Crystallite sizes were estimated using the Scherrer equation, yielding average sizes of 32.6 nm for pure ZnO and 34.8 nm for 10% Co-doped ZnO, suggesting a slight grain growth upon doping. Optical properties, investigated using UV–Visible spectroscopy, demonstrate enhanced absorption in the visible region with increased cobalt doping. A redshift in the absorption edge was observed, indicating a narrowing of the optical bandgap. This phenomenon is attributed to increased defect states, particularly oxygen vacancies and donor levels induced by Co doping. Bandgap values were extracted from Tauc plots using the Kubelka–Munk function, revealing a decrease from 3.21 Ev (undoped ZnO) to 2.85 Ev (5% Co:ZnO) and further to 2.69 Ev (10% Co:ZnO). This bandgap narrowing is consistent with sp–d exchange interactions between localized Co^{2+} d-electrons and the conduction/valence band of ZnO, which modifies the band structure (Jellal et al., 2019; Martínez Ruano, Taimbu de la Cruz, Orrego Alzate, & Cardona Alzate, 2019). The presence of such defect-mediated states is advantageous for visible-light photocatalysis, as it enables enhanced light harvesting in the solar spectrum.

Morphological analysis using SEM shows that pure ZnO films possess relatively uniform and compact structures, while cobalt doping introduces more irregularity, with particles appearing more randomly distributed. The surface becomes rougher and more granular with increased Co content, possibly due to dopant-induced strain and altered nucleation/growth dynamics during deposition. These morphological changes can influence photocatalytic performance by affecting surface area and light scattering. Finally, EDX analysis confirms the elemental composition of the films and verifies the successful incorporation of cobalt into the ZnO matrix without contamination by extraneous elements.

9.6 THEORETICAL ANALYSIS OF STRUCTURAL AND ELECTRONIC PROPERTIES

The optimized lattice parameters of ZnO and cobalt-doped ZnO, calculated using density functional theory with a generalized gradient approximation plus Hubbard U correction (Figure 9.3). The theoretical values for pure ZnO ($a = 3.25$ Å, $c = 5.22$ Å) show excellent agreement with experimental measurements, confirming the reliability of the computational approach adopted in this study. Upon Co doping, both lattice parameters decrease slightly, which is attributed to the substitution of Zn^{2+}

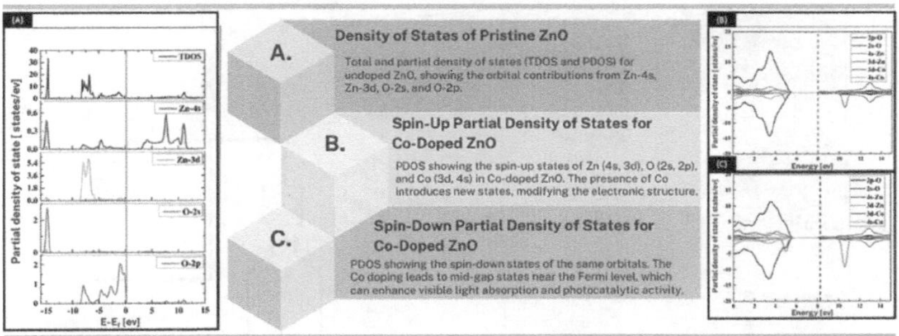

FIGURE 9.3 Theoretical study of the optical properties of ZnO:Co thin films.

(ionic radius 0.74 Å) with the smaller Co^{2+} ions (ionic radius 0.65 Å). This structural contraction is consistent with experimental XRD findings and further supports the successful incorporation of cobalt into the ZnO matrix. Moreover, the magnetic properties of the doped systems were evaluated. All Co-doped ZnO structures exhibited a stable ferromagnetic ground state. Notably, increasing the Co concentration from 6.25% to 12.5% results in a larger energy difference between the ferromagnetic and non-magnetic states, rising from 0.28 Ry to 0.57 Ry. This energy difference suggests enhanced magnetic ordering at higher doping levels, which can potentially influence charge separation and electron–hole recombination dynamics in photocatalysis.

To further elucidate the electronic structure, the partial density of states was computed. For undoped ZnO, the valence band is primarily composed of O (2p) orbitals, while the conduction band is mainly Zn (4s) in character. Upon doping with Co, Co (3d) orbitals contribute significantly to the upper valence band, resulting in a higher density of states near the Fermi level. This suggests that the introduction of Co creates new electronic states, enhancing light absorption in the visible region. In addition, increasing the Co concentration further amplifies the density of states in the valence band, which aligns with the observed optical redshift and improved absorption behavior in experimental UV–vis measurements. Finally, the calculated bandgap values (3.38 eV for undoped ZnO, 2.81 eV for 6.25% Co:ZnO, and 2.75 eV for 12.5% Co:ZnO) closely match the experimental data, reinforcing the validity of the theoretical approach and confirming the observed trend of bandgap narrowing with increasing Co content.

9.7 RHB PHOTODEGRADATION OPTIMIZATION, CONVERSION UNDER OPTIMAL CONDITIONS, CATALYST REUSABILITY, AND MECHANISTIC INSIGHT

The surface plots using Box-Behnken Design as a response surface methodology optimization indicate that cobalt doping has the highest effect on conversion efficiency, with RhB degradation reaching a maximum at high Co content (10%) and an extended contact time (120 min), regardless of the initial RhB concentration (Figure 9.4).

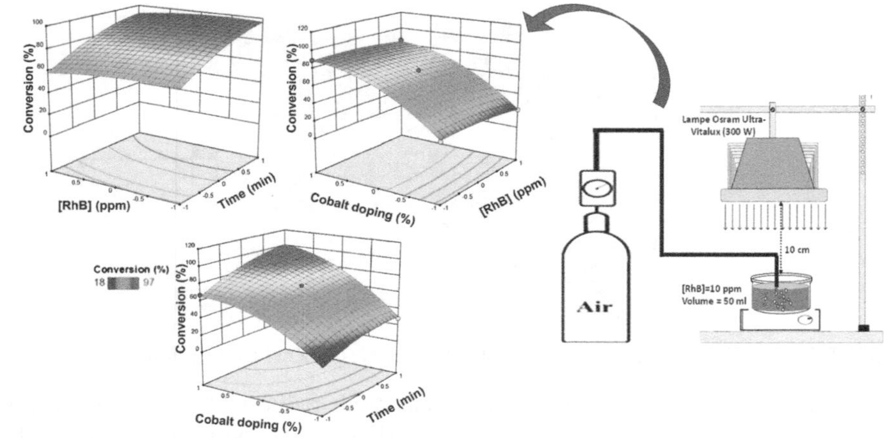

FIGURE 9.4 Response surface methodology optimization of RhB photodegradation using ZnO:Co thin films.

To validate the efficiency of cobalt-doped ZnO films, RhB photodegradation was carried out under optimal conditions identified by the Box-Behnken Design (10% Co-doping, 10 ppm RhB, 120 min reaction time), where the experimental RhB conversion reached 93% (Figure 9.5). In contrast, pristine ZnO exhibited only 25% conversion under identical conditions, underscoring the enhancement due to cobalt incorporation. Catalyst stability was assessed over four reuse cycles, with conversion efficiency slightly declining from 98% to 87%, and a minor increase in COD levels indicating minimal leaching or deactivation, hence demonstrating good reusability. Radical scavenger tests confirmed that superoxide radicals $\left(^{\bullet}O_2^{-}\right)$ played the dominant role in the photocatalytic process, while hydroxyl radicals (•OH) and photogenerated holes (h^+) had limited contributions (Boughelout, Macaluso, Kechouane, & Trari, 2020). The proposed mechanism involves visible-light-induced excitation of electrons from the valence to the conduction band of Co-doped ZnO, generating e^-/h^+ pairs that interact with surface-adsorbed O_2 and H_2O to form $\bullet O_2^-$ and •OH, respectively, which in turn oxidize RhB into CO_2 and H_2O. Kinetic studies further confirmed that RhB degradation followed a pseudo-first-order reaction model with rate constants increasing from 0.004 min^{-1} (ZnO) to 0.03 min^{-1} (10% Co–ZnO), demonstrating superior photocatalytic activity ($R^2 = 0.99$). Finally, the performance of this catalyst compared to those from the literature showed that the 10% Co-doped ZnO films synthesized herein offer competitive, if not superior, degradation efficiency, making them promising candidates for wastewater treatment under visible-light activation.

9.8 PROCESS MODELING AND ENGINEERING ASSESSMENT

To evaluate the practical feasibility and scale-up potential of the RhB degradation process, a comprehensive process simulation was conducted using Aspen Plus software (Figure 9.6). The process flowsheet was constructed based on experimental results and kinetic parameters derived from batch tests. The reactor was modeled using a Plug Flow Reactor block, which simulates a long tubular reactor and assumes

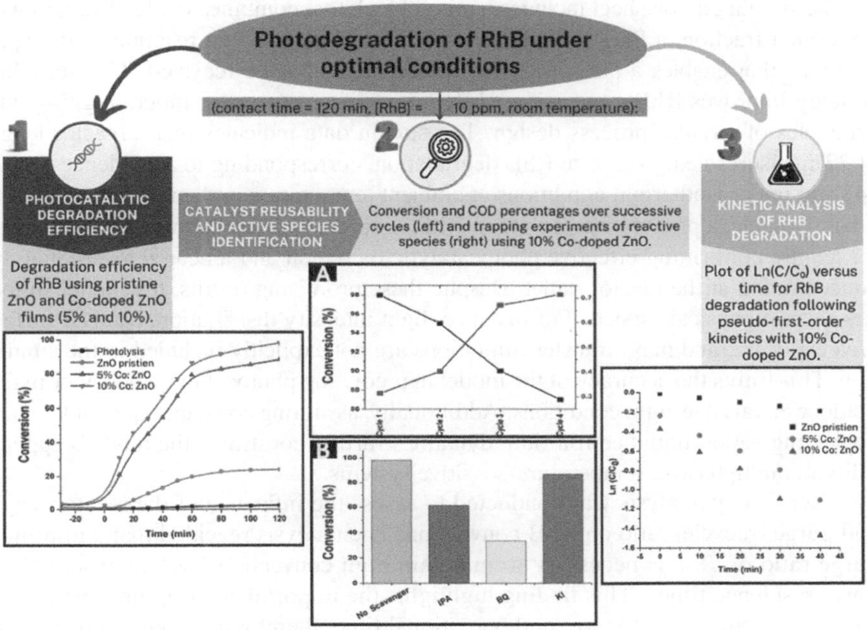

FIGURE 9.5 RhB photodegradation tests under optimum conditions.

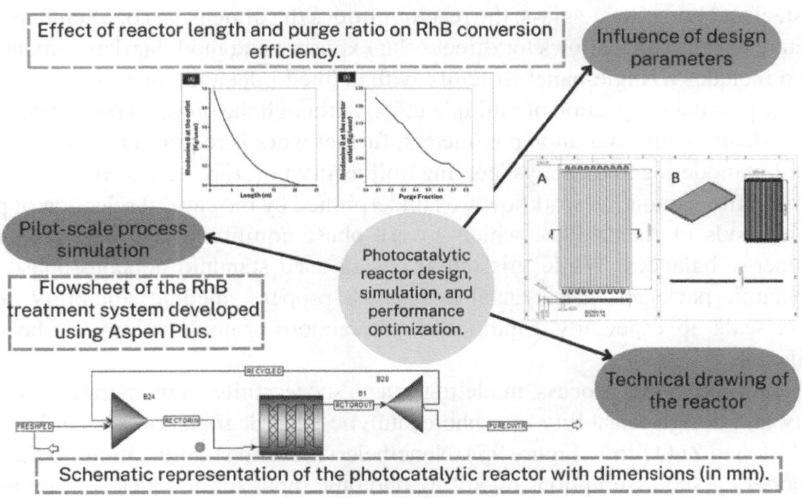

FIGURE 9.6 Schematic representation of the photocatalytic reactor and its simulation flowsheet using Aspen Plus.

ideal plug flow behavior with no axial mixing and constant radial composition. Although pressure drops across pipe bends and elbows were neglected, this assumption is reasonable for a preliminary design focused primarily on quantifying chemical conversion rather than hydraulic performance.

The simulated flowsheet includes a mixer block that combines the feed stream with a recycled fraction, a Plug Flow Reactor unit for photocatalytic reaction, and a splitter block that enables a portion of the treated effluent to be recycled. This recycling strategy improves RhB conversion while minimizing raw water input, aligning with principles of circular process design. The stream data indicates that a reactor length of 22 m ensures near-complete RhB degradation, corresponding to a residence time of 143 min under isothermal conditions at ambient temperature and an initial pressure of 2 bar. Notably, a progressive decline in RhB concentration is observed along the reactor length, confirming effective photocatalytic oxidation and a near-zero residual dye concentration at the reactor outlet. Despite these promising results, the model simplifies several physical aspects. For instance, light intensity distribution, catalyst surface area exposure, and mass transfer limitations are not explicitly included in the simulation. This limits the accuracy of the model in predicting photocatalytic behavior in real outdoor or variable-light conditions. Additionally, assuming constant liquid holdup and neglecting vapor–liquid equilibrium dynamics further constrains the model's applicability to multiphase or temperature-sensitive systems.

A sensitivity analysis was conducted to assess the influence of the reactor length and purge (recycle) ratio on RhB conversion. The analysis revealed that a minimum purge ratio of 78% is necessary to maintain high conversion levels, particularly at lower residence times. This finding highlights the importance of optimizing recycle streams to reduce reactor size and operational time, thereby improving process economics and footprint efficiency. To facilitate the experimental validation of the model at a pilot scale, a 3D module prototype reactor was designed. Constructed from stainless steel and borosilicate glass, the reactor mimics the dimensions of a standard photovoltaic panel, which allows for direct solar exposure and modular deployment. The design includes a single-panel structure with defined inlet and outlet ports, allowing serial or parallel integration of multiple units. Although the physical prototype aligns geometrically with simulation parameters, further work is required to integrate radiation flux modeling and catalyst coating uniformity in future iterations.

The entire simulation workflow was underpinned by the careful selection of property methods in Aspen Plus, which govern phase equilibria, transport properties, and energy balances. While this initial design used standard aqueous-phase thermodynamic packages, it is crucial to reassess property method suitability during further scale-up, especially if multiphase interactions or non-ideal mixing becomes significant.

In summary, the process modeling stage successfully demonstrates the scalability and design feasibility of a photocatalytic RhB degradation system based on cobalt-doped ZnO films (Figure 9.7). Nonetheless, several simplifications in the reactor modeling and assumptions regarding photonic input and kinetics warrant refinement. Future studies should focus on integrating CFD-based light transport models, reactor hydrodynamics, and real sunlight variation into the design process to fully capture the performance and stability of the proposed photocatalytic reactor under real-world conditions.

This case study presents a promising integration of experimental and computational approaches for the photocatalytic degradation of RhB using cobalt-doped ZnO films, demonstrating high degradation efficiencies up to 93% under optimized

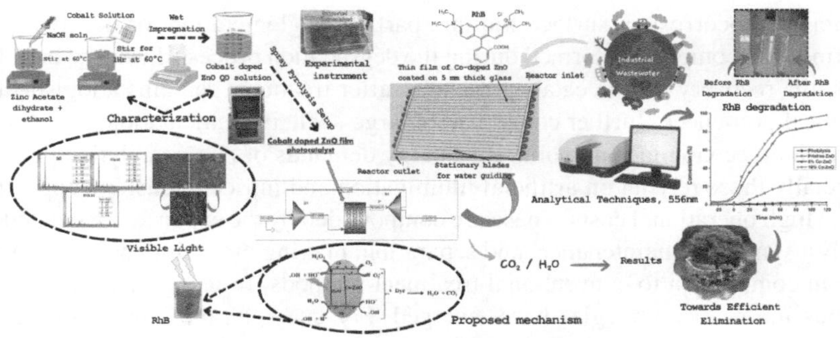

FIGURE 9.7 Summary of RhB photodegradation using ZnO:Co thin films and its simulation in a continuous reactor.

conditions. The combination of Box-Behnken Design and Aspen Plus modeling provides a structured framework for both process optimization and scale-up simulation. The work is particularly valuable in its attempt to bridge lab-scale performance with pilot-scale feasibility, notably through the design of a modular reactor system compatible with solar exposure. However, several limitations temper the overall impact. The process simulation, while robust in mass balance terms, does not account for critical photocatalytic variables such as light intensity gradients, photon flux, and catalyst surface heterogeneity, factors that are central to photocatalytic performance. Furthermore, the assumption of ideal plug flow and neglect of hydrodynamic effects limit the reactor model's predictive accuracy in real-world applications. The catalyst reuse study, although positive, would benefit from longer-term cycling and stability assessments under environmental conditions. In addition, a life cycle assessment should be applied to examine the environmental impacts associated with the entire process, from catalyst powder synthesis to film development, to degradation tests in batch and continuous systems. Overall, this work lays a solid foundation for photocatalytic water treatment using doped semiconductors, but future investigations must integrate reactor photophysics, outdoor variability, and long-term catalyst performance to fully validate its practical deployment.

9.9 CHALLENGES, MITIGATION STRATEGIES, AND FUTURE PERSPECTIVES

Despite promising progress in the development of photocatalytic materials and reactor configurations, several key limitations continue to hinder the large-scale deployment of photocatalytic technologies for wastewater treatment. These limitations are primarily linked to material inefficiencies, operational complexity, and economic feasibility.

One major constraint arises from the intrinsic properties of widely used photocatalysts such as ZnO and TiO_2. Their relatively large band gaps restrict photoactivity to the ultraviolet region, which accounts for only a small fraction of the solar spectrum. As a result, their ability to harness solar energy efficiently remains limited. Additionally, many photocatalysts exhibit reduced activity over time due to factors such as photocorrosion, surface fouling, particle agglomeration, or deactivation by intermediate compounds formed during the degradation process. In slurry-based systems, the recovery of fine catalyst particles after treatment remains a logistical and technical bottleneck, further complicating large-scale implementation.

From an economic standpoint, the energy demands of photocatalytic operations, especially those relying on artificial illumination and intricate reactor setups, translate to high operational costs. These are compounded by the expenses associated with catalyst synthesis, maintenance, and separation, making the technology less competitive in comparison to conventional treatment methods. In response to these issues, various innovative strategies have emerged. For example, the integration of magnetic components such as Fe_3O_4 or $CoFe_2O_4$ into photocatalytic materials allows for straightforward recovery through external magnetic fields, offering a cost-effective and efficient alternative to conventional separation methods. Similarly, the incorporation of transition metal co-catalysts, like the studied case, Co doped in ZnO matrix, enhances photocatalytic efficiency by facilitating charge carrier separation and providing additional active sites for redox reactions. These modifications help reduce electron–hole recombination and improve quantum efficiency. Hybrid systems that combine photocatalysis with membrane filtration have also shown promise. In these configurations, photocatalysis breaks down complex organic pollutants, while membranes provide a physical barrier for residual particles and degraded by-products, resulting in a more robust and comprehensive treatment process.

Looking forward, the future of photocatalytic wastewater treatment lies in smarter, more adaptable systems. Artificial intelligence and machine learning tools are increasingly being used to guide catalyst design, optimize process parameters, and predict performance under real-world conditions, particularly in the visible-light range. The development of solar-driven photoreactors offers a pathway toward reducing energy consumption and environmental impact. Moreover, aligning photocatalytic treatment with circular economy principles, such as recovering valuable by-products during effluent processing, can enhance the sustainability and economic attractiveness of the technology.

Nonetheless, several systemic barriers persist. The lack of regulatory standards, high initial investment costs, limited industrial uptake, and the scarcity of pilot-scale validations under actual operating conditions remain significant hurdles. Addressing these will require a concerted effort involving supportive policy frameworks, academic–industry partnerships, financial incentives, and the design of scalable, modular systems that can be easily integrated into existing treatment infrastructures.

9.10 CONCLUSION

In the examined case study, Aspen Plus software was applied to simulate the RhB degradation process and analyze its sensitivity based on the findings received from the batch experiences. In summary, the successfully synthesized cobalt-doped ZnO films have high photocatalytic activity and good recycling results, and the efficient scaling-up design of the photo-reactor could promote a wide use for a variety of potential environmental and industrial applications.

Indeed, photocatalytic technologies for water treatment have witnessed significant advancements, driven by innovations in material design, reactor engineering, and process integration. The development of nanostructures, transition metals doping strategies (i.e., ZnO:Co in the case study), composite and hybrid systems such as heterojunctions and Z-scheme photocatalysts, and the emergence of novel materials like MOFs, perovskites, and carbon-based catalysts have greatly expanded the operational window for photocatalysis, especially under visible light. Simultaneously, engineering advances in reactor design, including immobilized systems, fluidized beds, and solar-driven configurations, have improved the scalability and efficiency of photocatalytic processes. However, unresolved challenges persist. Low solar-to-chemical conversion efficiencies, catalyst deactivation, difficulties in catalyst recovery, and high operational costs continue to hinder full-scale deployment. Moreover, there remain significant gaps in translating laboratory successes into reliable, cost-effective, and sustainable industrial applications. Issues related to policy regulation, environmental impact assessments, and economic viability must also be addressed to ensure real-world application. Moving forward, the following key points are recommended: photocatalysis demands interdisciplinary collaboration. Materials scientists, chemists, reactor engineers, environmental scientists, and policy makers must work together to bridge the gaps between innovative material synthesis and practical reactor deployment. Only through an integrated approach, combining advances in catalyst development with intelligent reactor engineering and system optimization, can the photocatalysis process achieve its full potential as a sustainable solution for wastewater treatment. In fact, during the development of a new photocatalytic process, the scale-up is a crucial step, which allows the valorization of the experimental findings; however, it often proves to be challenging and costly. The existing approaches of scale-up could be organized into: (i) test and error, in which experimental process data are utilized for the building of each scale's empirical relations; (ii) thumb rules, considering as a general rule during scale-up a constant value of a specific operational parameter; (iii) basic approach or fundamental approach of the modeling process.

Successfully scaling up is key to turning photocatalytic technologies from promising laboratory discoveries into practical, efficient, and affordable solutions for large-scale water treatment. But to make this a reality, an ongoing scientific effort is needed to tackle the technical and operational challenges that still stand in the way.

REFERENCES

Ajiboye, T. O., Oyewo, O. A., & Onwudiwe, D. C. (2021). The performance of bismuth-based compounds in photocatalytic applications. *Surfaces and Interfaces*, *23*, 100927. https://doi.org/10.1016/J.SURFIN.2021.100927

Ariza-Tarazona, M. C., Villarreal-Chiu, J. F., Hernández-López, J. M., Rivera De la Rosa, J., Barbieri, V., Siligardi, C., & Cedillo-González, E. I. (2020). Microplastic pollution reduction by a carbon and nitrogen-doped TiO_2: Effect of pH and temperature in the photocatalytic degradation process. *Journal of Hazardous Materials*, *395*. https://doi.org/10.1016/j.jhazmat.2020.122632

Behera, R., Patel, S. R., & Parikh, J. K. (2023). Hydrodynamics and transport mechanism of microfluidic mixing in precipitation of nanodrugs: A review. *Crystal Research and Technology*, *58*(7), 2300007. https://doi.org/10.1002/CRAT.202300007

Boughelout, A., Macaluso, R., Kechouane, M., & Trari, M. (2020). Photocatalysis of rhodamine B and methyl orange degradation under solar light on ZnO and Cu_2O thin films. *Reaction Kinetics, Mechanisms and Catalysis*, *129*(2), 1115–1130. https://doi.org/10.1007/S11144-020-01741-8

Chakravorty, A., & Roy, S. (2024). A review of photocatalysis, basic principles, processes, and materials. *Sustainable Chemistry for the Environment*, *8*, 100155. https://doi.org/10.1016/J.SCENV.2024.100155

Daher, E. A., Riachi, B., Chamoun, J., Laberty-Robert, C., & Hamd, W. (2023). New approach for designing wrinkled and porous ZnO thin films for photocatalytic applications. *Colloids and Surfaces A: Physicochemical and Engineering Aspects*, *658*, 130628. https://doi.org/10.1016/J.COLSURFA.2022.130628

Das, S., & Mahalingam, H. (2019). Exploring the synergistic interactions of TiO_2, rGO, and g-C_3N_4 catalyst admixtures in a polystyrene nanocomposite photocatalytic film for wastewater treatment: Unary, binary and ternary systems. *Journal of Environmental Chemical Engineering*, *7*(4), 103246. https://doi.org/10.1016/J.JECE.2019.103246

Dimitropoulos, M., Aggelopoulos, C. A., Sygellou, L., Tsantis, S. T., Koutsoukos, P. G., & Yannopoulos, S. N. (2024). Unveiling the photocorrosion mechanism of zinc oxide photocatalyst: Interplay between surface corrosion and regeneration. *Journal of Environmental Chemical Engineering*, *12*(2), 112102. https://doi.org/10.1016/J.JECE.2024.112102

El Gaidoumi, A., Arrahli, A., Loqman, A., Baragh, F., El Bali, B., & Kherbeche, A. (2022). Efficient sol-gel nanocomposite TiO_2-clay in photodegradation of phenol: Comparison to labe-made and commercial photocatalysts. *Silicon*, *14*(10), 5401–5414. https://doi.org/10.1007/S12633-021-01275-1

El Hajam, M., Kandri, N. I., Zerouale, A., Wang, X., Gustafsson, J., Wang, L., … Xu, C. (2022). Lignocellulosic nanocrystals from Sawmill waste as biotemplates for free-surfactant synthesis of photocatalytically active porous silica. *ACS Applied Materials and Interfaces*, *14*(17), 19547–19560. https://doi.org/10.1021/ACSAMI.2C02550

Erum, J. K. E., Alamgir, Yu, Y., & Gao, J. (2025). Recent progress in Zr-MOFs for antibiotic adsorption and photocatalysis. *Separation and Purification Technology*, *363*, 132138. https://doi.org/10.1016/J.SEPPUR.2025.132138

Fahoul, Y., Tanji, K., González Díaz, O. M., Quesada-Cabrera, R., Naciri, Y., El Mrabet, I., … Kherbeche, A. (2023). Development of a new CoS-Supported $ZnAl_2O_4$ catalyst for the visible photodegradation of a basic textile dye from water. *Optical Materials*, *143*, 114148. https://doi.org/10.1016/J.OPTMAT.2023.114148

Fahoul, Y., Zouheir, M., Tanji, K., & Kherbeche, A. (2021). Synthesis of a novel $ZnAl_2O_4$/CuS nanocomposite and its characterization for photocatalytic degradation of acid red 1 under UV illumination. *Journal of Alloys and Compounds*, *889*, 161708. https://doi.org/10.1016/J.JALLCOM.2021.161708

Gao, G., Zhou, P., & Chen, C. qi. (2024). Facile construction of magnetic M-ZnO-Ni-RSA composite as a functional photocatalyst for tetracycline degradation: Performance and mechanism. *Journal of Photochemistry and Photobiology A: Chemistry*, *448*, 115249. https://doi.org/10.1016/J.JPHOTOCHEM.2023.115249

Gao, W., Lu, J., Zhang, S., Zhang, X., Wang, Z., Qin, W., … Sang, Y. (2019). Suppressing photoinduced charge recombination via the Lorentz force in a photocatalytic system. *Advanced Science*, *6*(18), 1901244. https://doi.org/10.1002/ADVS.201901244

González-González, R. B., Sharma, A., Parra-Saldívar, R., Ramirez-Mendoza, R. A., Bilal, M., & Iqbal, H. M. N. (2022). Decontamination of emerging pharmaceutical pollutants using carbon-dots as robust materials. *Journal of Hazardous Materials*, *423*, 127145. https://doi.org/10.1016/J.JHAZMAT.2021.127145

Hachhach, M., Akram, H., Hanafi, M., Achak, O., & Chafik, T. (2019). Simulation and sensitivity analysis of molybdenum disulfide nanoparticle production using Aspen Plus. *International Journal of Chemical Engineering*, *2019*(1), 3953862. https://doi.org/10.1155/2019/3953862

Hakki, H. K., & Sillanpää, M. (2024). Comprehensive analysis of photocatalytic and photoreactor challenges in photocatalytic wastewater treatment: A case study with ZnO photocatalyst. *Materials Science in Semiconductor Processing*, *181*, 108592. https://doi.org/10.1016/J.MSSP.2024.108592

Hamdan, M. A. H., Madon, R. H., Hairom, N. H. H., Makhtar, S. N. N. M., Ahmad, M. K., Hamed, N. K. A., … Setiabudi, H. D. (2024). High-efficiency river water treatment via pilot-scale low-pressure hybrid membrane photocatalytic reactor (MPR) utilizing ZnO-Kaolin photocatalyst. *Journal of Water Process Engineering*, *68*, 106543. https://doi.org/10.1016/J.JWPE.2024.106543

Huang, B., Xu, K., Zhao, Y., Li, B., Jiang, S., Liu, Y., … Li, Y. (2025). Review of the versatility and application potentials of g-C_3N_4-based S-scheme heterojunctions in photocatalytic antibiotic degradation. *Molecules*, *30*(6), 1240. https://doi.org/10.3390/MOLECULES30061240

Jellal, I., Ahmoum, H., Khaaissa, Y., Nouneh, K., Boughrara, M., Fahoume, M., … Naja, J. (2019). Experimental and ab-initio investigation of the microstructure and optoelectronic properties of FCM–CVD-prepared Al-doped ZnO thin films. *Applied Physics A: Materials Science and Processing*, *125*(9), 1–7. https://doi.org/10.1007/S00339-019-2947-4

Jellal, I., Nouneh, K., Jedryka, J., Chaumont, D., & Naja, J. (2020). Non-linear optical study of hierarchical 3D Al doped ZnO nanosheet arrays deposited by successive ionic adsorption and reaction method. *Optics & Laser Technology*, *130*, 106348. https://doi.org/10.1016/J.OPTLASTEC.2020.106348

Khader, E. H., Muslim, S. A., Saady, N. M. C., Ali, N. S., Salih, I. K., Mohammed, T. J., … Zendehboudi, S. (2024). Recent advances in photocatalytic advanced oxidation processes for organic compound degradation: A review. *Desalination and Water Treatment*, *318*, 100384. https://doi.org/10.1016/J.DWT.2024.100384

Khan, H., & Shah, M. U. H. (2023). Modification strategies of TiO_2 based photocatalysts for enhanced visible light activity and energy storage ability: A review. *Journal of Environmental Chemical Engineering*, *11*(6), 111532. https://doi.org/10.1016/J.JECE.2023.111532

Laeim, H., Molahalli, V., Prajongthat, P., Pattanaporkratana, A., Pathak, G., Phettong, B., … Chattham, N. (2025). Porosity Tunable Metal-Organic Framework (MOF)-based composites for energy storage applications: Recent progress. *Polymers*, *17*(2), 130. https://doi.org/10.3390/POLYM17020130

Liu, J., Du, H., Du, X., & Jiang, Y. (2025). Construction of highly dispersive Fe-MOFs-induced Fe doped C_3N_5 photocatalyst for simultaneous efficient photocatalytic degradation of co-existing multiple antibiotics in water. *Journal of Environmental Chemical Engineering*, *13*(2), 116152. https://doi.org/10.1016/J.JECE.2025.116152

Liu, X., Chen, X., Xu, L., Wu, B., Tu, X., Luo, X., … Lin, J. (2020). Non-noble metal ultrathin MoS_2 nanosheets modified $Mn_{0.2}Cd_{0.8}S$ heterostructures for efficient photocatalytic H_2 evolution with visible light irradiation. *International Journal of Hydrogen Energy*, *45*(51), 26770–26784. https://doi.org/10.1016/J.IJHYDENE.2020.07.086

Ma, Z., Wei, Y., Ding, J., Chang, P., Guo, J., & Peng, Z. (2022). Investigation of electronic and optical properties of Al/Ag and Al/N Co-implanted ZnO thin films. *Coatings*, *12*(6), 733. https://doi.org/10.3390/COATINGS12060733

Martínez Ruano, J. A., Taimbu de la Cruz, C. A., Orrego Alzate, C. E., & Cardona Alzate, C. A. (2019). Techno–economic analysis of chitosan-based hydrogels production. In M. Mondal (Ed.), *Cellulose-Based Superabsorbent Hydrogels. Polymers and Polymeric Composites: A Reference Series* (pp. 1769–1790). Springer, Cham. https://doi.org/10.1007/978-3-319-77830-3_58

Mohod, A. V., Momotko, M., Shah, N. S., Marchel, M., Imran, M., Kong, L., & Boczkaj, G. (2023). Degradation of rhodamine dyes by Advanced Oxidation Processes (AOPs) – Focus on cavitation and photocatalysis - A critical review. *Water Resources and Industry*, *30*, 100220. https://doi.org/10.1016/J.WRI.2023.100220

Mondal, A., Prabhakaran, A., Gupta, S., & Subramanian, V. R. (2021). Boosting photocatalytic activity using Reduced Graphene Oxide (RGO)/semiconductor nanocomposites: Issues and future scope. *ACS Omega*, *6*(13), 8734–8743. https://doi.org/10.1021/ACSOMEGA.0C06045

Monsalve-Bravo, G. M., Moscoso-Vasquez, H. M., & Alvarez, H. (2014). Scaleup of batch reactors using phenomenological-based models. *Industrial and Engineering Chemistry Research*, *53*(22), 9439–9453. https://doi.org/10.1021/IE500587R

Morsy, M., Hassan, S. S., Shaker, H. M., & Ismail, K. M. (2025). Synthesis and characterization of MIL-88B(Fe) MOF for high-performance humidity sensing. *Sensors and Actuators B: Chemical*, *438*, 137806. https://doi.org/10.1016/J.SNB.2025.137806

Naciri, Y., Hsini, A., Bouziani, A., Djellabi, R., Ajmal, Z., Laabd, M., … Albourine, A. (2022). Photocatalytic oxidation of pollutants in gas-phase via Ag_3PO_4-based semiconductor photocatalysts: Recent progress, new trends, and future perspectives. *Critical Reviews in Environmental Science and Technology*, *52*(13), 2339–2382. https://doi.org/10.1080/10643389.2021.1877977

Navidpour, A. H., Xu, B., Ahmed, M. B., & Zhou, J. L. (2024). Immobilization of TiO_2 and ZnO by facile surface engineering methods to improve semiconductor performance in photocatalytic wastewater treatment: A review. *Materials Science in Semiconductor Processing*, *179*, 108518. https://doi.org/10.1016/J.MSSP.2024.108518

Nazir, A., Huo, P., Wang, H., Weiqiang, Z., & Wan, Y. (2023). A review on plasmonic-based heterojunction photocatalysts for degradation of organic pollutants in wastewater. *Journal of Materials Science*, *58*(15), 6474–6515. https://doi.org/10.1007/S10853-023-08391-W

Nguyen, T. B., & Doong, R. A. (2017). Heterostructured $ZnFe_2O_4/TiO_2$ nanocomposites with a highly recyclable visible-light-response for bisphenol A degradation. *RSC Advances*, *7*(79), 50006–50016. https://doi.org/10.1039/C7RA08271A

Nkwachukwu, O. V., & Arotiba, O. A. (2025). A perovskite/spinel structured $LaFeO_3/Co_3O_4$ heterojunction for photoelectrocatalytic applications in degradation of tetracycline and water splitting. *Journal of Power Sources*, *640*, 236790. https://doi.org/10.1016/J.JPOWSOUR.2025.236790

Ojha, S., Tripathi, S. M., Vishwakarma, P. K., & Mishra, S. (2024). Pharmaceuticals in the water: Emerging concerns and innovative remediation solutions. *Current Green Chemistry*, *11*(1), 50–62. https://doi.org/10.2174/0122133461272526231012064151

Padmavathy, N., & Vijayaraghavan, R. (2008). Enhanced bioactivity of ZnO nanoparticles—An antimicrobial study. *Science and Technology of Advanced Materials*, *9*(3), 35004–35011. https://doi.org/10.1088/1468-6996/9/3/035004

Palanivel, B., Lallimathi, M., Arjunkumar, B., Shkir, M., Alshahrani, T., Al-Namshah, K. S., … Ramalingam, G. (2021). rGO supported g-$C_3N_4/CoFe_2O_4$ heterojunction: Visible-light-active photocatalyst for effective utilization of H_2O_2 to organic pollutant degradation and OH radicals production. *Journal of Environmental Chemical Engineering*, *9*(1), 104698. https://doi.org/10.1016/J.JECE.2020.104698

Phoon, B. L., Lai, C. W., Juan, J. C., Show, P. L., & Chen, W. H. (2019). A review of synthesis and morphology of $SrTiO_3$ for energy and other applications. *International Journal of Energy Research*, *43*(10), 5151–5174. https://doi.org/10.1002/ER.4505

Qian, X., Xu, Y., & Xu, Y. (2024). Bacterial cellulose based TiO_2-CdS nanocomposite gel with enhanced photocatalytic activity for adsorptive degradation of cationic dye. *International Journal of Biological Macromolecules*, *259*, 127873. https://doi.org/10.1016/J.IJBIOMAC.2023.127873

Rahal, A., Bouchama, I., Ghebouli, M. A., Alanazi, F. K., Ghebouli, B., Fatmi, M., … Khettab, K. (2025). Experimental investigation of structural and optical properties of Mn-doped ZnO thin films deposited by pneumatic spray technique. *Scientific Reports*, *15*(1), 1–13. https://doi.org/10.1038/S41598-025-90425-1

Tanji, K., Zouheir, M., Hachhach, M., Ahmoum, H., Jellal, I., El Masaoudi, H., … Kherbeche, A. (2021). Design and simulation of a photocatalysis reactor for rhodamine B degradation using cobalt-doped ZnO film. *Reaction Kinetics, Mechanisms and Catalysis*, *134*(2), 1017–1038. https://doi.org/10.1007/s11144-021-02116-3

Tatarchuk, T., Soltys, L., & Macyk, W. (2023). Magnetic adsorbents for removal of pharmaceuticals: A review of adsorption properties. *Journal of Molecular Liquids*, *384*, 122174. https://doi.org/10.1016/J.MOLLIQ.2023.122174

Tuama, A. N., Alzubaidi, L. H., Jameel, M. H., Abass, K. H., bin Mayzan, M. Z. H., & Salman, Z. N. (2024). Impact of electron–hole recombination mechanism on the photocatalytic performance of ZnO in water treatment: A review. *Journal of Sol-Gel Science and Technology*, *110*(3), 792–806. https://doi.org/10.1007/S10971-024-06385-X

Verma, S., Mezgebe, B., Hejase, C. A., Sahle-Demessie, E., & Nadagouda, M. N. (2024). Photodegradation and photocatalysis of per- and polyfluoroalkyl substances (PFAS): A review of recent progress. *Next Materials*, *2*, 100077. https://doi.org/10.1016/J.NXMATE.2023.100077

Wang, B., Qian, K., Yang, W., An, W., Lou, L. L., Liu, S., & Yu, K. (2023). $ZnFe_2O_4/BiVO_4$ Z-scheme heterojunction for efficient visible-light photocatalytic degradation of ciprofloxacin. *Frontiers of Chemical Science and Engineering*, *17*(11), 1728–1740. https://doi.org/10.1007/S11705-023-2322-Z

Wang, N., Li, X., Yang, Y., Guo, T., Zhuang, X., Ji, S., … Zhou, Z. (2019). Enhanced photocatalytic degradation of sulfamethazine by Bi-doped TiO_2 nano-composites supported by powdered activated carbon under visible light irradiation. *Separation and Purification Technology*, *211*, 673–683. https://doi.org/10.1016/J.SEPPUR.2018.10.040

Wudil, Y. S., Ahmad, U. F., Gondal, M. A., Al-Osta, M. A., Almohammedi, A., Sa'id, R. S., … Mohamed, M. J. S. (2023). Tuning of graphitic carbon nitride (g-C_3N_4) for photocatalysis: A critical review. *Arabian Journal of Chemistry*, *16*(3), 104542. https://doi.org/10.1016/J.ARABJC.2023.104542

Xin, Z., He, Q., Wang, S., Han, X., Fu, Z., Xu, X., & Zhao, X. (2022). Recent progress in ZnO-based nanostructures for photocatalytic antimicrobial in water treatment: A review. *Applied Sciences*, *12*(15), 7910. https://doi.org/10.3390/APP12157910

Zhang, L., Li, X., Chen, S., Guan, J., Guo, Y., & Yu, W. (2023). 3D chitosan/GO/ZnO hydrogel with enhanced photocorrosion-resistance and adsorption for efficient removal of typical water-soluble pollutants. *Catalysis Communications*, *176*, 106627. https://doi.org/10.1016/J.CATCOM.2023.106627

Zhao, Y., Linghu, X., Shu, Y., Zhang, J., Chen, Z., Wu, Y., … Wang, B. (2022). Classification and catalytic mechanisms of heterojunction photocatalysts and the application of titanium dioxide (TiO_2)-based heterojunctions in environmental remediation. *Journal of Environmental Chemical Engineering*, *10*(3), 108077. https://doi.org/10.1016/J.JECE.2022.108077

Zheng, F., Queirós, J. M., Martins, P. M., de Luis, R. F., Fidalgo-Marijuan, A., Vilas-Vilela, J. L., … Reguera, J. (2023). Au-sensitised TiO_2 and ZnO nanoparticles for broadband pharmaceuticals photocatalytic degradation in water remediation. *Colloids and Surfaces A: Physicochemical and Engineering Aspects*, *671*, 131594. https://doi.org/10.1016/J.COLSURFA.2023.131594

Zhou, S., Gao, J., Zhu, J., Peng, D., Zhang, Y., & Zhang, Y. (2020). Self-cleaning, antibacterial mixed matrix membranes enabled by photocatalyst Ti-MOFs for efficient dye removal. *Journal of Membrane Science*, *610*, 118219. https://doi.org/10.1016/J.MEMSCI.2020.118219

Zhu, C., & Wang, X. (2025). Nanomaterial ZnO synthesis and its photocatalytic applications: A review. *Nanomaterials*, *15*(9), 682. https://doi.org/10.3390/NANO15090682

10 Microalgal Biomass

Applications and Mechanisms in Removing Conventional and Industrial Pollutants

Radouane El Amri, Rajaa Zahnoune, Otmane Boudouch, Noureddine Kamil, Saad ALjlil, Mounir Akhatar, Nour-Eddine El Mansouri, and Reda Elkacmi

10.1 INTRODUCTION

Water pollution is a major threat to society and the environment. Industrial, domestic, and agricultural effluent discharges are the main sources of water contamination, containing hazardous substances such as HMs, dyes, pharmaceuticals, organic matter, nutrients, and pathogens. When these effluents are discharged into natural ecosystems without undergoing appropriate treatment, this leads to the destruction of ecosystems, affects public health, and alters water resources (Gonçalves et al., 2017; Elkacmi & Bennajah, 2019; Rawat et al., 2011). Consequently, it has become crucial to treat this wastewater before discharging it in order to protect the environment and preserve water resources.

Biological treatment using microalgae appears to be a promising alternative for integrating or replacing conventional treatment systems used to purify wastewater. Microalgae are microorganisms capable of photosynthetic activity (Tomaselli, 2004). They have a faster growth rate and use less surface area than terrestrial plants (Clarens et al., 2010). Several studies have demonstrated the ability of microalgae to remove organic matter, nutrients, HMs, synthetic dyes, and pharmaceutical pollutants from wastewater (El Amri et al., 2022; Spain et al., 2021; Fan et al., 2020; Fito et al., 2019; Chin et al., 2020). Other recent studies have reported on the potential of microalgae–bacteria consortia in several applications, such as algal biomass production and nutrient removal (Bélanger-lépine et al., 2018; Delgadillo-Mirquez, 2016; Su et al., 2012).

A symbiotic interaction has been found in microalgae–bacteria systems. Bacteria oxidize organic matter with the oxygen released by microalgae during photosynthesis in exchange for the carbon dioxide required for photosynthetic reactions (Gonçalves et al., 2017). In general, microalgae-based liquid effluent treatment is considered a

DOI: 10.1201/9781003624905-10

relatively cost-effective method compared to other wastewater treatment technologies, especially when microalgae are grown from free nutrient sources, as in the case of using municipal wastewater, which is very rich in nitrogen, phosphorus, and elements essential for microalgal growth (Plöhn et al., 2022). In addition, sunlight can be used as a source of illumination, particularly in countries with long periods of sunshine, thus further reducing cultivation costs.

This chapter aims to explore the potential of microalgae for the removal of conventional and industrial pollutants from wastewater, highlighting the biological, biochemical, and physicochemical mechanisms involved in these processes.

10.2 REMOVAL OF CONVENTIONAL POLLUTANTS

Microalgae-based wastewater treatment offers superior nutrient removal, meeting increasingly stringent discharge and reuse standards, as microalgae depend mainly on P and N for protein synthesis. In addition, microalgae play an indirect role in the degradation of organic carbon and nutrients. Through photosynthesis, they use CO_2 and produce the oxygen needed to degrade organic forms of C, N, and P, converting them into CO_2 and inorganic N and P through the action of heterotrophic bacteria. Microalgae can then use these inorganic substances in a new phase of photosynthesis to synthesize biomass (Chai et al., 2021).

The growing interest in microalgae-based wastewater treatment is due to their low energy requirements, their ability to thrive in different environmental conditions, and their ability to recover and transform nutrients present in wastewater into valuable bioactive compounds. A wide range of microalgal strains have developed well in municipal wastewater, but a limited range can be incorporated into agricultural and industrial wastewater. In general, different types of wastewater have varying concentrations of nutrients and pollutants, and each strain of microalgae has its own ideal growth conditions. Consequently, it is necessary to formulate nutrients in such a way as to provide optimal conditions for microalgal growth (Abdelfattah et al., 2022).

Microalgal species frequently used in wastewater treatment and nutrient and organic matter removal include *Scenedesmus obliquus*, *Chlorella vulgaris*, and *Chlorella sorokiniana* (Abdelfattah et al., 2022). A symbiotic relationship between microalgae and aerobic bacteria present in wastewater has been identified as the main driver of chemical oxygen demand (COD) removal. In this relationship, microalgae provide aerobic bacteria with essential oxygen to oxidize organic matter (Mujtaba and Lee, 2016). Microalgae can use various organic compounds as carbon sources in addition to CO_2. Some *Chlorella* species are mixotrophic, meaning they can simultaneously use light and organic sources as energy sources, and assimilate CO_2 and glucose or simple acids as carbon sources. It has been found that when *Chlorella pyrenoidosa* is grown under moderate light conditions, an exponential mixotrophic phase initially occurs. When the organic substrate is completely consumed, an autotrophic phase follows after a period of adaptation (Li et al., 2011). Gonzalez-Camejo et al. (2018) evaluated the effect of light in a microalgae–bacteria hybrid consortium on NH_4^+ removal and microalgal growth. In the case of NH_4^+ removal, light intensity favored NH_4^+ uptake by microalgae over NH_4^+ oxidation by nitrifying bacteria due to microalgal proliferation (Gonzalez-Camejo et al., 2018). Arcila and Buitron (2017) demonstrated that the maximum values for biomass productivity, dissolved oxygen, PO_4^{3-} removal (92%), and pH were obtained under high irradiation conditions with a temperature of 26°C, and that these conditions hindered the nitrification process. The improvement in P removal was attributed to the P precipitation mechanism at higher pH values. Conversely, with low irradiation levels, higher removal efficiencies for total nitrogen (TN) (60%) and COD (89%) were observed with moderate PO_4^{3-} removal (28%) (Arcila and Buitrón, 2017).

The effect of light/dark regime and initial sludge/microalgae ratio for domestic wastewater treatment using a hybrid consortium of *Chlorella sp* and activated sludge was investigated. The hybrid culture showed superior nutrient and COD removal, better settleability, and lower oxygen consumption, compared to activated sludge alone under conditions of 12 h light and 12 h dark, with a culture period of five days and an initial sludge/microalgae ratio of 1:2, highlighting the economic and practical feasibility of cooperation between *Chlorella sp* and activated sludge as a less light-dependent system for domestic wastewater treatment (Abdelfattah et al., 2022).

Nutrient concentrations and C/N and N/P ratios can affect microalgal growth and thus treatment performance. Nutrient concentrations fluctuated remarkably between sampling points at the municipal wastewater treatment plant. Wang et al (2010) concluded that growing microalgae in nutrient-rich sewage sludge was more practical in terms of microalgal growth and COD, PO_4^{3-}, NO_3^-, and NH_4^+ removal compared to primary and secondary treated effluent (Wang et al., 2010). Zheng et al (2018) mixed piggery wastewater with brewery wastewater to determine the optimal C/N ratio for nutrient removal and microalgal growth. At an optimal C/N ratio of 7.9, NH_4^+, TN, total phosphorus (TP), and COD removal efficiencies reached 100%, 96%, 90%, and 93%, respectively (Zheng et al., 2018). Culturing *Scenedesmus sp* in molasses wastewater with a C/N ratio of 15 resulted in better treatment with removal efficiencies of 87.2%, 90.5%, and 88.6% for COD, TN, and TP, respectively, at a temperature of 10°C (Ma et al., 2017). AlMomani and Ormeci (2016) investigated the feasibility of using *Chlorella vulgaris*, *Neochloris oleoabundans*, and mixed native microalgae in terms of microalgal growth and organic carbon, N, and P removal from primary effluent, secondary effluent, and centrate (effluent containing high levels of nutrients). The results showed that the appropriate culture for growth rate and nutrient removal varied according to wastewater type, and that no single wastewater type was the most favorable for culturing all microalgal species. The highest growth rate and corresponding maximum C, N, and P removal for *Chlorella vulgaris* and *Neochloris oleoabundans* were observed in primary wastewater with a C/N/P ratio of 24:5:1 and for mixed microalgal culture in secondary sludge with a C/N/P ratio of 4.4:1:1.5. Therefore, nutrient composition and ratios in different municipal effluents in microalgal treatment systems need to be fully evaluated (AlMomani and Örmeci, 2016).

10.3 REMOVAL OF INDUSTRIAL POLLUTANTS

Several treatment technologies, such as activated sludge, membrane processes, advanced oxidation, and electrochemical treatments, have been developed to remove industrial pollutants. However, most of these approaches have shortcomings, such as incomplete removal of HMs, high maintenance and operating costs, and the production of sludge or other toxic by-products. Biological treatment, in particular microalgae-based wastewater treatment, may be one of the most promising, environmentally friendly, and sustainable alternatives. Microalgae are capable of metabolizing hazardous materials into non-toxic substances. This is an affordable process that offers the dual benefit of wastewater purification and simultaneous biomass harvesting, due to their efficient ability to absorb inorganic nutrients and HMs, as well as sorbing many toxic pollutants present in wastewater. This section discusses the application of microalgae for the removal of toxic substances present in industrial wastewater, in particular the removal of HMs, dyes, and pharmaceuticals.

10.3.1 Removal of HMs

HMs are a class of metals with an atomic density greater than 4,000 $kg{\cdot}m^{-3}$. Due to their non-biodegradable properties, abundant sources, toxicity, and cumulative behavior, HM pollution is a critical global problem. HMs are naturally toxic and cause serious illness in humans and animals, even at low concentrations. They fall into three categories: radionuclides such as uranium (U), radium (Ra), americium (Am), and thorium (Th), precious metals such as gold (Au), palladium (Pd), platinum (Pt), and ruthenium (Ru), and toxic metals such as copper (Cu), chromium (Cr), arsenic (As), zinc (Zn), nickel (Ni), silver (Ag), tin (Sn), cobalt (Co), and lead (Pb). In addition to their natural presence, HMs enter aquatic systems via industrial discharges and agricultural runoff (Abdelfattah et al., 2022). Various treatment methods are available to remove HMs from the aquatic environment, with varying degrees of success. In all cases, these conventional treatment methods entail high operating and maintenance costs and produce secondary waste (Vardhan et al., 2019). It is therefore essential to develop robust, environmentally friendly, and cost-effective treatments.

Site characteristics, the extent of HM contamination, and the regulatory limits for the HMs concerned in this regulatory area all have an impact on the treatment strategy chosen. Chemical, physical, and biological approaches are the three basic categories of treatment techniques for removing HMs from wastewater using various types of microalgae (chemisorption, physisorption, bioaccumulation, biodegradation…etc.). The main binding sites for the removal of HMs from wastewater are the amino, hydroxyl, carboxyl, and sulfate functional groups present in the cell walls of microalgae. Due to the presence of these negatively charged groups, which enable ions to be bound from the surrounding environment, the outermost layer of the cell wall is the primary participant in HM removal. The removal of HMs by microalgae is mainly influenced by their properties, the type of HM, and the characteristics of the effluent. Microalgae (in freshwater) are the species most commonly used to remove various HMs from wastewater. A few green microalgae, including *Chlorella miniata*, *Chlorella vulgaris*, *Chlorella reinhardtii*, and *Sphaeroplea*, have been successfully used to remove toxic HMs from seawater (Abdelfattah et al., 2022). Table 10.1 summarizes the majority of previous studies on HM removal by microalgae.

TABLE 10.1
Overview of Previous Studies on HM Biosorption by Microalgae

HMs	[HM] ($mg \cdot L^{-1}$)	Treatment Method	Microalga Used	[Biomass] ($g \cdot L^{-1}$)	Biomass Type	References
Cadmium	50	Biosorption	*Spirulina platensis*	1	Non-living	Gunasundari (2017)
Cadmium	50	Biosorption	*Spirulina platensis*	0.6	Non-living	Gunasundari (2017)
Chromium	50	Biosorption	*Spirulina platensis*	1	Non-living	Gunasundari (2017)
Chromium	50	Biosorption	*Spirulina sp.*	3	Non-living	Rezaei (2016)
Chromium	50	Biosorption	*Chlorella vulgaris*	2	Non-living	Sibi (2016)
Cadmium	50	Biosorption	*Anabaena sphaerica*	0.25	Non-living	Abdel-Aty et al. (2013)
Lead	50	Biosorption	*Anabaena sphaerica*	0.25	Non-living	Abdel-Aty et al. (2013)
Lead	100	Biosorption	*Pseudochlorococcum typicum*	4.52 $mg \cdot mL^{-1}$ (chlorophyll a)	Living	Shanab et al. (2012)
Mercury	100	Biosorption	*Pseudochlorococcum typicum*	4.52 $mg \cdot mL^{-1}$ (chlorophyll a)	Living	Shanab et al. (2012)
Lead	-	Biosorption	*Chlorella vulgaris*	2	Non-living	Ferreira et al. (2011)
Nickel	-	Biosorption	*Chlorella vulgaris*	2	Non-living	Ferreira et al. (2011)
Zinc	-	Fixed-bed column	*Chlorella vulgaris*	2	Non-living	Ferreira et al. (2011)
Arsenic	6	Biosorption	*Maugeotia genuflexa*	4	Non-living	Sarı et al. (2011)

10.3.2 Removal of Dyes

Due to their many applications in diverse industries such as textiles, paper, and food manufacturing, the use of dyes is constantly increasing. Dyes discharged into ecosystems have unattractive aesthetic effects and serious environmental repercussions. Phytoremediation technology used to treat dye-contaminated water has gained importance in recent years. The cell walls of microalgae contain a variety of functional groups that contribute to the dye removal process. Many variables are essential for dye removal, including contact time, pH, temperature, initial dye concentration, and adsorbent dosage. Consequently, these criteria are taken into account to determine the efficiency of dye treatment by microalgae (Ayele et al., 2021).

Treatment of dye-contaminated wastewater is generally carried out using traditional techniques such as chemical, physicochemical, and biological approaches. Numerous chemical and physical procedures, including adsorption, advanced oxidation processes, ozonation, ultrafiltration, photo-catalytic oxidation, flocculation coagulation, and chemical and electrochemical coagulation, have been used in the past to remove dyes (Abdelfattah et al., 2022). Numerous limitations have been identified with all the above methods, including inefficient dye removal, sludge production, short O_3 half-life (20 min), generation of by-products, release of aromatic amines, high electricity costs, ineffectiveness against dispersant dyes and vat dyes (insoluble dyes rendered soluble with reducing agents in strongly alkaline media), and repetitive or frequent use of chemicals leading to secondary pollution problems. Consequently, an alternative technique for dye reduction is considered necessary. Bioremediation using microalgae is considered the best alternative to traditional physicochemical treatment. The uptake of dyes by microalgae has several advantages, including being environmentally safe, having a high adsorption efficiency, being a simple process, using readily available materials, and being inexpensive (Ayele et al., 2021). Several investigations have been carried out to examine the ability of microalgae to remove various types of dyes, as summarized in Table 10.2.

TABLE 10.2
Dye Removal by Microalgae

Dye	Microalgae	Dye Concentration ($mg \cdot L^{-1}$)	% Removal	References
Methylene blue	*Chlorella vulgaris*	100	83.04	Chin et al. (2020)
Reactive Black 5	*Chlorella vulgaris*	200	80	Ishchi and Sibi (2020)
Disperse Red 1	*Chlorella vulgaris*	300	84	Ishchi and Sibi (2020)
Malachite green	*Oscillatoria sp.*	5	93	Gelebo et al. (2020)
Safranin	*Oscillatoria sp.*	52	—	Gelebo et al. (2020)
Congo red	*Chlorella vulgaris*	50	100	Naji and Salman (2019)
Yellow dye	*Chlorella vulgaris*	10	3.12	Raymond and Omo (2019)
Aniline blue	*Chlorella vulgaris*	25	58	Arteaga et al. (2018)
Malachite green	*Haematococcus sp.*	100	67	Liu et al. (2018)
Acid Black 210	*Spirulina platensis*	125	98.55	Al Hamadi et al. (2017)
Acid Blue 7	*Spirulina platensis*	—	97.05	Al Hamadi et al. (2017)
Blue dye	*Oscillatoria sp.*	—	76.48	Brahmbhatt and Jasrai (2016)
Red dye	*Oscillatoria sp.*	—	62.63	Brahmbhatt and Jasrai (2016)
Blue dye	*Spirogyra sp.*	—	78.29	Brahmbhatt and Jasrai (2016)
Red dye	*Spirogyra sp.*	—	64.21	Brahmbhatt and Jasrai (2016)
Methylene blue	*Desmodesmus sp.*	20	98.6	Al-Fawwaz and Abdullah (2016)
Rhodamine B	*Coelastrella sp.*	100	80	Baldev et al. (2013)
Malachite green	*Chlorella vulgaris*	6	91.61	Kousha et al. (2013)
Malachite green	*Nostoc sp.*	100	97.13	Gajare and Menghani (2012)
Malachite green	*Vaucheria sp.*	—	85.9	Khataee et al. (2010)
Monoazo and diazo dyes	*Scenedesmus bijuga*	—	68	Omar (2008)
Malachite green	*Chlorella sp.*	—	80.7	Daneshvar et al. (2007)
Malachite green	*Cosmarium sp.*	10	87.1	Daneshvar et al. (2007)
Synazol	*Spirogyra sp.*	—	85	Khalaf (2008)
Methyl orange	*Chlorella sp.*	30	98.53	El Amri et al. (2023)

10.3.3 Removal of Pharmaceutical Compounds (PCs)

PCs constitute a vast category of chemical substances widely used throughout the world. Farming operations, hospital effluents, industrial pollution, and household waste all pollute aquatic environments with PC residues. Although most PC concentrations in the environment range from $ng{\cdot}L^{-1}$ to a few $mg{\cdot}L^{-1}$, there is growing evidence that they have negative ecological effects on target and non-target microorganisms, such as inhibiting microbial growth, modifying microbial communities, reducing soil microbial activity, and affecting denitrification rates. *N,N*-Nitrosodimethylamine is formed when amine-based PCs (e.g., doxylamine, nizatidine, carbinoxamine, and ranitidine) interact with disinfectants used to disinfect drinking water. *N,N*-Nitrosodimethylamine poses a significant risk to human health due to its carcinogenic properties (Kelly et al., 2007).

The scientific community is becoming increasingly interested in the biotreatment of PCs using microalgae. The advantages of these microalgae are that they run on solar energy, have low operating costs, are environmentally friendly, and contribute to carbon fixation and renewal. Microalgae have the ability to overcome many of the constraints associated with bacteria and fungi, which require a stoichiometric balance of carbon and other nutrients for growth and pollutant decomposition. Because of their adaptability, mixotrophic microalgae can be used to remove pharmaceutical pollutants from wastewater (Abdelfattah et al., 2022). The ability of mixotrophic microalgae to extract and absorb PCs from wastewater has been reported using a variety of microalgal species. The growth of microalgae in wastewater reduces the demand for chemical nutrients, which are required to prepare artificial growth media, and thus reduces the impact on the environment (Abdelfattah et al., 2022). Mechanisms such as biosorption, bioaccumulation, and biodegradation are all involved in the removal of PCs from the environment by microalgae. It has been reported that the removal mechanisms of various PCs by microalgae vary according to the physicochemical properties of the targeted PCs. For example, a study by De Godos et al. (2012) showed that *Chlorella vulgaris* can remove more than 50% of tetracycline in a High-Rate Algal Pond by adsorption due to the interaction of positively charged molecules on the microalgal cell surface (de Godos et al., 2012). Sulfamethoxazole, carbamazepine, trimethoprim, and florfenicol were bioaccumulated by passive diffusion using *Chlorella sp* (Song et al., 2019). Zhang et al (2014) demonstrated that *Scenedesmus dimorphus* can biodegrade around 85% of 17 α-estradiol in seven days (Zhang et al., 2014).

Extracellular enzymes produced by microalgae break down certain antibiotics into less toxic or even non-toxic intermediates. These intermediates are then bioaccumulated and biodegraded by cellular enzymes. In addition, the degree of structural complexity determines the degree of biodegradability. Ketoprofen and ibuprofen, classified as non-steroidal anti-inflammatory drugs, are both eliminated from wastewater in different ways. Using vascular plants and a microalgae-based treatment, ibuprofen was eliminated more efficiently than ketoprofen. Both paracetamol and ibuprofen were completely destroyed by *Chlorella sorokiniana* (Abdelfattah et al., 2022). The matrix of extracellular polymeric substances (EPS) facilitates the adsorption, accumulation, and, ultimately, biodegradation of various substances. The presence of carboxyl, hydroxyl, and phosphoryl functional groups is responsible for the predominant negative charge of microalgae, which has been shown to favor the uptake of positively charged PC molecules. The rapid uptake of PCs onto the cell wall using solid phase extractors (SPEs) has been demonstrated by numerous recent studies. In general, studies suggest that PCs are eliminated by several different mechanisms, with bioadsorption being the most widespread. The contribution of bioadsorption varies according to the microalgae used and the micropollutant absorbed (Abdelfattah et al., 2022). Some PCs adsorbed to the surface of microalgae can enter the cell and bioaccumulate. As an active process, bioaccumulation requires energy. The first line of defense is provided by the EPS and the cell wall. Quantifying bioaccumulation and bioadsorption is a difficult task, as both processes are ongoing (Abdelfattah et al., 2022). Table 10.3 summarizes various studies on the removal of various pharmaceutical pollutants from different matrixes using microalgae.

10.4 MECHANISMS INVOLVED IN MICROALGAL BIOTREATMENT

The elimination of pollutants through the use of microalgae involves various mechanisms, such as biosorption, bioaccumulation, and biodegradation, particularly in the case of industrial pollutants (pharmaceutical substances, dyes, HMs, etc.), and assimilation in the case of conventional pollutants (carbon, nitrogen, phosphorus, etc.).

TABLE 10.3
Removal of Pharmaceutical Pollutants by Microalgae

Type of Wastewater	Pollutants	Microalgae	% Removal	References
Lake water	Ibuprofen	*Nannochloris sp.*	40	Bai and Acharya (2017)
	Trimethoprim		10	
	Ciprofloxacin		100	
	Carbamazepine		20	
	Triclosan		100	
Urine, anaerobically treated blackwater, and synthetic urine	Diclofenac	*Chlorella sorokiniana*	40–60	de Wilt et al. (2016)
	Ibuprofen		100	
	Paracetamol		100	
	Metoprolol		100	
	Carbamazepine		30	
	Trimethoprim		40	
Wastewater digestate and growth medium	Estradiol	*Selenastrum capricornutum*	88–100	Hom-Diaz et al. (2015)
Urban wastewater	Acetaminophen	Microalgal consortia in high-rate algal ponds	99	Matamoros et al. (2015)
	Ibuprofen		99	
	Carbamazepine		62	
Urban or synthetic wastewater	Carbamazepine	Microalgal consortia in high-rate algal ponds dominated by *Chlorella sp.* and *Scenedesmus sp.*	20	Matamoros et al. (2016)
Urban wastewater	Hydroxychloroquine	*Chlorella sp.*	88.96	El Amri et al. (2022)

10.4.1 Microalgal Pathways for Conventional Pollutant Removal

Microalgae possess specific mechanisms for the elimination of each conventional pollutant, which may involve nutrient storage pathways in the biomass or their assimilation into molecules essential for microalgal growth. Microalgae are mainly photoautotrophs, using light as an energy source and CO_2 from simulated or real pure gas streams or bicarbonates $\left(HCO_3^-\right)$ in the culture medium as a carbon source, converting it into energy-storing molecules such as carbohydrates through photosynthesis and CO_2 fixation (Singh et al., 2019). Most microalgal species possess carbon-concentrating mechanisms that involve the reversible intracellular or extracellular conversion of HCO_3^- to CO_2. These mechanisms enable efficient CO_2 fixation by obtaining inorganic carbon, even when the atmospheric CO_2 concentration is low (Singh et al., 2019).

Microalgal growth occurs mainly at pH values between 7.0 and 8.4, at which HCO_3^- generally accounts for most of the inorganic carbon available in the culture medium. Inorganic carbon can enter the microalgal cell by three processes: (i) direct passive diffusion of CO_2; (ii) conversion of HCO_3^- to CO_2, catalyzed by extracellular carbonic anhydrase, followed by passive diffusion of CO_2; and (iii) direct active transport of HCO_3^-, aided by inorganic carbon uptake transporters. However, the conversion of HCO_3^- to CO_2 produces hydroxyl radicals, which can contribute to increasing the pH of the culture. Inside the cell, CO_2 is fixed in a reaction that triggers the Calvin-Benson cycle (Figure 10.1), catalyzed by ribulose biphosphate carboxylase oxygenase (Rubisco), ultimately yielding carbohydrates. Although CO_2 is the preferred carbon source, in the presence of certain organic molecules (e.g., sugars, acetate, and

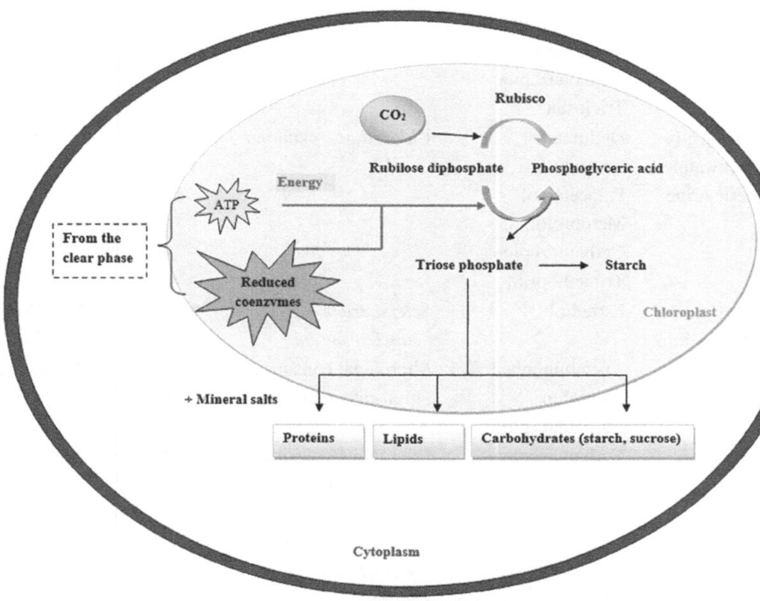

FIGURE 10.1 Calvin-Benson cycle.

alcohols) and in the absence of light, some microalgae, such as *Chlorella sp*, can adopt a heterotrophic metabolism, using these molecules as a carbon source (Amaro et al., 2023). In addition, mixotrophic metabolism is common in some of these microorganisms, in which CO_2 fixation coincides with photosynthesis and organic compounds are optionally used as an energy source for respiration (Amaro et al., 2023).

Nitrogen is a crucial nutrient for microalgae; it is found in peptides and proteins, including enzymes, chlorophylls, and nucleic acids, and is therefore essential for the growth of microalgae and the regulation of their metabolism. Microalgae can assimilate this nutrient from wastewater in the form of inorganic nitrogen (Figure 10.2), such as NH_4^+, NO_2^-, NO_3^-, and organic nitrogen, such as urea, purines, amino acids, and pyrimidines (Amaro et al., 2023). Inorganic nitrogen species are actively transported into the microalgal cell and assimilated by the enzymatic glutamine synthetase/glutamate synthase (also known as glutamine oxoglutarate aminotransferase) GS/GOGAT pathway. NH_4^+ can be directly incorporated into 2-oxoglutarate, leading to glutamate production and, consequently, amino acid synthesis via the GS/GOGAT cycle. However, NO_3^- and NO_2^- must be reduced to NH_4^+ before they can be assimilated; NO_3 is mainly converted to NO_2^- in the cytosol by the action of nitrate reductase, and NO_2^- is then reduced to $NH_4{}^+$ in the chloroplast, in a reaction catalyzed by nitrite reductase. As a result, NH_4^+ is the preferred and most readily absorbed form of nitrogen for most microalgae, as it does not involve redox reactions and therefore has a lower energy demand than other nitrogen sources (Amaro et al., 2023). In addition, some cyanobacteria, formerly described as blue-green algae, can also fix molecular nitrogen, converting it to ammonia (NH_3) in a reaction catalyzed by a nitrogenase

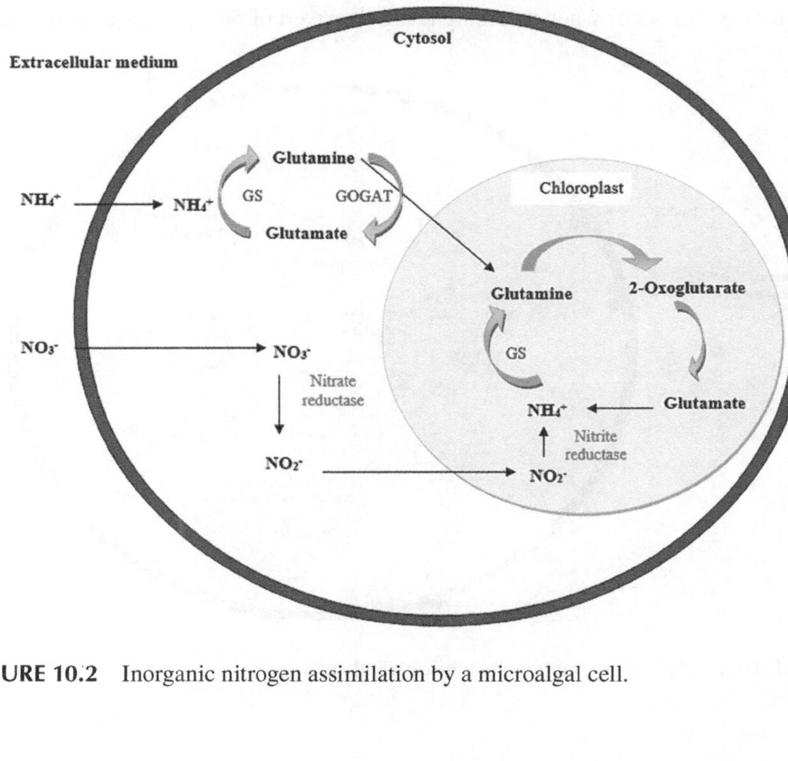

FIGURE 10.2 Inorganic nitrogen assimilation by a microalgal cell.

enzyme complex, which can be converted to NH_4^+ for assimilation. In addition to the production of hydroxyl radicals, the positive uptake of H^+ via the co-transport of nutrients such as N and P across the microalgal cell membrane can lead to an increase in the pH of the medium. As a result of this increase in pH or elevated temperatures, indirect removal of $NH_4{}^+$ from wastewater by microalgae can occur via its conversion to NH_3 gas and subsequent volatilization into water. However, it should be noted that this is not a desirable event, as NH_3 is a precursor of the greenhouse gas N_2O (Amaro et al., 2023). Ammonia reacts in the presence of water, according to Eq. (10.1), forming an ammonia/ammonium system, which is in equilibrium as a function of pH.

$$NH_4^+ + OH^- \Leftrightarrow NH_3 + H_2O \tag{10.1}$$

In wastewater, phosphorus can exist in the form of organic or inorganic compounds. However, the main source of phosphorus for microalgae is inorganic phosphate in the form of $H_2PO_4^-$, HPO_4^{2-} or PO_4^{3-}. Phosphorus uptake by a microalgal cell is shown in Figure 10.3. These phosphate species are transferred across the cell membrane by active transport using inorganic P transporters. Once inside the cell, they can be used directly to synthesize nucleic acids, phospholipids, and adenosine triphosphate by phosphorylation (Amaro et al., 2023). Furthermore, in the presence of light, phosphates can be converted to acid-soluble polyphosphates in a reaction catalyzed by polyphosphate kinase and used for the synthesis of deoxyribonucleic acid and proteins. When phosphorus is in excess, some microalgae can also convert phosphates into acid-insoluble polyphosphates and store them as granules inside the vacuole. This process can occur when microalgae are deprived of phosphorus and then re-fed

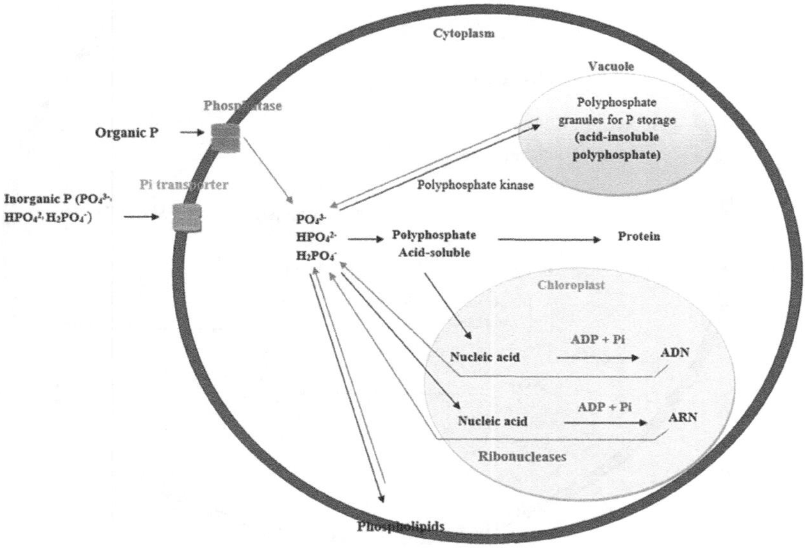

FIGURE 10.3 Phosphorus uptake by a microalgal cell.

(starvation uptake) or when exposed to excess phosphorus (luxury uptake), consuming more than is required for growth (Amaro et al., 2023). On the contrary, under conditions of low inorganic P availability, some microalgae possess phosphatases to consume external organic phosphorus present in compounds, such as phosphate esters, to support essential cellular processes. In addition to uptake and storage by microalgae, phosphorus can also be indirectly removed from wastewater through precipitation by complexing with calcium, magnesium, and iron, due to the high pH (>8) resulting from microalgal growth (Amaro et al., 2023).

10.4.2 Factors Influencing the Removal of Conventional Pollutants

Several abiotic and biotic factors influence the effectiveness of microalgal systems in removing conventional pollutants. Among abiotic factors, pH, temperature, light intensity, hydraulic retention time, wastewater composition, and water turbidity are the most critical. With regard to biotic factors, the microalgae–bacteria ratio and microalgal species are the most studied (Amaro et al., 2023). As for abiotic factors, pH can significantly affect N and P removal, mainly due to phosphorus precipitation and NH_3 volatilization under alkaline conditions. The availability of free CO_2 increases with higher pH, which favors its uptake. However, extremely high pH values can inhibit microalgal metabolism and reduce nutrient uptake (Amaro et al., 2023).

Overall, the removal of conventional pollutants and biomass productivity are improved with increasing temperature and light intensity. The ideal temperature for microalgal growth is generally between 15°C and 30°C. Lower or higher temperatures can reduce the activity of certain enzymes, such as Rubisco, inhibiting carbon sequestration, and temperatures above 35°C can be lethal for many species. This has a direct impact on conventional nutrient removal (Amaro et al., 2023). Furthermore, if light intensity is too high, exceeding the species-specific saturation point, oxidative damage can occur due to photoinhibition of microalgae, resulting in reduced nutrient removal. High light intensity can also affect nitrite-oxidizing bacteria. Light intensities above 180 $\mu mol \cdot m^{-2} \cdot s^{-1}$ induce inhibition of nitrite-oxidizing bacteria, leading to a decrease in biotreatment efficiency and an accumulation of nitrate nitrogen (NO_2^-) in the medium (Amaro et al., 2023). Furthermore, low hydraulic retention time values can lead to a washout of the microbial population composition in the photobioreactor, reducing the number of microalgal cells and altering the composition of the bacterial community, particularly in terms of dominance. This event can lead to a reduction in the efficiency of microalgal systems in removing specific conventional pollutants (Amaro et al., 2023).

The mechanisms described for nitrogen, phosphorus, and carbon removal by microalgae appear to be linked; consequently, external nutrient concentration is also a key factor, with N/P mass ratio values between 5 and 30 being considered optimal. For example, low nitrogen concentrations can limit phosphorus uptake, as nitrogen is needed to synthesize the proteins that enable phosphorus assimilation. In addition, phosphorus limitation can reduce adenosine triphosphate synthesis, which is essential for microalgal metabolism, such as carbon concentration mechanisms (Amaro et al., 2023). In some cases, wastewater pretreatment is required to ensure effective bioremediation by microalgae due to high total suspended solids concentration (and therefore turbidity) and NH_3 content, which negatively affect microalgal growth. For example, pretreatment by sedimentation can enhance bioremediation performance in wastewater (Amaro et al., 2023). However, depending on the microalgal species, sedimentation may not be sufficient, and filtration or dilution pre-treatments are sometimes necessary due to the ammonium concentration tolerance of many microalgal species, which should ideally be between 25 and 1,000 $\mu mol{\cdot}L^{-1}$. On the other hand, industrial wastewater has very different compositions depending on the type of industry, whether food (dairy, palm oil mill...), textile, pharmaceutical, or paper. So, to ensure effective treatment, dilution or mixing of different types of wastewaters can also be useful to achieve optimal conditions for microalgal growth, depending on the elemental stoichiometry of the microalgae, particularly with regard to the ratio of conventional pollutants (Amaro et al., 2023).

The effects of biotic factors are linked to the presence of microorganisms other than microalgae and to their interactions, particularly with bacteria, due to their cooperative or inhibitory effects. Several authors have reported that interactions between microalgae and bacteria can change depending on the species used, the ratio between these two types of organisms, and the culture phase (Amaro et al., 2023). In a microalgae–bacteria consortium, in particular, competition for nutrients and space between these two types of microorganisms can lead to the suppression of the growth of one type of microorganism. This was observed in a system composed of *Chlorella sp* and *Bacillus firmus*, where, in an advanced phase of the culture, the growth of the microalgae was inhibited, even though there were still nutrients in the culture medium, probably due to the increased concentration of bacteria in the culture (Huo et al., 2020). However, this dynamic between the concentration of microalgae and bacteria is highly species-specific. When the pair *Chlorella sp* and *Beijerinckia fluminensis* was used, the opposite situation occurred, with the number of microalgae increasing at an advanced stage of culture, while the number of bacteria decreased significantly. This phenomenon is probably due to an increase in the metabolites released by the microalgae, which may have a bactericidal effect (Huo et al., 2020).

10.4.3 Industrial Pollutant Removal Mechanisms

10.4.3.1 Biosorption

Biosorption is considered a passive process in which the sorbent is a biological material that can bind and concentrate water pollutants. In other words, biosorption is a mass transfer process in which a material is displaced from the liquid phase and bound to the surface of a solid. Biosorption comprises physical, chemical, and metabolism-independent processes, supported by different mechanisms including precipitation, ion exchange, surface complexation, electrostatic interaction, absorption, and adsorption (Mustafa et al., 2021). The mechanisms governing this phenomenon require a biosorbent (solid phase) and a target sorbate that is dissolved or dispersed in water. The biomaterial involved can be living or dead microorganisms and even their components. The immense affinity of the biosorbent for the target sorbent is the driving force behind the attraction of sorbent species, and the total capacity is related to the number of sorbent molecules. The process continues until an equilibrium is reached between the substance adsorbed by the biosorbent and its remaining concentration in the liquid. In addition, the degree of affinity of the biosorbent for a specific sorbent is responsible for its distribution between the solid and liquid phases (Abdelfattah et al., 2022).

Biosorption was first observed in several microalgae in the early 1970s, when radionuclides and Mixed Liquor Suspended Solids (MLs) released by a nuclear reactor were concentrated in microalgae. The cell wall of microalgae is directly responsible for biosorption, and its chemical composition plays an essential role in the process and determsines the mechanism by which this phenomenon occurs. In addition, pores are present on the surface of microalgae, and their surface charge promotes biosorption. Various chemical groups present in the cell wall of microalgae, such as carboxyls, hydroxyls, and sulfates, serve as binding sites and are also effective ion exchangers, contributing to the complexation of metal ions and the adsorption of organic substances present in polluted waters (Soto-Ramírez et al., 2021). In addition, other molecules such as lipids, proteins, and nucleic acids can be deposited on the cell surface; however, they are mainly present in the cytoplasm and plasma membrane and are able to bind metal cations thanks to their different functional groups (carboxylic, aminic, thiolic, imidazolic, thioesteric, nitrogen, and oxygen in peptide bonds). In general, the cell wall structure of microalgae comprises a fibrillar matrix that provides high mechanical strength, while its flexibility is due to the amorphous fraction, and both these elements, together with the intercellular spaces present on the cell wall, can enhance the biosorption process (Kumar et al., 2021). The porosity and surface roughness of microalgae can also influence the adsorption of HMs. Smaller cells increase adsorbent surface area by providing more contact surface per unit of biomass. The greater the number of biosorbent particles, the greater the number of active sites for biosorption; however, a greater amount of biomass does not necessarily indicate better biosorption properties, as metal cations captured in an aqueous medium may be reduced due to the repulsion of active sites between microalgal surfaces, decreasing removal efficiency (Abdelfattah et al., 2022).

HMs, such as manganese (Mn), molybdenum (Mo), cobalt (Co), copper (Cu), zinc (Zn), iron (Fe), and boron (B), are trace elements essential to cellular metabolism and enzymatic processes. Some of these are toxic in high concentrations; other elements, such as chromium (Cr), cadmium (Cd), mercury (Hg), lead (Pb), and arsenic (As), are generally toxic to microalgae. HMs stimulate the growth and metabolism of microalgae when traces of toxic metal ions are present, through the phenomenon of hormesis. Hormesis is a generally favorable biological defense response to low-dose exposures to toxins or other stress-generating agents or phenomena (Erofeeva, 2022). In addition, active binding sites on the cell surface can form complexes with specific pollutants in water, stimulating flocculation and leading to a reduction in total dissolved and suspended matter (Al-Tohamy et al., 2022).

Biosorption of HM ions by microalgae takes place according to a two-stage mechanism. A metabolism-independent process characterized by rapid, reversible adsorbate binding to active sites on the microalgal surface is followed by a slow phase that refers to positive intracellular diffusion and mainly involves the metabolic activity of microalgae. In addition to polymeric substances such as exopolysaccharides with uronic groups and peptides, the cell wall of microalgae can consist of cellulose, alginates, lipids, and proteins, which feature numerous functional groups, including hydroxyls, phosphates $\left(PO_4^{3-}\right)$, amines, sulfonates, carboxyls, and thiols. Microalgae capture both cationic and anionic species of HMs as they possess a large number of deprotonated carboxyl and sulfate groups as well as monomeric alcohols that stimulate biosorption (Matamoros et al., 2015).

Algae-derived EPS can accelerate metal biosorption, and their effectiveness depends on EPS structure, solution properties, metal species, and operating conditions. SPEs released by *Cyanothece sp* can remove MLs from contaminated water through organic functional groups, namely carboxyl and hydroxyl, rather than ion exchange. At low pH, *Nostoc linckia* SPEs showed biosorption capacity for Co(II) and Cr(IV), attributed to interactions between metal ions and the negatively charged functional groups of the SPEs. The interaction of *Chlorella vulgaris* SPEs with cadmium decreased intracellular Cd^{2+} concentration, enhancing the ability of *Chlorella vulgaris* to remove NH_4^+ and PO_4^{3-} (Abdelfattah et al., 2022).

Studying the chemical structure of a biosorbent provides substantial and accurate information for predicting its affinity and capacity for specific HM ions. Nuclear magnetic resonance spectroscopy and Fourier transform infrared spectroscopy are two valuable analytical techniques that can help scientists define the particular functional groups responsible for complexation with HMs and determine the exact basic mechanisms (Abdelfattah et al., 2022).

The optimum pH for biosorption varies according to metal and microalgal species. In general, pH ranges from 3 to 6.5. Values below 3 lead to competition between dissociated H^+ ions for active sites on the cell surface, while metals can precipitate in their hydroxide form at values above 6.5, rendering biosorbents useless and leading to the accumulation of highly toxic sludge (Nateras-Ramírez et al., 2022). However, some microalgae, such as *Nostoc linckia* and *Nannochloris oculata*, were able to adsorb Cr^{4+} at pH 2 (Abdelfattah et al., 2022). Due to the limited number of active sites on the cell surface, metal concentration and competition are two major factors in biomass saturation (Nateras-Ramírez et al., 2022). Other studies have shown that low atomic number metal cations compete with target actions for active sites on the cell surface, reducing the efficiency of HM removal (Andrès and Gérente, 2011). With regard to temperature, several studies have shown that temperature influences the uptake of HMs, which can be favorable or unfavorable in wastewater treatment. In other studies, however, temperature had a negligible impact on metal sorption (Abdelfattah et al., 2022).

10.4.3.2 Bioaccumulation

Unlike biosorption, bioaccumulation is considered an active process, and living biomass is used in this process to detoxify effluents. In this process, biomass absorbs substances through its surfaces and nutrients. The compounds are then accumulated or metabolized, depending on the type of biomass. It can also be defined as the process by which inorganic and organic pollutants (sulfates, nitrates, phosphates, HMs, and pesticides) are transferred within the cells of living organisms and eliminated from aqueous solutions. Although the mechanisms of bioaccumulation and biosorption are substantially different, it is difficult to quantify biosorbed and bioaccumulated pollutants because both mechanisms evolve dynamically. Microalgae accumulate different pollutants along with the nutrients and microelements present. Microalgae can resist pollutants at low concentrations thanks to their ability to adapt to the environment. What's more, microalgae have immense tolerance to a broad class of pollutants from the domestic, agricultural, and industrial sectors, enhancing their bioremediation capacity (Mustafa et al., 2021).

The bioaccumulation process is characterized by the bioconcentration factor, which is the ratio between the concentration of a pollutant in the adsorbent and that in the medium. Bioadsorption is the prerequisite for bioaccumulation. However, not all adsorbed molecules can be bioaccumulated. For example, *Spirogyra* can absorb 850,000 times more radio-phosphorus than the surrounding water. *Cladophora sp* can accumulate the lipophilic organic pollutants triclosan and triclocarban, widely used as antimicrobial agents, up to 50–400 $ng{\cdot}g^{-1}$ of microalgae. In contrast, the concentration in the environment does not exceed 200 $ng{\cdot}kg^{-1}$ (Abdelfattah et al., 2022).

Pollutants move along a concentration gradient from an area of high concentration (external) to an area of low concentration (internal) without requiring energy. Due to the hydrophobicity of the cell membrane, low-molecular-weight materials that are non-polar and lipid-soluble can probably penetrate the cell membrane by passive diffusion. In contrast, polar compounds, high-molecular-weight molecules, and ions cannot cross the cell membrane. Bioaccumulation of sulfamethoxazole, carbamazepine, trimethoprim, and florfenicol has already been reported, with the aforementioned pollutants penetrating microalgal cells by passive diffusion. In addition, changes in cell membrane permeability caused by pollutants or a stressful environment can lead to passive diffusion, which is ensured by depolarization or hyperpolarization of the membrane. For example, when the cyanobacterium *Anabaena CPB4337* was exposed to perfluorinated alkyl acid compounds, such as perfluorooctane sulfonate and perfluorooctanoic acid, its sensitivity to herbicides changed, with some herbicides becoming more harmful and others less toxic. Furthermore, the bioaccumulation of levofloxacin by *Chlorella vulgaris* was significantly enhanced from 34 to 101 $mg{\cdot}g^{-1}$ by the addition of 1% (w/v) NaCl. Assisted passive diffusion is the process by which pollutants cross the cell membrane with the help of transport proteins, which support polar molecules entering the cell. The third process is active transport across the cell membrane, which requires energy as the molecules diffuse against a concentration gradient (Abdelfattah et al., 2022).

Regardless of the mechanism, external and internal physicochemical parameters such as pH, contact time, temperature, and pollutant concentration have a significant impact on bioaccumulation. For example, temperature has been found to significantly influence the bioaccumulation of antibiotics in the planktonic food web. The relationship between temperature and bioaccumulation is biphasic. Carbamazepine bioaccumulation in *Chlamydomonas mexicana* and *Scenedesmus obliquus* improved with increasing culture time and carbamazepine concentration. One of the problems of bioaccumulation is that some bioaccumulated substances stimulate reactive oxygen species. Free radicals (alcoholic radicals, hydroxyl radicals (OH), superoxide radicals ($\bullet O^{2-}$), perhydroxyl radicals ($\bullet HO_2$), singlet oxygen ($^1O_2$), and non-radicals (hydrogen peroxide (H_2O_2)) form inside cells, fundamentally controlling cellular metabolism and potentially causing oxidative damage to biomolecules, cellular dysfunction, and ultimately cell death. To improve the efficiency of microalgae in bioremediation practice, the physicochemical environment must be optimized, as the rate and capacity of microalgae depend on these parameters in the bioaccumulation process. In addition, the selection of microalgal species tolerant to high pollutant concentrations is an effective strategy for achieving higher bioaccumulation capacities and rates (Abdelfattah et al., 2022).

10.4.3.3 Biodegradation

One of the most effective processes for removing pollutants from effluents is biodegradation, in which complex compounds break down into simple, safe chemical elements. Unlike bioaccumulation and biosorption, which involve microorganisms acting as biological filters to concentrate pollutants and separate them from the surrounding water, bioremediation involves the breakdown of target pollutants either by complete mineralization of the parent molecules into CO_2 and H_2O, or by biotransformation, which contains a series of enzymatic reactions to produce various metabolic intermediates (Mustafa et al., 2021).

The fundamental mechanisms of biodegradation can be divided into two categories: (i) metabolic degradation, in which pollutants serve as the electron donor/acceptor and carbon source for microalgae, and (ii) cometabolism, in which pollutants serve as both the electron donor and carbon source for non-living matter. Biodegradation by microalgae takes place extracellularly or intracellularly, or in a combination of both locations, with extracellular degradation followed by intracellular breakdown of degradation intermediates. Extracellular degradation is based on EPS excreted mainly by microalgae. They remain confined to the outer cell walls, enabling microalgae to mineralize pollutants in a dissolved state and function as an external digestive system. In addition, EPS function as emulsifiers in the form of biosurfactants that increase the bioavailability of pollutants in the environment, enabling future bioaccumulation by microalgae. The exact mechanisms governing the interaction between pollutants and EPS emitted by microalgae are still unknown, requiring further research (Abdelfattah et al., 2022).

The biodegradation of organic pollutants by microalgae involves three complex enzymatic processes. Stage I requires cytochrome P450 enzymes, including hydroxylase, carboxylase, decarboxylase, and mono-oxygenase. The function of these enzymes is to increase the hydrophilicity of the pollutant by adding or unmasking a hydroxyl group through redox or hydrolysis reactions (Sutherland and Ralph, 2019). In stage II, the presence of enzymes such as glutathione-S-transferases and glucosyltransferases can stimulate the conjugation of glutathione with various compounds with electrophilic centers (amide, epoxide ring, and carboxylic acid) to protect the cell from oxidative damage (Xiong et al., 2019). Detoxification step III comprises various enzymes (such as dehydrogenase, glutamyl-tRNA reductase, carboxylase, mono(di) oxygenase, laccases, transferase, hydrolases, pyrophosphatase, and dehydratase) to biotransform molecules into less or non-toxic intermediates (Ding et al., 2017).

Cometabolism is indicated by the conversion of a non-living element in the obligatory presence of organisms that biodegrade other compounds. Organic substrates serve as electron donors for the cometabolism of non-living matter and the production of biomass (Abdelfattah et al., 2022). To stimulate the biodegradation of recalcitrant pollutants by microalgae, some studies have reported an effective strategy of adding other nutrients or organic substrates to form a cometabolic system (Vo et al., 2020). For example, the addition of sodium acetate was found to increase the efficiency of ciprofloxacin elimination by *Ceratozamia mexicana* from 13% to 56%. The increase in the elimination efficiency of parent molecules, as well as the regulation of transformation products and pathways, is mainly linked to acetate concentration. In particular, the type of nutrient and its concentration have the strongest effect among the other factors (Rodea-Palomares et al., 2015). The effect of carbon sources on the cometabolism of pollutants by microalgae has been systematically investigated. Although results have shown that the highest EPS, enzyme concentrations, and optimal removal efficiency are obtained with sugar-based carbon sources, the removal efficiency of micropollutants can be decreased in the presence of certain organic substrates, probably due to the repression of catabolites (Vo et al., 2020). Consequently, it is essential to evaluate the effects of different carbon sources in order to select the most suitable source and verify that it can be used for the biodegradation of pollutants in wastewater. Furthermore, this invention will increase biomass recovery and gains for various sectors. Generally speaking, it has been reported that the three mechanisms can be combined in microalgal wastewater treatment, as biodegraded particles can be adsorbed to their surface and then transferred into cells by active processes. Finally, enzymatic reactions promote precipitation, biotransformation, and bioaccumulation within cells (Abdelfattah et al., 2022). Figure 10.4 summarizes the mechanisms involved in the removal of industrial pollutants by microalgae.

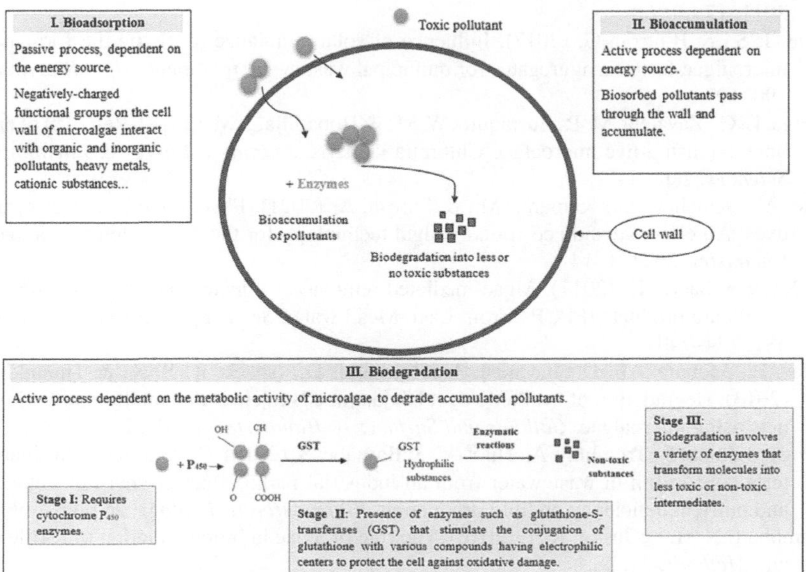

FIGURE 10.4 Mechanisms of industrial pollutant elimination by microalgae.

REFERENCES

Abdel-Aty, A. M., Ammar, N. S., Ghafar, H. H. A., & Ali, R. K. (2013). Biosorption of cadmium and lead from aqueous solution by fresh water alga Anabaena sphaerica biomass. *Journal of Advanced Research*, 4(4), 367–374.

Abdelfattah, A., Ali, S. S., Ramadan, H., El-Aswar, E. I., Eltawab, R., Ho, S. H., ... & Sun, J. (2022). Microalgae-based wastewater treatment: Mechanisms, challenges, recent advances, and future prospects. *Environmental Science and Ecotechnology*, 13, 100205.

Al-Fawwaz, A. T., & Abdullah, M. (2016). Decolorization of methylene blue and malachite green by immobilized Desmodesmus sp. isolated from North Jordan. *International Journal of Environmental Science and Development*, 7(2), 95.

Al Hamadi, A., Güven, U. R. A. Z., Katircioğlu, H., & Osmanağaoğlu, Ö. (2017). Adsorption of azo dyes from textile wastewater by Spirulina Platensis. *Eurasian Journal of Environmental Research*, 1(1), 19–27.

AlMomani, F. A., & Örmeci, B. (2016). Performance of Chlorella vulgaris, Neochloris oleoabundans, and mixed indigenous microalgae for treatment of primary effluent, secondary effluent and centrate. *Ecological Engineering*, 95, 280–289.

Al-Tohamy, R., Ali, S. S., Li, F., Okasha, K. M., Mahmoud, Y. A. G., Elsamahy, T., ... & Sun, J. (2022). A critical review on the treatment of dye-containing wastewater: Ecotoxicological and health concerns of textile dyes and possible remediation approaches for environmental safety. *Ecotoxicology and Environmental Safety*, 231, 113160.

Amaro, H. M., Salgado, E. M., Nunes, O. C., Pires, J. C., & Esteves, A. F. (2023). Microalgae systems-environmental agents for wastewater treatment and further potential biomass valorisation. *Journal of Environmental Management*, 337, 117678.

Andrès, Y., & Gérente, C. (2011). Removal of rare earth elements and precious metal species by biosorption. In: *Microbial Biosorption of Metals*, Dordrecht: Springer Netherlands, 2011. 179–196.

Arcila, J. S., & Buitrón, G. (2017). Influence of solar irradiance levels on the formation of microalgae-bacteria aggregates for municipal wastewater treatment. *Algal Research*, 27, 190–197.

Arteaga, L. C., Zavaleta, M. P., Eustaquio, W. M., & Bobadilla, J. M. (2018). Removal of aniline blue dye using live microalgae Chlorella vulgaris. *Journal of Energy & Environmental Sciences*, 2(1), 6–12.

Ayele, A., Getachew, D., Kamaraj, M., & Suresh, A. (2021). Phycoremediation of synthetic dyes: An effective and eco-friendly algal technology for the dye abatement. *Journal of Chemistry*, 2021, 1–14.

Bai, X., & Acharya, K. (2017). Algae-mediated removal of selected pharmaceutical and personal care products (PPCPs) from Lake Mead water. *Science of the Total Environment*, 581, 734–740.

Baldev, E., MubarakAli, D., Ilavarasi, A., Pandiaraj, D., Ishack, K. S. S., & Thajuddin, N. (2013). Degradation of synthetic dye, Rhodamine B to environmentally non-toxic products using microalgae. *Colloids and Surfaces B: Biointerfaces*, 105, 207–214.

Bélanger-Lépine, F., Tremblay, A., Huot, Y., & Barnabé, S. (2018). Cultivation of an algae-bacteria consortium in wastewater from an industrial park: Effect of environmental stress and nutrient deficiency on lipid production. *Bioresource Technology*, 267, 657–665.

Brahmbhatt, N. H., & Jasrai, R. T. (2016). The role of algae in bioremediation of textile effluent. *Methods*, 21, 28.

Chai, W. S., Tan, W. G., Munawaroh, H. S. H., Gupta, V. K., Ho, S. H., & Show, P. L. (2021). Multifaceted roles of microalgae in the application of wastewater biotreatment: A review. *Environmental Pollution*, 269, 116236.

Chin, J. Y., Chng, L. M., Leong, S. S., Yeap, S. P., Yasin, N. H. M., & Toh, P. Y. (2020). Removal of synthetic Dye by Chlorella vulgaris microalgae as natural adsorbent. *Arabian Journal for Science and Engineering*, 45, 7385–7395.

Clarens, A.F., Resurreccion, E.P., White, M.A., Colosi, L.M., 2010. Environmental life cycle comparison of algae to other bioenergy feedstocks. *Environmental Science & Technology*, 44(5), 1813–1819.

Daneshvar, N., Ayazloo, M., Khataee, A. R., & Pourhassan, M. (2007). Biological decolorization of dye solution containing Malachite Green by microalgae Cosmarium sp. *Bioresource Technology*, 98(6), 1176–1182.

de Godos, I., Muñoz, R., & Guieysse, B. (2012). Tetracycline removal during wastewater treatment in high-rate algal ponds. *Journal of Hazardous Materials*, 229, 446–449.

de Wilt, A., Butkovskyi, A., Tuantet, K., Leal, L. H., Fernandes, T. V., Langenhoff, A., & Zeeman, G. (2016). Micropollutant removal in an algal treatment system fed with source separated wastewater streams. *Journal of Hazardous Materials*, 304, 84–92.

Delgadillo-Mirquez, L., Lopes, F., Taidi, B., & Pareau, D. (2016). Nitrogen and phosphate removal from wastewater with a mixed microalgae and bacteria culture. *Biotechnology Reports*, 11, 18–26.

Ding, T., Yang, M., Zhang, J., Yang, B., Lin, K., Li, J., & Gan, J. (2017). Toxicity, degradation and metabolic fate of ibuprofen on freshwater diatom Navicula sp. *Journal of Hazardous Materials*, 330, 127–134.

El Amri, R., Elkacmi, R., & Boudouch, O. (2023). Removal of methyl orange from water using microalgae: Effect of operating parameters, equilibrium, kinetic and thermodynamic studies. *Chemistry Africa*, 6(5), 2569–2580.

El Amri, R., Elkacmi, R., Hasib, A., & Boudouch, O. (2022). Removal of hydroxychloroquine from an aqueous solution using living microalgae: Effect of operating parameters on removal efficiency and mechanisms. *Water Environment Research*, 94(9), e10790.

Elkacmi, R., & Bennajah, M. (2019). Advanced oxidation technologies for the treatment and detoxification of olive mill wastewater: A general review. *Journal of Water Reuse and Desalination*, 9(4), 463–505.

Erofeeva, E. A. (2022). Environmental hormesis: From cell to ecosystem. *Current Opinion in Environmental Science & Health*, 29, 100378.

Fan, H., Wang, K., Wang, C., Yu, F., He, X., Ma, J., & Li, X. (2020). A comparative study on growth characters and nutrients removal from wastewater by two microalgae under optimized light regimes. *Environmental Technology & Innovation*, 19, 100849.

Ferreira, L. S., Rodrigues, M. S., De Carvalho, J. C. M., Lodi, A., Finocchio, E., Perego, P., & Converti, A. (2011). Adsorption of Ni2+, Zn2+ and Pb2+ onto dry biomass of Arthrospira (Spirulina) platensis and Chlorella vulgaris. I. Single metal systems. *Chemical Engineering Journal*, 173(2), 326–333.

Fito, J., & Alemu, K. (2019). Microalgae–bacteria consortium treatment technology for municipal wastewater management. *Nanotechnology for Environmental Engineering*, 4, 1–9.

Gajare, S. M., & Menghani, S. (2012). Biosorption of malachite green by naturally grown algal biomass from Girna river, Jalgaon District, Maharashtra. *Journal of Algal Biomass Utilization*, 3(4), 60–65.

Gelebo, G. G., Tessema, L. H., Kehshin, K. T., Gebremariam, H. H., Gebremikal, E. T., Motuma, M. T., ... & Suresh, A. (2020). Phycoremediation of synthetic dyes in an aqueous solution using an indigenous Oscillatoria sp., from Ethiopia. *Ethiopian Journal of Science and Sustainable Development*, 7(2), 14-20. https://doi.org/10.1080/15226514.2021.1984387

Gonçalves, A. L., Pires, J. C., & Simões, M. (2017). A review on the use of microalgal consortia for wastewater treatment. *Algal Research*, 24, 403–415.

Gonzalez-Camejo, J., Barat, R., Pachés, M., Murgui, M., Seco, A., & Ferrer, J. (2018). Wastewater nutrient removal in a mixed microalgae–bacteria culture: Effect of light and temperature on the microalgae–bacteria competition. *Environmental Technology*, 39(4), 503–515.

Gunasundari, E. (2017). Adsorption isotherm, kinetics and thermodynamic analysis of Cu (II) ions onto the dried algal biomass (Spirulina platensis). *Journal of Industrial and Engineering Chemistry*, 56, 129–144.

Hom-Diaz, A., Llorca, M., Rodríguez-Mozaz, S., Vicent, T., Barceló, D., & Blánquez, P. (2015). Microalgae cultivation on wastewater digestate: β-estradiol and 17α-ethynylestradiol degradation and transformation products identification. *Journal of Environmental Management*, 155, 106–113.

Huo, S., Kong, M., Zhu, F., Qian, J., Huang, D., Chen, P., & Ruan, R. (2020). Co-culture of Chlorella and wastewater-borne bacteria in vinegar production wastewater: Enhancement of nutrients removal and influence of algal biomass generation. *Algal Research*, 45, 101744.

Ishchi, T., & Sibi, G. (2020). Azo dye degradation by Chlorella vulgaris: Optimization and kinetics. *International Journal of Biological and Chemical Science*, 14(1), 1–7.

Kelly, B. C., Ikonomou, M. G., Blair, J. D., Morin, A. E., & Gobas, F. A. (2007). Food web specific biomagnification of persistent organic pollutants. *Science*, 317(5835), 236–239.

Khalaf, M. A. (2008). Biosorption of reactive dye from textile wastewater by non-viable biomass of Aspergillus niger and Spirogyra sp. *Bioresource Technology*, 99(14), 6631–6634.

Khataee, A. R., Dehghan, G., Ebadi, E., & Pourhassan, M. (2010). Central composite design optimization of biological dye removal in the presence of macroalgae Chara sp. *CLEAN–Soil, Air, Water*, 38(8), 750–757.

Kousha, M., Farhadian, O., Dorafshan, S., Soofiani, N. M., & Bhatnagar, A. (2013). Optimization of malachite green biosorption by green microalgae—Scenedesmus quadricauda and Chlorella vulgaris: Application of response surface methodology. *Journal of the Taiwan Institute of Chemical Engineers*, 44(2), 291–294.

Kumar, V., Jaiswal, K. K., Verma, M., Vlaskin, M. S., Nanda, M., Chauhan, P. K., ... & Kim, H. (2021). Algae-based sustainable approach for simultaneous removal of micropollutants, and bacteria from urban wastewater and its real-time reuse for aquaculture. *Science of the Total Environment*, 774, 145556.

Li, Y., Chen, Y. F., Chen, P., Min, M., Zhou, W., Martinez, B., ... & Ruan, R. (2011). Characterization of a microalga Chlorella sp. well adapted to highly concentrated municipal wastewater for nutrient removal and biodiesel production. *Bioresource Technology*, 102(8), 5138–5144.

Liu, J. H., Zhang, L., Zha, D. C., Chen, L. Q., Chen, X. X., & Qi, Z. M. (2018). Biosorption of malachite green onto Haematococcus pluvialis observed through synchrotron Fourier-transform infrared microspectroscopy. *Letters in Applied Microbiology*, 67(4), 348–353.

Ma, C., Dong, M., & Shen, D. (2017, October). A printed magneto-electric dipole based on substrate integrated gap waveguide. In *2017 Sixth Asia-Pacific Conference on Antennas and Propagation (APCAP)* (pp. 1–3). IEEE.

Matamoros, V., Gutiérrez, R., Ferrer, I., García, J., & Bayona, J. M. (2015). Capability of microalgae-based wastewater treatment systems to remove emerging organic contaminants: A pilot-scale study. *Journal of Hazardous Materials*, 288, 34–42.

Matamoros, V., Uggetti, E., García, J., & Bayona, J. M. (2016). Assessment of the mechanisms involved in the removal of emerging contaminants by microalgae from wastewater: A laboratory scale study. *Journal of Hazardous Materials*, 301, 197–205.

Mujtaba, G., & Lee, K. (2016). Advanced treatment of wastewater using symbiotic co-culture of microalgae and bacteria. *Applied Chemistry for Engineering*, 27(1), 1–9.

Mustafa, S., Bhatti, H. N., Maqbool, M., & Iqbal, M. (2021). Microalgae biosorption, bioaccumulation and biodegradation efficiency for the remediation of wastewater and carbon dioxide mitigation: Prospects, challenges and opportunities. *Journal of Water Process Engineering*, 41, 102009.

Naji, N. S., & Salman, J. M. (2019). Effect of temperature variation on the efficacy of Chlorella vulgaris in decolorization of Congo red from aqueous solutions. *Biochemical and Cellular Archives*, 19(2), 4169–4174.

Nateras-Ramírez, O., Martínez-Macias, M. R., Sánchez-Machado, D. I., López-Cervantes, J., & Aguilar-Ruiz, R. J. (2022). An overview of microalgae for Cd^{2+} and Pb^{2+} biosorption from wastewater. *Bioresource Technology Reports*, 17, 100932.

Omar, H. H. (2008). Algal decolorization and degradation of monoazo and diazo dyes. *Pakistan Journal of Biological Science*, 11(10), 1310–1316.

Plöhn, M., Scherer, K., Stagge, S., Jönsson, L. J., & Funk, C. (2022). Utilization of different carbon sources by nordic microalgae grown under mixotrophic conditions. *Frontiers in Marine Science*, 9, 830800.

Rawat, I., Kumar, R. R., Mutanda, T., & Bux, F. (2011). Dual role of microalgae: Phycoremediation of domestic wastewater and biomass production for sustainable biofuels production. *Applied Energy*, 88(10), 3411–3424.

Raymond, E. S., & Omo, K. M. (2019). Biodecolourization of yellow dye using chlorella vulgaris and sphaerocystics schroeteri. *American Journal of Biological and Environmental Statistics*, 5(1), 1–6.

Rezaei, H. (2016). Biosorption of chromium by using Spirulina sp. *Arabian Journal of Chemistry*, 9(6), 846–853.

Rodea-Palomares, I., Makowski, M., Gonzalo, S., González-Pleiter, M., Leganés, F., & Fernández-Piñas, F. (2015). Effect of PFOA/PFOS pre-exposure on the toxicity of the herbicides 2, 4-D, Atrazine, Diuron and Paraquat to a model aquatic photosynthetic microorganism. *Chemosphere*, 139, 65–72.

Sarı, A., Uluozlü, Ö. D., & Tüzen, M. (2011). Equilibrium, thermodynamic and kinetic investigations on biosorption of arsenic from aqueous solution by algae (Maugeotia genuflexa) biomass. *Chemical Engineering Journal*, 167(1), 155–161.

Shanab, S., Essa, A., & Shalaby, E. (2012). Bioremoval capacity of three heavy metals by some microalgae species (Egyptian Isolates). *Plant Signaling & Behavior*, 7(3), 392–399.

Sibi, G. (2016). Biosorption of chromium from electroplating and galvanizing industrial effluents under extreme conditions using Chlorella vulgaris. *Green Energy & Environment*, 1(2), 172–177.

Singh, J., Sharma, S., Soni, S., Sharma, S., & Singh, R. C. (2019). Influence of different milling media on structural, morphological and optical properties of the ZnO nanoparticles synthesized by ball milling process. *Materials Science in Semiconductor Processing*, 98, 29–38.

Song, C., Wei, Y., Qiu, Y., Qi, Y., Li, Y., & Kitamura, Y. (2019). Biodegradability and mechanism of florfenicol via Chlorella sp. UTEX1602 and L38: Experimental study. *Bioresource Technology*, 272, 529–534.

Soto-Ramírez, R., Lobos, M. G., Córdova, O., Poirrier, P., & Chamy, R. (2021). Effect of growth conditions on cell wall composition and cadmium adsorption in Chlorella vulgaris: A new approach to biosorption research. *Journal of Hazardous Materials*, 411, 125059.

Spain, O., Plöhn, M., & Funk, C. (2021). The cell wall of green microalgae and its role in heavy metal removal. *Physiologia Plantarum*, 173(2), 526–535.

Su, Y., Mennerich, A., & Urban, B. (2012). Synergistic cooperation between wastewater-born algae and activated sludge for wastewater treatment: Influence of algae and sludge inoculation ratios. *Bioresource Technology*, 105, 67–73.

Sutherland, D. L., & Ralph, P. J. (2019). Microalgal bioremediation of emerging contaminants-Opportunities and challenges. *Water Research*, 164, 114921.

Tomaselli, L. (2004). The microalgal cell. In: Richmon, A. (Ed.). *Handbook of microalgal culture: Biotechnology and applied phycology*, 1, 3–19. Blackwell Publishing Ltd.

Vardhan, K. H., Kumar, P. S., & Panda, R. C. (2019). A review on heavy metal pollution, toxicity and remedial measures: Current trends and future perspectives. *Journal of Molecular Liquids*, 290, 111197.

Vo, H. N. P., Ngo, H. H., Guo, W., Nguyen, K. H., Chang, S. W., Nguyen, D. D., ... & Bui, X. T. (2020). Micropollutants cometabolism of microalgae for wastewater remediation: Effect of carbon sources to cometabolism and degradation products. *Water Research*, 183, 115974.

Wang, L., Min, M., Li, Y., Chen, P., Chen, Y., Liu, Y., ... & Ruan, R. (2010). Cultivation of green algae Chlorella sp. in different wastewaters from municipal wastewater treatment plant. *Applied Biochemistry and Biotechnology*, 162, 1174–1186.

Xiong, J. Q., Kim, S. J., Kurade, M. B., Govindwar, S., Abou-Shanab, R. A., Kim, J. R., ... & Jeon, B. H. (2019). Combined effects of sulfamethazine and sulfamethoxazole on a freshwater microalga, Scenedesmus obliquus: Toxicity, biodegradation, and metabolic fate. *Journal of Hazardous Materials*, 370, 138–146.

Zhang, Y., Habteselassie, M. Y., Resurreccion, E. P., Mantripragada, V., Peng, S., Bauer, S., & Colosi, L. M. (2014). Evaluating removal of steroid estrogens by a model alga as a possible sustainability benefit of hypothetical integrated algae cultivation and wastewater treatment systems. *ACS Sustainable Chemistry & Engineering*, 2(11), 2544–2553.

Zheng, H., Liu, M., Lu, Q., Wu, X., Ma, Y., Cheng, Y., ... & Ruan, R. (2018). Balancing carbon/nitrogen ratio to improve nutrients removal and algal biomass production in piggery and brewery wastewaters. *Bioresource Technology*, 249, 479–486.

11 Microalgae as a Natural Resource

Economic Potential and Applications

Radouane El Amri, Rajaa Zahnoune, Otmane Boudouch, Maryeme Hamdouch, Ikrame Hattab, Aziz Hasib, Sudip Chakraborty, and Reda Elkacmi

11.1 INTRODUCTION

Microalgae, photosynthetic microscopic organisms, represent an exceptionally rich natural resource, offering promising prospects in a variety of economic and industrial fields (Hachicha et al., 2022). Their ability to produce a variety of high-value-added biochemical compounds makes them key players in sectors as diverse as food, health, energy, and the environment. These compounds include natural pigments, lipids, polyunsaturated fatty acids (such as omega-3), polysaccharides, and bioactive substances with therapeutic properties (Dolganyuk et al., 2020). These unique characteristics position microalgae as a renewable and sustainable source, capable of meeting today's challenges linked to the growing demand for biological resources and the need to reduce the environmental impact of human activities (Alam et al., 2020). With the emergence of new cultivation technologies and bioprocesses, the potential of microalgae continues to expand. Recent advances in production systems, such as photobioreactors and open cultures, have made it possible to optimize the growth of these microorganisms and explore new economically viable applications. These innovations pave the way for the discovery of previously untapped products, reinforcing the role of microalgae in the green and circular economy (Anyaoha et al., 2024; Kazakis, 2025).

In the field of nutrition, microalgae are exploited for their richness in proteins, vitamins, and minerals, for both human and animal consumption (Gaurav et al., 2024; Fal et al., 2023).

The commercial applications of microalgae (Figure 11.1) are vast and diverse. They include the production of third-generation biofuels, offering a sustainable alternative to fossil fuels (Gaurav et al., 2024), as well as their use in wastewater treatment and CO_2 sequestration, helping to combat climate change (Fal et al., 2023). In the field of nutrition, microalgae are exploited for their richness in proteins, vitamins,

DOI: 10.1201/9781003624905-11

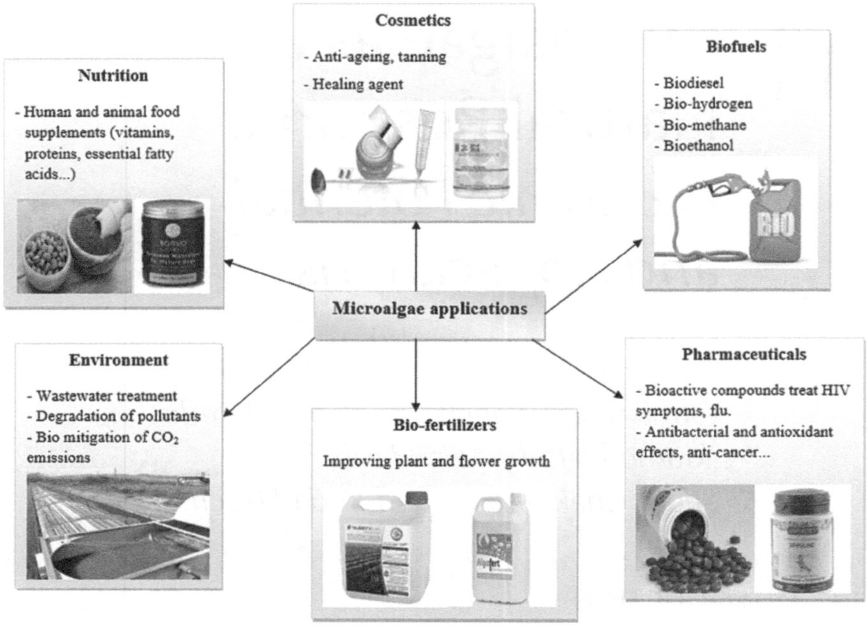

FIGURE 11.1 Microalgae applications.

and minerals, for both human and animal consumption (Silva et al., 2025). They also have applications in the cosmetics industry, where their extracts are valued for their antioxidant and moisturizing properties, as well as in the synthesis of biofertilizers and drugs with antimicrobial, antiviral, and anticancer properties (Sathasivam et al., 2019; Vázquez-Romero et al., 2022). Among the microalgal species most commonly used in these applications are *Arthrospira (Spirulina), Chlorella, Dunaliella, Chaetoceros (diatoms)*, and *Isochrysis*. Each of these species has specific biochemical characteristics that make them ideal candidates for particular uses, underlining the diversity and versatility of this natural resource (Lee, 1997).

This chapter explores in detail the economic potential of microalgae and their multiple applications, highlighting technological advances, current challenges, and future prospects. It aims to provide a comprehensive vision of how microalgae can contribute to a more sustainable and environmentally friendly economy, while meeting society's growing needs for health, energy, and nutrition.

11.2 MICROALGAE IN THE ENVIRONMENTAL FIELD

Microalgae can be used in the environmental field, particularly for wastewater treatment, because of their ability to absorb nutrients and transform them into biomass (Chinnasamy et al., 2010). Microalgae assimilate nutrients contained in wastewater for growth, reducing the quantity of nutrients in wastewater and lowering chemical oxygen demand and biochemical oxygen demand (El Amri, Elkacmi, Benadda, et al., 2023). The small size of microalgae offers a large specific surface area, which increases the rate of assimilation of nutrients and the various pollutants present in wastewater (Mohd Udaiyappan et al., 2017). As a result, overall treatment costs and environmental impact are reduced. From an environmental point of view, microalgae are a sustainable option, as they have an immense capacity to fix CO_2 from the surrounding environment through photosynthesis. The primary photosynthetic products are then converted into high-value compounds, making this approach attractive from an economic point of view too (Abdelfattah et al., 2023).

A number of studies carried out on wastewater purification have demonstrated the considerable potential of microalgae in the removal of nutrients, notably ammonium and phosphorus, and organic matter, and the absorption of heavy metals, synthetic dyes, and pharmaceutical pollutants (El Amri, Elkacmi, & Boudouch, 2023; El Amri et al., 2022; Fan et al., 2020; Spain et al., 2021). The use of high-rate algal ponds (HRAP) is considered an efficient way to treat wastewater (Phang et al., 2000). In addition, the biomass generated in HRAP can be used for animal feed and biofuel production (Rizwan et al., 2018a). *Spirulina platensis*, grown in HRAP, eliminated 99% of the phosphate from the wastewater of an anaerobically digested starch plant (Phang et al., 2000). *Chlorella vulgaris* was effective in the final treatment of textile wastewater in the HRAP (Lim et al., 2010). Some studies have considered the use of mixed species of microalgae for wastewater treatment. A consortium of five microalgal species in HRAP has been shown to be effective in treating landfill leachate (Mustafa et al., 2012). The removal efficiency of environmentally toxic substances can be further enhanced by the use of immobilized microalgae (de-Bashan & Bashan, 2010). It has been reported that *Chlorella vulgaris* immobilized in alginate beads can be used to remove color from textile dyes (Chu et al., 2009). It has also been reported that immobilized *Chlorella vulgaris* and *Scenedesmus obliquus* effectively removed N and P from domestic wastewater in semi-continuous mode (Ruiz-Marin et al., 2010).

11.3 MICROALGAE AND BIOFUEL PRODUCTION

Global demand for energy is growing, with fossil fuels accounting for around 88% of the total. However, these fuels are increasing greenhouse gas emissions into the atmosphere (Cea-Barcia et al., 2014). Biofuel produced by microalgae can significantly reduce greenhouse gas emissions compared with fossil fuels. The term biofuel refers to biodiesel, biohydrogen, biomethane, and bioethanol produced by microalgae. Biofuels have attracted a great deal of interest because of the depletion of fossil fuels due to overexploitation and the high CO_2 (Huang et al., 2010). Microalgae have favorable properties for biofuels, as they have exceptionally high growth rates, high photosynthetic efficiencies, and the ability to grow on non-arable land, in fresh and salt water, as well as in wastewater (Mahdy et al., 2015).

Biodiesel and fatty acid methyl esters are derived from biomass oils, mainly vegetable oils (Naik et al., 2010). The first generation of biofuels raised food concerns, as they consume part of the world's food supply (Melis et al., 2000). At present, oilseed crops such as rapeseed, palm, and soya are interesting, but they require the use of large areas of land (Caporgno et al., 2015). The second generation of biofuels uses the Jatropha plant, but its slow growth rate and heavy use of arable land make it unprofitable (Mohd Udaiyappan et al., 2017). Microalgae are therefore known as the third generation of biofuels based on lipid production. Microalgae, however, do not pose feed-related problems and have the capacity to store large quantities of lipids, which can in turn be converted into biofuel (Keffer & Kleinheinz, 2002). It has been reported that the production of biofuels from microalgal biomass is a hundred times greater than using higher plants as feedstock (Rodionova et al., 2017). Some species of microalgae can accumulate up to 20%–50% (by dry weight) of triglycerols, which are considered the main component essential for biodiesel production (Zullaikah et al., 2019). There are various methods for converting microalgal biomass into biofuels. These include biochemical conversion, thermochemical conversion, chemical reaction, and direct combustion. Transesterification is a chemical reaction process commonly used to convert lipids into biodiesel (Wang et al., 2008).

A great deal of research was carried out in the early 1970s to produce hydrogen from microalgae using sunlight and water. Microalgae commonly used for hydrogen production include *Chlamydomonas reinhardtii*, *Chlorella fusca*, *Chlorella sorokiniana*, *Scenedesmus obliquus*, *Chlorococcum littorale*, and *Platymonas subcordiformis* (Nagarajan et al., 2017). *Scenedesmus obliquus*, a green microalga, was the first microalga to produce a small amount of hydrogen in the dark. Hydrogen production was increased when the microalgae were exposed to light (Benemann, 2000). *Chlamydomonas reinhardtii* and *Chlorella sorokiniana* are also able to switch from oxygen production to hydrogen production under sulfur-starved conditions (Chader et al., 2011). The various processes involved in biohydrogen production from microalgae include direct biophotolysis, indirect biophotolysis, dark fermentation, and photo-fermentation (Saifuddin & Priatharsini, 2016). For ethanol production, microalgae are receiving a great deal of attention as a raw material for bioethanol production (Nigam & Singh, 2011). The microalgae most commonly used for bioethanol production are *Chlorella sp, Chlamydomonas reinhardtii, Chlorococcum sp, Scenedesmus sp, and Spirulina fusiformis*. Studies on these microalgae have revealed their potential as feedstocks for bioethanol production, as they have demonstrated their ability to accumulate large quantities of starch (e.g., *Chlorella vulgaris* can accumulate 37% starch) (Hirano et al., 1997). Bioethanol production from microalgae can therefore be achieved by one of the following methods: from microalgal metabolites, dark fermentation, and direct production by modified microalgae. Consequently, bioethanol from microalgal biomass is considered a suitable alternative to other traditional technologies, but commercialization has not yet begun (John et al., 2011).

11.4 MICROALGAE AND ANIMAL FEED

The use of microalgal species as feed supplements has developed rapidly over the last few decades. The biomass obtained from microalgae has been used as a raw material for animals, both fish and farm animals (Gunasundari & Senthil Kumar, 2017). Around 30% of the algal biomass produced worldwide is sold as animal feed (Spolaore et al., 2006). Microalgae-based feeds can provide vitamins, essential amino acids, polysaccharides, polyunsaturated fatty acids, minerals, and pigments (Priyadarshani & Rath, 2012). In fact, a number of nutritional and toxicological tests have shown that microalgal biomass can be used as an animal feed supplement

(Richmond, 2004). For example, *Arthrospira* is used as a feed for a variety of animals, such as aquarium fish, breeding bulls, cows, cats, dogs, horses, and ornamental birds (Richmond, 2004). Over 50% of *Arthrospira* is grown for the production of dietary supplements (Rizwan et al., 2018a). *Chlorella, Isochrysis, Pavlova, Phaeodactylum, Chaetoceros, Nannochloropsis, Skeletonema, Thalassiosira, Haematococcus, and Tetraselmis* are the most important microalgal species for fish farming, especially for feeding fish larvae (Enzing et al., 2014). Given the large number of microalgal species, in-depth biochemical analysis is required to select the most suitable microalgae for specific applications in animal feed technology, notably as supplements in new animal diets (Batista et al., 2013).

11.5 MICROALGAE AND HUMAN NUTRITION

Microalgae are rich sources of carbohydrates, proteins, and lipids. They have been used as a food source for humans in China and Japan, as well as in Africa and Mexico, due to their abundant composition in proteins (up to 70% of dry weight), vitamins, and essential fatty acids. To this day, *Chlorella* and *Spirulina* are the best-selling food microalgae, as they grow more rapidly (Varfolomeev & Wasserman, 2011). Japanese scientists have used *Chlorella ellipsoidea* in the development of various food products (breads and rolls, ice creams, soups, noodle powders, green tea, cookies, and soy sauce). *Dunaliella sp* (particularly *D. salina*) is also considered a food source. They are rich in lipids and proteins, glycerol, and β-carotene, and have the particular ability to thrive in brackish water (Dewapriya & Kim, 2014).

11.6 MICROALGAE AND PHARMACEUTICALS

Microalgae are known for their secondary metabolites, which produce value-added pharmaceuticals and food supplements such as astaxanthin, omega-3 fatty acids, sterols, proteins, enzymes, vitamins, and pigments (Maadane et al., 2015). Primary metabolites are components used for growth and include carbohydrates, proteins, and lipids. Secondary metabolites, on the other hand, are components that are not used for growth; they generally serve as defense mechanisms for organisms. These metabolites, also known as bioactive compounds, can be manipulated to produce useful drugs or supplements. The most commonly used microalgae for the production of bioactive compounds of pharmaceutical interest are *Arthrospira* (*Spirulina*), *Chlorella*, *Dunaliella*, *Haematococcus*, and *Nostoc*. According to advances in microalgae research, *Chlamydomonas reinhardtii* has proved to be one of the most important microalgae that have been employed in the manufacture of pharmaceutical proteins (erythropoietin, interferon β, insulin, and immunoglobulin A) (Scaife et al., 2015; Yan et al., 2016). Glycerol, a compound widely used in the pharmaceutical industry, was produced by *Chlamydomonas reinhardtii* when it was deprived of sulfur (Skjånes et al., 2013). Cyanovirin, a bioactive compound obtained from *Nostoc*'s biomass, has shown positive results in treating the symptoms of HIV and influenza A (H1N1). *Nostoc* also produces essential fatty acids and polyunsaturated fatty acids. The biomass of the *Chlorella* species is known to be rich in vitamin B complex, particularly B12, α-carotene, β-carotene, lutein, ascorbic acid, and α-tocopherol, and these compounds may help reduce the onset of certain cancers and prevent macular degeneration. *Dunaliella* has also been considered a rich source of bioactive compounds such as enzymes and vitamins. The crude extract obtained from this microalga strongly limits the growth of harmful bacteria, thus acting as an effective antimicrobial (Bhattacharjee, 2016). The valuable compounds produced by microalgae also include antioxidants (Millao & Uquiche, 2016). Microalgae are exposed to radical stress and high oxygen levels due to their phototrophic nature. As a result, they have developed a number of effective protection systems against radical and oxidative stress factors. These systems enable them to prevent the accumulation of reactive oxygen species and free radicals, and to protect themselves from cell-damaging activities. However, it should be noted that these bioactive compounds obtained from microalgae represent only a small proportion of algal biomass (Mohd Udaiyappan et al., 2017).

11.7 MICROALGAE AND COSMETICS

Microalgae have received a great deal of attention because they contain bioactive compounds that can be used effectively in the treatment of skin disorders such as aging, tanning, and pigmentation-related problems (Wang et al., 2015). Microalgae commonly used in the cosmetics industry include *Spirulina, Chlorella, Dunaliella salina, and Nannochloropsis oculata* (Rizwan et al., 2018a). Conventional methods such as maceration, aqueous extraction, and Soxhlet extraction are still the most widely used techniques for extracting bioactive compounds from microalgae (Azmir et al., 2013). Fucoxanthin and phloroglucinol derivatives obtained from marine msicroalgae have been used as skin-whitening agents by inhibiting the enzyme tyrosinase, resulting in a reduction in melanin pigment. This is considered the most common approach to skin whitening, as melanin is the compound responsible for skin, hair, and eye color (Wang et al., 2011). Skin wrinkles are due to the formation of matrix metalloproteinase, which leads to the degradation of skin collagen and is generally caused by reactive oxygen species due to oxidative stress. Phenolic compounds obtained from marine microalgae have been shown to inhibit the formation of matrix metalloproteinase, thus preventing skin aging (Thomas & Kim, 2013). Polysaccharides obtained from microalgae have a wide range of cosmetic functions. Fucoidans obtained from brown algae, carrageenans obtained from red algae, and ulvans obtained from green algae are the best examples. They are used as rheology modifiers, hair conditioners, suspension agents, and wound-healing agents (Aditya et al., 1026). Extracts obtained from *Chlorella* and *Spirulina* biomass have been used in the production of creams, lotions, shampoos, and sun protection products (Bhalamurugan et al., 2018). On a commercial level, there are several microalgae-based products on the market, for example, Pepha-Tight based on *Nannochloropsis oculata* and Pepha-Ctive based on *Dunaliella salina* from the Swiss company Pentapharm, which have exceptional properties for refining the skin and improving its energy metabolism, Protulines from the French company Exsymol S. A.M, Monaco, which contains a protein-rich *Arthrospira* extract used as an anti-aging agent, and Dermochlorella from the French company Codif, St. Malo, which contains a *Chlorella vulgaris* extract that stimulates collagen production in the skin, leading to tissue regeneration and reduced wrinkle formation (Stolz & Obermayer, 2005).

11.8 MICROALGAE AND BIOFERTILIZERS

Fertilizers contain nutrients that are essential for plant growth. Fertilizers play an essential role in improving agriculture, thereby increasing production. Arable land is constantly being used to grow crops, resulting in a loss of nutrients essential for optimal plant growth. The most common nutrients incorporated in fertilizers are nitrogen and phosphorus. The cost of producing chemical fertilizers is rising, so a suitable alternative is needed. One possible solution is the use of biofertilizers produced from microalgae (Nur et al., 2025). Biofertilizers are inexpensive products containing natural compounds derived from microorganisms such as microalgae, bacteria, and fungi that can help improve soil fertility and stimulate plant growth (Samoraj et al., 2025). The majority of cyanobacteria are capable of fixing atmospheric nitrogen and can be used effectively as biofertilizers. They can help boost soil fertility and thus improve crop yields, as demonstrated with rice (Song et al., 2005). A study showed that dry biomass obtained from *Acutodesmus dimorphus* applied as a biofertilizer was able to increase plant growth and flower production in tomato plants (Garcia-Gonzalez & Sommerfeld, 2016).

A study carried out on lettuce growth using *Chlorella vulgaris* as a biofertilizer revealed that *Chlorella vulgaris* dry powder was considered a suitable substrate for the germination of *Lactuca sativa* seeds, as it significantly improved the germination process compared with the unfertilized control. It also significantly increased the amount of pigments (chlorophyll a, chlorophyll b, and carotenoids) in lettuce seedlings. As a result, the study concluded that dry microalgae can be used as plant nutrients for ideal growth, as they can improve soil fertility (Faheed & El-Fattah, 2008). Another study on wheat cultivation revealed that filamentous microalgae increase nitrogen, phosphorus, and potassium in the soil, improving crop yields (Renuka et al., 2016). The impact of two microalgae, *Chlorella vulgaris* and *Spirulina platensis*, on corn cultivation was also studied with different treatments. The results concluded that microalgal treatment with cow manure + *Spirulina platensis* or *Chlorella vulgaris* improved maize yield and growth by up to 51.1% 60 days after planting (Dineshkumar et al., 2019).

REFERENCES

Abdelfattah, A., Ali, S. S., Ramadan, H., El-Aswar, E. I., Eltawab, R., Ho, S. H., Elsamahy, T., Li, S., El-Sheekh, M. M., Schagerl, M., Kornaros, M., & Sun, J. (2023). Microalgae-based wastewater treatment: Mechanisms, challenges, recent advances, and future prospects. In *Environmental Science and Ecotechnology* (Vol. 13). Editorial Board, Research of Environmental Sciences. https://doi.org/10.1016/j.ese.2022.100205

Aditya, T., Bitu, G., & Mercy Eleanor, G. (1026). The role of algae in pharmaceutical development. *Journal of Pharmaceutical Nanotechnology*, 4, 82–89.

Alam, M. A., Xu, J. L., & Wang, Z. (2020). Microalgae biotechnology for food, health and high value products. In *Microalgae Biotechnology for Food, Health and High Value Products*. Springer Singapore. https://doi.org/10.1007/978-981-15-0169-2

Anyaoha, K. E., Krujatz, F., Hodgkinson, I., Maletz, R., & Dornack, C. (2024). Microalgae contribution in enhancing the circular economy drive of biochemical conversion systems – A review. In *Carbon Resources Conversion* (Vol. 7, Issue 2). KeAi Publishing Communications Ltd. https://doi.org/10.1016/j.crcon.2023.10.003

Azmir, J., Zaidul, I. S. M., Rahman, M. M., Sharif, K. M., Mohamed, A., Sahena, F., Jahurul, M. H. A., Ghafoor, K., Norulaini, N. A. N., & Omar, A. K. M. (2013). Techniques for extraction of bioactive compounds from plant materials: A review. *Journal of Food Engineering*, 117(4), 426–436. https://doi.org/10.1016/J.JFOODENG.2013.01.014

Batista, A. P., Gouveia, L., Bandarra, N. M., Franco, J. M., & Raymundo, A. (2013). Comparison of microalgal biomass profiles as novel functional ingredient for food products. *Algal Research*, 2(2), 164–173. https://doi.org/10.1016/J.ALGAL.2013.01.004

Benemann, J. R. (2000). Hydrogen production by microalgae. *Journal of Applied Phycology*, 12(3), 291–300.

Bhalamurugan, G. L., Valerie, O., & Mark, L. (2018). Valuable bioproducts obtained from microalgal biomass and their commercial applications: A review. *Environmental Engineering Research*, *23*(3), 229–241. https://doi.org/10.4491/eer.2017.220

Bhattacharjee, M. (2016). Pharmaceutically valuable bioactive compounds of algae. *Asian Journal of Pharmaceutical and Clinical Research*, 9(6), 43–47). https://doi.org/10.22159/ajpcr.2016.v9i6.14507

Caporgno, M. P., Taleb, A., Olkiewicz, M., Font, J., Pruvost, J., Legrand, J., & Bengoa, C. (2015). Microalgae cultivation in urban wastewater: Nutrient removal and biomass production for biodiesel and methane. *Algal Research*, 10, 232–239. https://doi.org/10.1016/J.ALGAL.2015.05.011

Cea-Barcia, G., Buitrón, G., Moreno, G., & Kumar, G. (2014). A cost-effective strategy for the bio-prospecting of mixed microalgae with high carbohydrate content: Diversity fluctuations in different growth media. *Bioresource Technology*, 163, 370–373. https://doi.org/10.1016/J.BIORTECH.2014.04.079

Chader, S., Mahmah, B., Chetehouna, K., Amrouche, F., & Abdeladim, K. (2011). Biohydrogen production using green microalgae as an approach to operate a small proton exchange membrane fuel cell. *International Journal of Hydrogen Energy*, 36(6), 4089–4093. https://doi.org/10.1016/J.IJHYDENE.2010.07.117

Chinnasamy, S., Bhatnagar, A., Hunt, R. W., & Das, K. C. (2010). Microalgae cultivation in a wastewater dominated by carpet mill effluents for biofuel applications. *Bioresource Technology*, 101(9), 3097–3105. https://doi.org/10.1016/j.biortech.2009.12.026

Chu, W. L., See, Y. C., & Phang, S. M. (2009). Use of immobilised Chlorella vulgaris for the removal of colour from textile dyes. *Journal of Applied Phycology*, 21(6), 641–648. https://doi.org/10.1007/s10811-008-9396-3

de-Bashan, L. E., & Bashan, Y. (2010). Immobilized microalgae for removing pollutants: Review of practical aspects. *Bioresource Technology*, 101(6), 1611–1627. https://doi.org/10.1016/j.biortech.2009.09.043

Dewapriya, P., & Kim, S. kwon. (2014). Marine microorganisms: An emerging avenue in modern nutraceuticals and functional foods. *Food Research International*, 56, 115–125. https://doi.org/10.1016/J.FOODRES.2013.12.022

Dineshkumar, R., Subramanian, J., Gopalsamy, J., Jayasingam, P., Arumugam, A., Kannadasan, S., & Sampathkumar, P. (2019). The impact of using microalgae as biofertilizer in maize (Zea mays L.). *Waste and Biomass Valorization*, 10(5), 1101–1110. https://doi.org/10.1007/s12649-017-0123-7

Dolganyuk, V., Belova, D., Babich, O., Prosekov, A., Ivanova, S., Katserov, D., Patyukov, N., & Sukhikh, S. (2020). Microalgae: A promising source of valuable bioproducts. *Biomolecules*, 10(8), 1–24. https://doi.org/10.3390/biom10081153

El Amri, R. El, Elkacmi, R., Benadda, B., Kacem, M., Hasib, A., & Boudouch, O. (2023). Treatment of wastewater from Azilal (Morocco) using microalgae: Comparing indoor and outdoor cultures and biomass harvesting by coagulation–flocculation. *Desalination and Water Treatment*, 286, 90–100. https://doi.org/10.5004/dwt.2023.29316

El Amri, R., Elkacmi, R., & Boudouch, O. (2023). Removal of methyl orange from water using microalgae: Effect of operating parameters, equilibrium, kinetic and thermodynamic studies. *Chemistry Africa*, 6(5), 2569–2580. https://doi.org/10.1007/s42250-023-00671-x

El Amri, R., Elkacmi, R., Hasib, A., & Boudouch, O. (2022). Removal of hydroxychloroquine from an aqueous solution using living microalgae: Effect of operating parameters on removal efficiency and mechanisms. *Water Environment Research*, *94*(9). https://doi.org/10.1002/wer.10790

Enzing, C., Ploeg, M., Barbosa, M., & Sijtsma, L. (2014). Microalgae-based products for the food and feed sector: an outlook for Europe. *JRC Scientific and Policy Reports*, 19–37.

Faheed, F. A., & El-Fattah, Z. A. (2008). Effect of Chlorella vulgaris as bio-fertilizer on growth parameters and metabolic aspects of lettuce plant. *Journal of Agriculture and Social Sciences*, 4(4), 165–169.

Fal, S., Smouni, A., & Arroussi, H. El. (2023). Integrated microalgae-based biorefinery for wastewater treatment, industrial CO_2 sequestration and microalgal biomass valorization: A circular bioeconomy approach. *Environmental Advances*, *12*. https://doi.org/10.1016/j.envadv.2023.100365

Fan, H., Wang, K., Wang, C., Yu, F., He, X., Ma, J., & Li, X. (2020). A comparative study on growth characters and nutrients removal from wastewater by two microalgae under optimized light regimes. *Environmental Technology & Innovation*, 19, 100849. https://doi.org/10.1016/J.ETI.2020.100849

Garcia-Gonzalez, J., & Sommerfeld, M. (2016). Biofertilizer and biostimulant properties of the microalga Acutodesmus dimorphus. *Journal of Applied Phycology*, 28(2), 1051–1061. https://doi.org/10.1007/s10811-015-0625-2

Gaurav, K., Neeti, K., & Singh, R. (2024). Microalgae-based biodiesel production and its challenges and future opportunities: A review. *Green Technologies and Sustainability*, 2(1), 100060. https://doi.org/10.1016/j.grets.2023.100060

Gunasundari, E., & Senthil Kumar, P., (2017). Adsorption isotherm, kinetics and thermodynamic analysis of Cu(II) ions onto the dried algal biomass (Spirulina platensis). *Journal of Industrial and Engineering Chemistry*, 56, 129–144. https://doi.org/10.1016/J.JIEC.2017.07.005

Hachicha, R., Elleuch, F., Hlima, H. Ben, Dubessay, P., de Baynast, H., Delattre, C., Pierre, G., Hachicha, R., Abdelkafi, S., Michaud, P., & Fendri, I. (2022). Biomolecules from microalgae and cyanobacteria: Applications and market survey. *Applied Sciences*, 12(4). https://doi.org/10.3390/app12041924

Hirano, A., Ueda, R., Hirayama, S., & Ogushi, Y. (1997). CO_2 fixation and ethanol production with microalgal photosynthesis and intracellular anaerobic fermentation. *Energy*, 22(2–3), 137–142. https://doi.org/10.1016/S0360-5442(96)00123-5

Huang, G. H., Chen, F., Wei, D., Zhang, X. W., & Chen, G. (2010). Biodiesel production by microalgal biotechnology. *Applied Energy*, 87(1), 38–46. https://doi.org/10.1016/j.apenergy.2009.06.016

John, R. P., Anisha, G. S., Nampoothiri, K. M., & Pandey, A. (2011). Micro and macroalgal biomass: A renewable source for bioethanol. *Bioresource Technology*, 102(1), 186–193. https://doi.org/10.1016/J.BIORTECH.2010.06.139

Kazakis, N. A. (2025). Green approaches to heavy metal removal from wastewater: Microalgae solutions in a circular economy framework. *Societal Impacts*, 5, 100103. https://doi.org/10.1016/j.socimp.2025.100103

Keffer, J. E., & Kleinheinz, G. T. (2002). Use of Chlorella vulgaris for CO_2 mitigation in a photobioreactor. *Journal of Industrial Microbiology and Biotechnology*, 29(5), 275–280. https://doi.org/10.1038/sj.jim.7000313

Lee, Y.-K. (1997). Commercial production of microalgae in the Asia-Pacific rim. *Journal of Applied Phycology*, 9, 403–411.

Lim, S. L., Chu, W. L., & Phang, S. M. (2010). Use of Chlorella vulgaris for bioremediation of textile wastewater. *Bioresource Technology*, 101(19), 7314–7322. https://doi.org/10.1016/j.biortech.2010.04.092

Maadane, A., Merghoub, N., Ainane, T., El Arroussi, H., Benhima, R., Amzazi, S., Bakri, Y., & Wahby, I. (2015). Antioxidant activity of some Moroccan marine microalgae: Pufa profiles, carotenoids and phenolic content. *Journal of Biotechnology*, 215, 13–19. https://doi.org/10.1016/J.JBIOTEC.2015.06.400

Mahdy, A., Mendez, L., Ballesteros, M., & González-Fernández, C. (2015). Algaculture integration in conventional wastewater treatment plants: Anaerobic digestion comparison of primary and secondary sludge with microalgae biomass. *Bioresource Technology*, 184, 236–244. https://doi.org/10.1016/J.BIORTECH.2014.09.145

Melis, A., Zhang, L., Forestier, M., Ghirardi, M. L., & Seibert, M. (2000). *Sustained Photobiological Hydrogen Gas Production upon Reversible Inactivation of Oxygen Evolution in the Green Alga Chlamydomonas reinhardtii 1*. www.plantphysiol.org

Millao, S., & Uquiche, E. (2016). Antioxidant activity of supercritical extracts from Nannochloropsis gaditana: Correlation with its content of carotenoids and tocopherols. *The Journal of Supercritical Fluids*, 111, 143–150. https://doi.org/10.1016/J.SUPFLU.2016.02.002

Mohd Udaiyappan, A. F., Abu Hasan, H., Takriff, M. S., & Sheikh Abdullah, S. R. (2017). A review of the potentials, challenges and current status of microalgae biomass applications in industrial wastewater treatment. *Journal of Water Process Engineering*, 20, 8–21. https://doi.org/10.1016/J.JWPE.2017.09.006

Mustafa, E. M., Phang, S. M., & Chu, W. L. (2012). Use of an algal consortium of five algae in the treatment of landfill leachate using the high-rate algal pond system. *Journal of Applied Phycology*, 24(4), 953–963. https://doi.org/10.1007/s10811-011-9716-x

Nagarajan, D., Lee, D. J., Kondo, A., & Chang, J. S. (2017). Recent insights into biohydrogen production by microalgae – From biophotolysis to dark fermentation. *Bioresource Technology*, 227, 373–387. https://doi.org/10.1016/J.BIORTECH.2016.12.104

Naik, S. N., Goud, V. V., Rout, P. K., & Dalai, A. K. (2010). Production of first and second generation biofuels: A comprehensive review. *Renewable and Sustainable Energy Reviews*, 14(2), 578–597. https://doi.org/10.1016/J.RSER.2009.10.003

Nigam, P. S., & Singh, A. (2011). Production of liquid biofuels from renewable resources. *Progress in Energy and Combustion Science*, 37(1), 52–68. https://doi.org/10.1016/J.PECS.2010.01.003

Nur, M. M. A., Mahreni, Murni, S. W., Setyoningrum, T. M., Hadi, F., Widayati, T. W., Jaya, D., Sulistyawati, R. R. E., Puspitaningrum, D. A., Dewi, R. N., Hadiyanto, & Hasanuzzaman, M. (2025). Innovative strategies for utilizing microalgae as dual-purpose biofertilizers and phycoremediators in agroecosystems. *Biotechnology Reports*, 45. https://doi.org/10.1016/j.btre.2024.e00870

Phang, S. M., Miah, M. S., Yeoh, B. G., & Hashim, M. A. (2000). Spirulina cultivation in digested sago starch factory wastewater. *Journal of Applied Phycology*, 12, 395–400.

Priyadarshani, I., & Rath, B. (2012). Commercial and industrial applications of micro algae-A review. *Research Article Journal of Algal Biomass Utilization*, 2012(4), 2229–6905.

Renuka, N., Prasanna, R., Sood, A., Ahluwalia, A. S., Bansal, R., Babu, S., Singh, R., Shivay, Y. S., & Nain, L. (2016). Exploring the efficacy of wastewater-grown microalgal biomass as a biofertilizer for wheat. *Environmental Science and Pollution Research*, 23(7), 6608–6620. https://doi.org/10.1007/s11356-015-5884-6

Richmond, Amos. (2004). *Handbook of microalgal culture: Biotechnology and applied phycology*. Blackwell Science.

Rizwan, M., Mujtaba, G., Memon, S. A., Lee, K., & Rashid, N. (2018a). Exploring the potential of microalgae for new biotechnology applications and beyond: A review. *Renewable and Sustainable Energy Reviews*, 92, 394–404. https://doi.org/10.1016/j.rser.2018.04.034

Rizwan, M., Mujtaba, G., Memon, S. A., Lee, K., & Rashid, N. (2018b). Exploring the potential of microalgae for new biotechnology applications and beyond: A review. *Renewable and Sustainable Energy Reviews*, 92, 394–404. https://doi.org/10.1016/J.RSER.2018.04.034

Rodionova, M. V., Poudyal, R. S., Tiwari, I., Voloshin, R. A., Zharmukhamedov, S. K., Nam, H. G., Zayadan, B. K., Bruce, B. D., Hou, H. J. M., & Allakhverdiev, S. I. (2017). Biofuel production: Challenges and opportunities. *International Journal of Hydrogen Energy*, 42(12), 8450–8461. https://doi.org/10.1016/J.IJHYDENE.2016.11.125

Ruiz-Marin, A., Mendoza-Espinosa, L. G., & Stephenson, T. (2010). Growth and nutrient removal in free and immobilized green algae in batch and semi-continuous cultures treating real wastewater. *Bioresource Technology*, 101(1), 58–64. https://doi.org/10.1016/j.biortech.2009.02.076

Saifuddin, N., & Priatharsini, P. (2016). Developments in bio-hydrogen production from Algae: A review. *Research Journal of Applied Sciences, Engineering and Technology*, 12(9), 968–982. https://doi.org/10.19026/rjaset.12.2815

Samoraj, M., Chojnacka, K., Skrzypczak, D., Mikula, K., Mironiuk, M., & Calis, D. (2025). Microalgae as biobased fertilizers for sustainable agriculture. *Advances in Sustainable Applications of Microalgae*, 263–281. https://doi.org/10.1016/B978-0-443-22127-9.00011-1

Sathasivam, R., Radhakrishnan, R., Hashem, A., & Abd_Allah, E. F. (2019). Microalgae metabolites: A rich source for food and medicine. *Saudi Journal of Biological Sciences*, 26(4), 709–722. https://doi.org/10.1016/j.sjbs.2017.11.003

Scaife, M. A., Nguyen, G. T. D. T., Rico, J., Lambert, D., Helliwell, K. E., & Smith, A. G. (2015). Establishing Chlamydomonas reinhardtii as an industrial biotechnology host. *Plant Journal*, 82(3), 532–546. https://doi.org/10.1111/tpj.12781

Silva, M., Geada, P., Pereira, R. N., & Teixeira, J. A. (2025). Microalgae biomass–A source of sustainable dietary bioactive compounds towards improved health and well-being. *Food Chemistry Advances*, 6. https://doi.org/10.1016/j.focha.2025.100926

Skjånes, K., Rebours, C., & Lindblad, P. (2013). Potential for green microalgae to produce hydrogen, pharmaceuticals and other high value products in a combined process. *Critical Reviews in Biotechnology*, 33(2), 172–215. https://doi.org/10.3109/07388551.2012.681625

Song, T., Mårtensson, L., Eriksson, T., Zheng, W., & Rasmussen, U. (2005). Biodiversity and seasonal variation of the cyanobacterial assemblage in a rice paddy field in Fujian, China. *FEMS Microbiology Ecology*, 54(1), 131–140. https://doi.org/10.1016/j.femsec.2005.03.008

Spain, O., Plöhn, M., & Funk, C. (2021). The cell wall of green microalgae and its role in heavy metal removal. *Physiologia Plantarum*, 173(2), 526–535. https://doi.org/10.1111/ppl.13405

Spolaore, P., Joannis-Cassan, C., Duran, E., & Isambert, A. (2006). Commercial applications of microalgae. *Journal of Bioscience and Bioengineering*, 101(2), 87–96. https://doi.org/10.1263/JBB.101.87

Stolz, P., & Obermayer, B. (2005). *Manufacturing microalgae for skin care* (Vol. 120, Issue 3). www.TheCosmeticSite.comCosmetics&Toiletries®magazine|99

Thomas, N. V., & Kim, S. K. (2013). Beneficial effects of marine algal compounds in cosmeceuticals. *Marine Drugs*, 11(1), 146–164. https://doi.org/10.3390/md11010146

Varfolomeev, S. D., & Wasserman, L. A. (2011). Microalgae as source of biofuel, food, fodder, and medicines. *Applied Biochemistry and Microbiology*, 47(9), 789–807. https://doi.org/10.1134/S0003683811090079

Vázquez-Romero, B., Perales, J. A., Pereira, H., Barbosa, M., & Ruiz, J. (2022). Techno-economic assessment of microalgae production, harvesting and drying for food, feed, cosmetics, and agriculture. *Science of the Total Environment*, *837*. https://doi.org/10.1016/j.scitotenv.2022.155742

Wang, B., Li, Y., Wu, N., & Lan, C. Q. (2008). CO_2 bio-mitigation using microalgae. *Applied Microbiology and Biotechnology*, 79(5), 707–718. https://doi.org/10.1007/s00253-008-1518-y

Wang, H. M., Chen, C. Y., & Wen, Z. H. (2011). Identifying melanogenesis inhibitors from Cinnamomum subavenium with in vitro and in vivo screening systems by targeting the human tyrosinase. *Experimental Dermatology*, 20(3), 242–248. https://doi.org/10.1111/j.1600-0625.2010.01161.x

Wang, H. M. D., Chen, C. C., Huynh, P., & Chang, J. S. (2015). Exploring the potential of using algae in cosmetics. *Bioresource Technology*, 184, 355–362. https://doi.org/10.1016/J.BIORTECH.2014.12.001

Yan, N., Fan, C., Chen, Y., & Hu, Z. (2016). The potential for microalgae as bioreactors to produce pharmaceuticals. *International Journal of Molecular Sciences*, 17(6). https://doi.org/10.3390/ijms17060962

Zullaikah, S., Utomo, A. T., Yasmin, M., Ong, L. K., & Ju, Y. H. (2019). Ecofuel conversion technology of inedible lipid feedstocks to renewable fuel. *Advances in Eco-Fuels for a Sustainable Environment*, 237–276. https://doi.org/10.1016/B978-0-08-102728-8.00009-7

12 Sustainability and Environmental Impact in Liquid Waste Treatment

Innovations, Strategies, and Future Perspectives

Anas Hafdi, Oumayma El Wafi, Mouad Mouhsin, Mohamed Amine Kasbaji, Meriem Kasbaji, Khaoula Khaless, Mohamed M'Barki, Mounir El Achaby, Saad ALjlil, and Mustapha Oubenali

12.1 INTRODUCTION TO SUSTAINABILITY IN LIQUID WASTE TREATMENT

Sustainability, as defined by the United Nations Brundtland report, is the ability to meet present needs without compromising the ability of future generations to meet their own needs [1]. In the context of liquid waste treatment, sustainability means practicing methods that can ensure resource management efficiency, reduce environmental hazards, and maintain ecological balance in the long run. Environmental issues have become a global concern in recent times. These challenges include waste management, water management, climate change, pollution, degradation, and deforestation. These are complex and interconnected issues and must be addressed through multidimensional approaches and dynamic solutions to mitigate their impacts to secure the future of our planet [2]. Water is a fundamental resource, essential to all life. In addition to its biological importance, water is essential to socio-economic development and environmental sustainability. It supports agriculture, industry, energy production, and healthy ecosystems, making it a cornerstone of global prosperity. According to statistics, the world contains approximately 1,400 million cubic kilometers of water, 97% of which is seawater, while a significant part of the freshwater is heavily polluted [3]. Global water scarcity is a critical issue that has intensified over recent decades. Due to economic development, population growth, and changing consumption patterns, global water consumption has increased sixfold over the past century and continues to grow steadily at a rate of around 1% per year [4]. Moreover, among the sustainable development goals agreed by the United Nations (UN), access to drinking water is a fundamental human

DOI: 10.1201/9781003624905-12

right. However, the UN World Water Development Report 2020 reveals that there is a serious water shortage. Nearly half of the world's population faces severe water scarcity at least once a year for at least one month.

Water availability, which is critical, is under increasing pressure. With over-exploitation, climate change impacts, population growth, and inefficiency in water use, water resource management is extremely challenging. Liquid waste is still the most serious issue nowadays. According to the Food and Agriculture Organization of the UN, the global annual water withdrawal amounts to 3,928 km^3, either consumed or discharged back into the environment as wastewater. Of this total, industrial wastewater accounts for 16%, while municipal wastewater represents 8% [5]. These liquid waste discharges not only contaminate surface water sources, such as lakes and rivers, but also infiltrate groundwater, threatening the quality and sustainability of global water supplies. Additionally, they are causing significant harm to the ecosystem and posing serious risks to human health [6]. The continuous discharge of liquid waste, primarily resulting from anthropogenic activities such as agricultural, urban, and industrial practices, represents a significant issue in the form of severe pollution [7].

Liquid waste is a broad category of waste materials that are mainly composed of liquids and produced in various residential, commercial, and industrial processes. Liquid waste includes wastewater, which also includes domestic, industrial, or agricultural wastewater, and fats, oils, and grease, which are by-products of cooking or industrial activities. Additionally, it includes used oils, such as lubricants and hydraulic fluids discarded after use in machinery or vehicles, which are typically toxic and challenging to treat. Liquid waste also contains suspended solids, including sludges or slurries with high liquid content, commonly produced during industrial or municipal processes. Liquid waste is classified into three main categories [8]. The first category is sewage, which is a liquid waste originating from domestic activities. The second category is trade liquid waste, which corresponds to non-human liquid waste generated by industrial, commercial, and business activities. The third category includes hazardous liquid waste released from industrial activities, which is particularly dangerous and harmful.

Liquid waste contains a wide range of harmful and toxic substances (Table 12.1), making its composition more complex to treat than that of domestic wastewater, which generally contains organic matter, nutrients, and microorganisms. Its composition is significantly influenced by the specific methods and raw materials employed in the industry. For instance, liquid effluents from the food processing industry are usually known to have a high concentration of biodegradable organic pollutants, mainly resulting from the processing of natural matrixes (fruits, vegetables, animal products, etc.). These effluents, however, are highly detrimental to the environment owing to the presence of heavy metals and high alkalinity [9]. Among the most hazardous pollutants commonly found in liquid waste are heavy metals (e.g., chromium, mercury, cadmium, nickel, etc.) [10,11] and synthetic dyes [12], which are particularly prevalent in wastewater from industries such as textiles and leather processing, along with other emerging pollutants such as perfluoroalkyl substances, pharmaceutical, and personal care products [13]. The presence of these compounds can result in a high chemical oxygen demand (COD), biochemical oxygen demand (BOD), and trace metals, which can harm water quality and various ecosystems [14]. The effects of water pollutants on health are diverse, leading to mild symptoms like headaches

TABLE 12.1
Main Categories of Liquid Waste Pollutants and Their Sources

Category	Main Pollutants	Sources	References
Inorganic pollutants	Heavy metals (Pb, Hg, Cd, Cr, As, etc.), nitrates, nitrites, phosphates, and sulfates.	Mining, metal plating, fertilizer runoff, chemical industries.	[16–18]
Organic pollutants	Phenols, polycyclic aromatic hydrocarbons, dyes and pigments, hydrocarbons, halogenated hydrocarbons, plasticizers, biphenyls, detergents, aromatic compounds, oils, and greases.	Food processing, textile industry, slaughterhouses, paper mills, oil refineries.	[19–22]
Chemical pollutants	Pesticides, herbicides, pharmaceutical residues, microplastics (fibers, fragments, and polymers).	Agriculture, pharmaceutical industries, household chemicals.	[7,22–24]
Biological pollutants	Bacteria, viruses, antibiotic resistance genes, protozoa, parasites, and microbial toxins.	Domestic sewage, hospitals, agriculture, and food industry.	[25–27]
Radioactive pollutants	Uranium, Radium, Thorium, Cesium-137, Iodine-131, and Strontium-90.	Nuclear power plants, medical waste, mining of radioactive ores, industrial and pilot plants, and processing of thorium and uranium nuclear fuels.	[28–30]

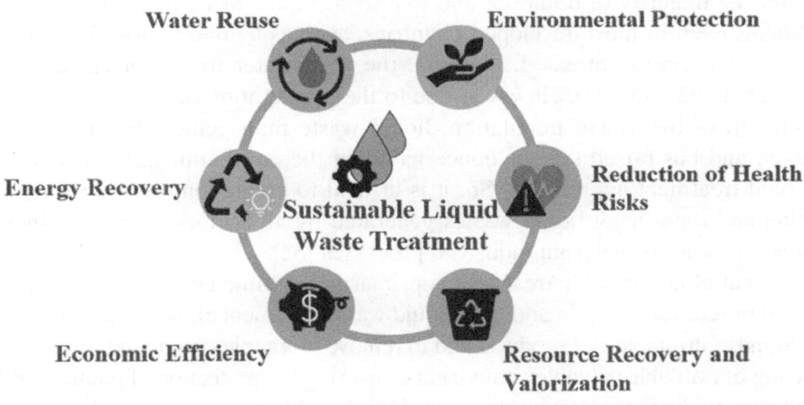

FIGURE 12.1 Main purposes of sustainable liquid waste treatment.

and skin rashes as well as serious respiratory, nervous, and cardiac disorders. Liquid waste pollution has environmental ramifications, leading to phenomena like acid rain, algal blooms, and disruptions to the aquatic food chain [15].

Therefore, treating liquid waste is essential to address water pollution and scarcity issues (Figure 12.1). The goal is to reduce BOD and COD levels as well as toxic particles before they are released into the environment. In addition, by implementing liquid waste treatment processes, we can effectively recycle water resources and reuse them in various sectors, including artificial recharge, agriculture, industries, and landscape irrigation [2]. It is therefore important to develop efficient techniques for treating liquid waste and to maximize the overall efficiency of the wastewater treatment plants (WWTPs) [31]. However, most liquid waste treatment systems used have many drawbacks, such as the use of harmful chemicals, excessive energy consumption, and excessive sludge production, which make them unsustainable. The biggest challenge is to use efficient treatment systems that are both capable of removing toxic pollutants from liquid waste and are sustainable. Moreover, there are several reasons why sustainability is important in liquid waste treatment. It addresses the pressing need to mitigate water scarcity, resource recovery, and reduce pollution. In addition, sustainable practices in this domain support world efforts to achieve the UN sustainable development goals, particularly clean water and sanitation (goal 6) and responsible consumption and production (goal 12). The integration of sustainability in liquid waste treatment will lead to a cleaner environment, healthier communities, and a more resilient future.

12.2 OVERVIEW AND ENVIRONMENTAL IMPACTS OF TRADITIONAL LIQUID WASTE TREATMENT METHODS

Liquid waste presents significant challenges; human societies have been facing these challenges generated mainly by anthropogenic activities for centuries, using established methods to address this essential need. Many low- and middle-income countries lack adequate infrastructure for wastewater collection and treatment [31]. As a result, the majority of domestic and industrial liquid waste is discharged without treatment. Even in most developed countries, such as China, almost 55% of wastewater is discharged untreated. In India, the wastewater treatment capacity deficit is as high as 62.46% [31]. However, due to the acceleration of industrialization and the growth of the global population, liquid waste management has become more complex and has raised serious concerns about the environmental impacts of conventional treatment approaches. So, it is critical to create long-term techniques for treating and recovering liquid wastes generated by all sectors of society, including sewage and wastewater from industrial processes [32].

The goal of liquid waste treatment is primarily to eliminate pollutants prior to safe disposal or recovery [32]. In addition, liquid waste treatment aims to transform organic compounds into stable end products and to remove inorganic compounds, as well as the recycling of available valuable components, ensuring the protection of public health and compliance with environmental standards [15]. This process can primarily be classified into biological, physical, and chemical techniques. These techniques can be applied in combination or individually, depending on the origin of the liquid waste, which influences the type, size, and volume of the contaminants [8]. The liquid waste treatment conventionally consists of four main stages (Figure 12.2): preliminary, primary, secondary, and tertiary, followed by a sludge treatment stage [15]. First, preliminary treatment aims to remove larger debris and solids through physical and mechanical processes. Second, primary treatment employs physicochemical methods to further reduce contaminants

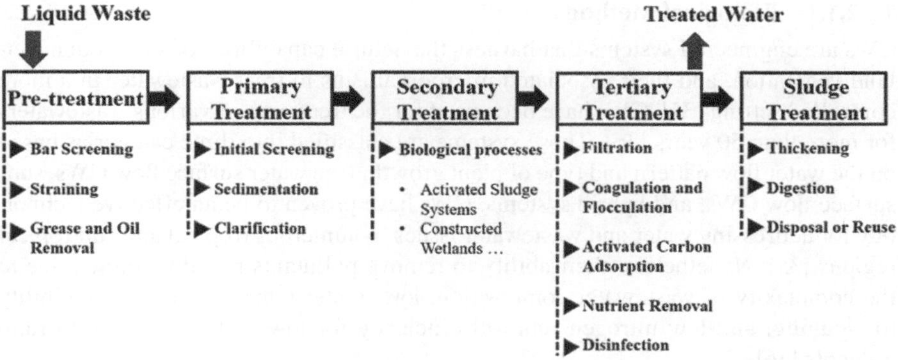

FIGURE 12.2 Schematic representation of the five stages of wastewater treatment and traditional methods used.

and suspended solids. The resulting waters then undergo secondary treatment, which focuses on the biological degradation of organic matter, often using microorganisms to break down the waste. Tertiary treatment is designed to remove additional contaminants, like pathogens and nutrients, using advanced physical and chemical processes. Finally, the process is left with sludge that will undergo treatment using methods such as controlled disposal, recycling, or incineration.

12.2.1 Overview of Traditional Liquid Waste Treatment Methods

In general, traditional liquid waste treatment comprises a combination of biological, chemical, and/or physical procedures and activities to eliminate and/or reduce various contaminants from the effluent [33]. A multitude of traditional techniques are used during the five stages of conventional liquid waste treatment (Figure 12.3); among these methods, the biological methods are the most used, such as constructed wetlands (CWs), moving bed biofilm reactors, activated sludge processes, up-flow anaerobic sludge blankets (UASB), and stabilization ponds [34]. These technologies have been widely implemented and integrated into WWTPs worldwide.

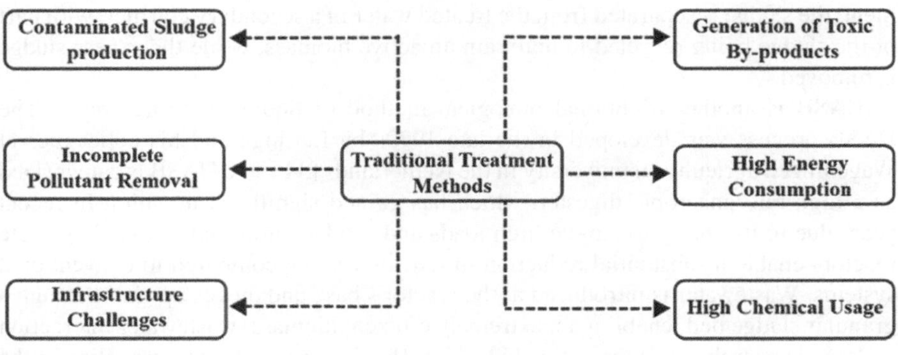

FIGURE 12.3 Effects of conventional treatments on the environment.

12.2.1.1 Biological Methods

CWs are engineered systems that harness the natural capabilities of soil media, wetland vegetation, and their associated microorganisms to treat wastewater in a more controlled setting [35]. CWs have been used for the treatment of various wastewaters for more than 50 years [36]. These systems are classified into three categories based on the water flow pattern and type of plant growth: free water surface flow CWs, subsurface flow CWs, and hybrid systems. CWs have proven to be an effective technology for addressing water and wastewater issues in numerous tropical and subtropical regions [35]. Nonetheless, their ability to remove pollutants remains limited due to the complexity of wastewater composition, low winter temperatures, susceptibility to clogging, and low nitrogen removal efficiency for low carbon-to-nitrogen ratio effluents [36].

A Moving Bed Biofilm Reactor is favored for its effectiveness in treating high organic and nutrient loads, as well as recalcitrant pollutants with increased removal efficiency [37]. The Moving Bed Biofilm Reactor process integrates both attached and suspended growth mechanisms. In this system, biofilms are grown on small sponge or plastic-based carriers with a density slightly less than water. These carriers remain in constant motion within the reactor through aeration in aerobic reactors or by mechanical stirring [38]. The Moving Bed Biofilm Reactor has gained significant popularity over the past four decades and is now widely implemented in WWTPs across the globe [39].

The Activated Sludge Process is one of the most widely used biological treatment technologies for municipal and industrial liquid waste. The Activated Sludge Process has played a crucial role in human development over the past century, recognized as one of the most significant engineering achievements [40]. It is based on the suspended growth of microorganisms in an aeration tank, where oxygen is supplied to promote the biodegradation of soluble organic pollutants. Phosphate and ammonium are eliminated mainly through microbial assimilation and dissimilation, including nitrification–denitrification, aerobic biological oxidation, etc. [40]. The infrastructure of a basic Activated Sludge Process starts with a single continuously operated bioreactor, where suspended microorganisms break down the dissolved and colloidal organic matter. The mixture is then transferred into a clarifier (settling tank), where activated sludge is separated from the treated wastewater by gravity [41]. After treatment, the sludge is separated from the treated water in a secondary clarifier, with part of the sludge being recycled to maintain an active biomass, while the excess sludge is removed.

UASB is another traditional biological method of liquid waste treatment. The UASB process was developed in the late 1970s by Lettinga and his colleagues at Wageningen agricultural university in the Netherlands [34]. The UASB is categorized as a high-rate anaerobic digester, which has gained significant attention in recent years due to its ability to handle high loads and produce minimal sludge. High-rate reactors enable a substantial reduction in reactor volume compared to conventional systems. Wastewater is introduced at the reactor's base and moves upward through a granular sludge bed, enabling an extremely efficient biomass–wastewater interaction and rapid anaerobic decomposition [42]. The effectiveness of the process relies on the quality of the biofilm and the intimate contact between sludge and wastewater, which are critical for pollutant removal. Unlike aerobic technologies, anaerobic processes like the UASB process can convert the high organic matter content of wastewater into methane gas, which can be used as an energy source, while reducing sludge production [43].

Waste Stabilization Pond (WSP) is also a traditional biological method of liquid waste treatment. The WSP is one of the simplest wastewater treatment methods, widely used worldwide, especially in countries with sufficient land and favorable climatic conditions (high temperature and sunlight). This technique is favored worldwide for its low technical demands, cost-effectiveness, and minimal energy usage [44]. It is even widely used in developed countries, representing over 50% of plants in the US and accounting for a significant share of treatment systems in France (20%) and Germany (30%) [45]. These large, shallow ponds rely on biological and physical processes to remove pollutants, organic matter, and pathogens from raw wastewater [46]. The WSP treats liquid waste through natural hydrodynamic, biochemical, physicochemical, photosynthetic, and microbiological processes. The treatment process takes place in a series of ponds where wastewater flows sequentially, allowing different microorganisms to break down contaminants under varying environmental conditions. Commonly referred to as lagoons, WSPs typically include three types of ponds: anaerobic, facultative, and maturation ponds. These ponds differ in their design, hydraulic flow, organic load, and biochemical processes [46]. Their efficiency in removing carbon, nutrients, and pathogens improves as the number of ponds increases. In some cases, raw liquid waste undergoes pretreatment, such as screening and grit removal, before flowing into the pond system [44].

12.2.1.2 Physicochemical Methods

Besides biological approaches, physicochemical processes (conventional treatments) are extensively used in sewage treatment plants. Such treatments consist of coagulation–flocculation, precipitation, activated carbon adsorption, and sand filtration [47]. Coagulation and flocculation are essential processes for removing pollutants from liquid waste. During effluent treatment, chemical coagulants, such as organic and inorganic polymers, as well as iron or aluminum-based inorganic coagulants (polyaluminum chloride, ferrous sulfate, and aluminum sulfate), are added to destabilize colloidal systems by charging phenomena, generating insoluble or bridge-forming species, thus promoting particle aggregation and the formation of heavy flocs. These flocs can then be separated through filtration, sedimentation, or flotation [48]. These processes significantly reduce turbidity and organic pollutants, improving overall treated water quality.

Chemical precipitation is a widely utilized technique and is one of the most preferred methods, particularly for the removal of phosphate compounds and toxic heavy metals from wastewaters [49]. The liquid waste is pH adjusted with an acid or base, resulting in precipitation of metal ions as insoluble solid particles, which can be separated from the treated wastewater by methods like sedimentation and filtration. This strategy is remarkably useful in decreasing toxic metal levels and mitigating eutrophication caused by excess phosphorus in aquatic ecosystems. In addition to precipitation, sand filtration and activated carbon adsorption are also widely employed techniques in liquid waste treatment to improve the treated wastewater quality by eliminating suspended solids and dissolved contaminants.

Sand filtration stands as one of the initial water treatment technologies and remains an important water purification process worldwide [50]. This process works by physical straining, which allows water to percolate through sand layers and trap both fine particles and microbial contaminants. Sand filtration can be classified into two main types based on the filtration rate: slow sand filtration or rapid sand filtration [50]. This process significantly reduces turbidity and bacterial content, which positions it as an essential final cleaning phase before wastewater release or recycling. In addition, sand filters are also widely used for drinking water production [51].

Activated carbon adsorption stands out as a highly effective method for eliminating dissolved organic pollutants from water because it removes dissolved organic pollutants, including pharmaceutical residues, pesticides, and industrial chemicals. Adsorption is a surface-based process in which pollutants, known as the adsorbate, adhere to the surface of a solid material, called the adsorbent, through physical or chemical interactions [52]. Activated carbon possesses a high retention capacity because of its extensive surface area and porous structure, which enhances adsorption and liquid waste decontamination. While sand filtration primarily addresses particulate and microbial removal, activated carbon adsorption specifically targets trace organic compounds that are resistant to conventional biological treatments. Using both these methods together offers an all-encompassing solution for eliminating contaminants.

12.2.2 Environmental Impacts of Traditional Treatment Methods

In response to rapid industrialization and increasing wastewater pollution, humanity has created many liquid waste management and treatment strategies with the objective of preserving ecosystems and protecting public health. Older methods of liquid waste treatment, which are based on a centralized approach, are proving less effective, resulting in the continuous accumulation of effluents [53]. WWTPs use a combination of biological, physicochemical, and mechanical processes [33]. This combination plays a key role in removing organic matter, suspended solids, and harmful contaminants before reuse or release into the environment. Liquid waste treatment is considered an effective way to reduce the human water footprint [54]. It aims to protect human health, thereby contributing to the world's natural water supply by returning treated water to the aquatic environment. In addition, properly treated wastewater can also be recycled for domestic and industrial uses to reduce the demand for fresh water. Different treatment methods can reduce the concentration of pollutants and suspended matter in wastewater and also degrade the content of biodegradable organic matter, measured by biological oxygen demand [53]. However, despite their widespread use, these traditional methods are not without drawbacks.

One of the main limitations of traditional methods lies in their inherent inability to handle the growing volumes of liquid waste produced. The rapid development of the industrial sector (textile industry, agricultural industry, paper and pulp industry, etc.) and the rapid demographic development are the reasons for the production of large quantities of liquid waste. This rapid increase poses two major challenges.

First, these high flows exceed the capacity of current WWTP infrastructures, which results in the direct discharge of untreated or insufficiently treated wastewater into the environment. On the other hand, the diversity of sources of these liquid wastes, coming from domestic wastewater and from different industries, results in a complex mixture of pollutants and hazardous substances. The traditional treatment methods adopted in WWTPs are not always able to effectively treat all types of contaminants, which leads to their release into the environment. Therefore, a controlled entry of liquid waste effluent [53] or a specific treatment adapted to each type of industrial discharge is necessary in order to limit their impact on water quality, human health, and aquatic ecosystems.

Despite their essential role in liquid waste management, WWTPs still have many side effects on the environment. In addition to requiring large amounts of chemical reagents, energy, and fresh water, treatment processes directly discharge treated effluent into the recipient streams. Moreover, their operations contribute to greenhouse gas emissions and waste sludge [55]. Traditional liquid waste treatment methods are often costly and less effective [56]. Among these, the physicochemical treatment techniques are widely accessible and commonly employed for liquid waste management, including processes such as neutralization/precipitation, solvent extraction, etc. [57]. However, the applicability of these methods is often not feasible due to their high chemical consumption and the significant production of sludge, which poses additional environmental and operational challenges.

Traditional liquid waste treatment processes rely heavily on chemical reagents to enhance operational efficiency [58]. These chemicals are utilized in WWTPs, particularly in physicochemical processes such as disinfection, precipitation, and coagulation–flocculation. For various purposes, including pH control, reducing BOD, eliminating odors, removing grease, etc. Sodium hydroxide and calcium hydroxide are pH adjustment agents used to neutralize acidic or alkaline wastewater to ensure effective treatment processes and facilitate subsequent treatment steps. Coagulants such as aluminum sulfate and ferric chloride help aggregate suspended matter, while disinfectants such as chlorine and ozone eliminate harmful microorganisms. While effective, the widespread usage of these chemicals poses significant environmental and health risks. Residual chemicals in treated water can cause secondary pollution, affecting aquatic ecosystems and human health. Chlorine and bromine gases, which are widely used as disinfectants for water treatment, can form toxic by-products such as trihaloacetic acids, trihalomethanes, and other compounds by reacting with organic matter present in the water [59]. In addition, these chemical reagents can interact with existing chemical compounds in liquid waste, create complex mixtures that enhance their persistence and toxicity, and finally be absorbed into the water cycle, thus aggravating water pollution [60]. Additionally, the handling and disposal of chemical sludge, which often contains heavy metals and other hazardous substances, poses serious risks of contamination of groundwater and soil.

In addition to the problem of excessive use of chemicals, WWTPs are large energy consumers [61,62]. The energy consumption of the water sector can account for 30%–40% of a municipality's energy costs [63]. In the year 2000, water supply and wastewater treatment constituted 3.7% of total electricity consumption in the United States. This demand is projected to increase by 63% by 2050 compared to

2000 levels [63]. Traditional liquid waste treatment processes require high energy consumption, contributing significantly to the carbon footprint of wastewater treatment systems. Energy operating costs of liquid waste treatment vary considerably, ranging from 9% to 50%, depending on the country and the size of WWTP [64]. Cardoso et al. [64] identified the main energy-consuming processes in WWTPs as aeration, sludge treatment, pumping, and other operations, with energy consumption ranging from 42% to 67%, 8% to 29%, 5% to 30%, and 8% to 35%, respectively, of the total plant energy consumption. This high energy demand is a major obstacle to the development of environmentally friendly and more sustainable solutions.

Another important challenge in traditional liquid waste treatment is the large-scale production of sludge, a by-product generated during both chemical and biological wastewater treatment processes [65]. This sludge contains a wide range of harmful substances such as phenols and their derivatives, dioxins, furans, pesticides, polychlorinated biphenyls, chlorine derivatives, phthalates, polycyclic aromatic hydrocarbons, and others [66]. Metals are an important residue of sludge inorganic matter, representing approximately 4%–8% of the total solids it contains [67], posing significant environmental and economic challenges in its management. In biological treatment systems, such as the Activated Sludge Process, excessive biomass accumulation necessitates frequent sludge extraction, dewatering, and disposal. Likewise, physicochemical processes, including coagulation and precipitation, generate considerable sludge volumes containing chemical residues, requiring additional disposal or treatment such as landfill, incineration, and composting [68]. However, even these methods have some drawbacks, such as contamination risk, energy consumption, and limited resource recovery [68].

Liquid waste treatment remains an important measure that has been used for many years; however, traditional treatment systems still have some drawbacks. Nonetheless, the growing volume and complexity of liquid discharges, associated with industrial development, population growth, and the introduction of novel toxic contaminants, increasingly challenge these traditional methods. Persistent pollutants and toxic contaminants in liquid waste are often challenging to remove through traditional physicochemical and biological treatment techniques. Furthermore, these processes require an enormous amount of energy and chemicals, leading to environmental pollution. Hence, there is an urgent need for a transition toward sustainable and innovative treatment strategies. This transition must prioritize the implementation of new sustainable technologies and promote resource recovery and energy efficiency. Unquestionably, the sustainability of wastewater treatment in the long run is dependent on putting the protection of natural ecosystems and the conservation of the environment first.

12.3 SUSTAINABLE TREATMENT TECHNOLOGIES

The environmental limitations of traditional treatment methods have catalyzed global optimizations in sustainable liquid waste management. These optimizations are continuing and will continue in the future, particularly as discharge rules become more rigorous and the ability to accept more wastewater increases [69]. Therefore, a massive scale-up of liquid waste treatment systems is needed to address the growing global sanitation crisis [70]. Worldwide, the sustainability of liquid waste treatment systems investments is a major concern, and many examples show that without addressing key sustainability aspects, WWTPs can fail during the operational phase [70]. To avoid these failures, sustainability is considered as the main parameter for the implementation of liquid waste treatment technology. Acceptable treatment technologies must be environmentally, economically, and socially sustainable [34].

12.3.1 Sustainability Criteria

As mentioned earlier, the operation of WWTPs using traditional liquid waste treatments presents significant sustainability challenges. Therefore, to achieve sustainable development, it is essential to ensure that the environmental and economic drawbacks associated with these treatments do not outweigh their benefits [64]. To assess the effectiveness and sustainability of treatment technologies, several key performance indicators must be considered (Figure 12.4).

First, environmental impact must be considered when evaluating treatment technologies. This involves assessing the process's impact on ecosystems, water quality, and biodiversity. Sustainable technologies should aim to minimize secondary pollution, avoid the excessive use of chemicals, and integrate with natural water cycles to enhance environmental resilience. The second most important thing is energy efficiency and emission reduction, which is also a key indicator used to evaluate the performance of a treatment system. Growing concern about climate change makes energy-positive or carbon-neutral treatment an important priority [71]. Sustainable technologies should minimize energy consumption while reducing greenhouse gas emissions to the lowest possible levels.

Additionally, resource recovery is one of the most promising aspects of sustainable liquid waste treatment. Treated wastewater and sludge could be valuable resources to exploit [15]. Resource recovery focuses on the ability to extract and reuse valuable components from treated waste. This includes the recovery of water and nutrients like nitrogen and phosphorus for use in agriculture, and the dried sludge can be used

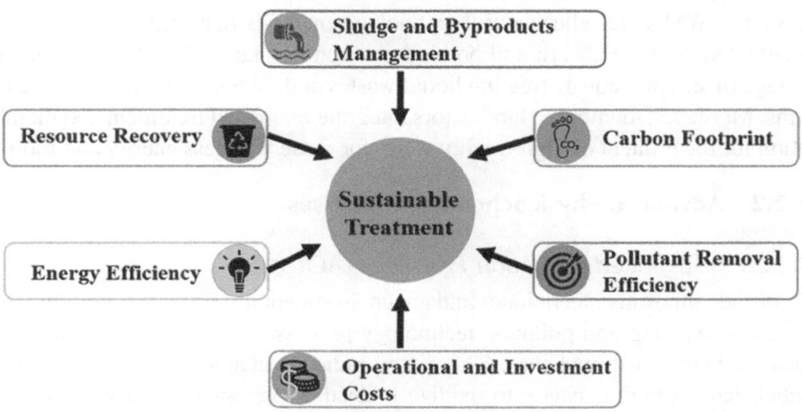

FIGURE 12.4 Sustainability criteria for liquid waste treatment methods.

as fuel, or other materials found in liquid waste can be repurposed for operations, as is commonly done in the pulp and paper sector [32]. Operating and capital costs are also key factors for determining the technology's feasibility. A key factor for the widespread adoption and scalability of innovative treatment solutions is their economic sustainability. Despite some advanced treatment methods requiring high start-up costs, they can ultimately save money in the long run by lowering disposal costs, increasing the efficiency of resource use, and reducing maintenance costs.

12.3.2 Overview of Sustainable Treatment Technologies

12.3.2.1 Biological and Natural Technologies

The development of sustainable liquid waste treatment systems should minimize energy and resource consumption and protect the environment. In this regard, innovative treatment options are being actively explored to produce net energy while also removing nutrients [72]. Anaerobic wastewater treatment has demonstrated its effectiveness, offering both energy and environmental benefits. Likewise, an alternative to the conventional nitrification–denitrification process involves the use of anaerobic ammonium-oxidizing bacteria, which significantly reduces energy and chemical consumption [72]. Additionally, under anaerobic conditions, these systems offer an efficient wastewater treatment while simultaneously generating usable energy in the form of clean electricity from organic substrates. The integration of these advanced technologies is crucial to optimizing energy recovery and advancing the development of truly sustainable wastewater treatment solutions.

Biological liquid waste treatment methods are considered among the most sustainable approaches. However, their effectiveness is often offset by significant sludge production and, in some cases, high energy consumption. Phytoremediation has emerged as one of the cost-effective, eco-friendly, and sustainable biological liquid waste treatment methods for the removal of several contaminants from contaminated water and soil [73,74]. Phytoremediation treats liquid waste using plants. It encompasses several techniques, including phytoextraction, phytovolatilization, phytodegradation, phytostabilization, and rhizofiltration [8].

CWs and WSPs are other natural technology methods that utilize plants and natural wetland systems to absorb and break down contaminants. They provide the added advantage of simultaneously treating liquid wastes and, in doing so, restoring local ecosystems. Moreover, membrane bioreactors integrate biological treatment and membrane filtration for the effluent of high-quality water for reuse with less energy and footprint.

12.3.2.2 Advanced Physicochemical Processes

12.3.2.2.1 Advanced Oxidation Processes (AOPs)

AOPs include innovative techniques and are environmentally friendly for treating wastewater and enhancing anti-pollution technology processes [75]. AOPs generate reactive species, including hydroxyl ions. These free radicals, characterized by a high redox potential, have a high capacity to oxidize molecules present in effluents. AOPs combine high pollutant removal efficiency through oxidation reactions with highly efficient hydroxyl radical generation, often assisted by photocatalytic and catalytic processes [76].

AOPs can be mainly categorized into Fenton and photo-Fenton, photocatalysis, UV–hydrogen peroxide processes, ozone-based processes, and sonolysis. The environmental consequences of these processes are considered to be of major importance due to the high reaction rates and yields, as well as the absence of highly dangerous by-products, which promote green chemistry characteristics [75].

12.3.2.2.2 Supercritical Water Oxidation (SCWO)

SCWO has been assumed to be a novel, efficient technology for the energy utilization and harmless treatment of organic wastes, resulting in rapid oxidation between oxidizers and organic matter, resulting in a product degraded into environmentally friendly green small molecules, such as inorganic salts, H_2O, and N_2 [77]. SCWO is a green chemistry technique for the degradation of organic contaminants using supercritical water ($Pc \geq 22.12$ MPa and $Tc \geq 374.15°C$) [78]. Operating under supercritical conditions, SCWO rapidly oxidizes organic matter, making it a viable option for treating high-strength waste streams with minimal harmful by-products.

In addition, SCWO is an exothermic process; when the mass concentration of organic matter is higher than 3% by weight, the heat released by the reaction can achieve self-sufficiency of the whole system, and the excess heat can even be recovered to produce steam or for electricity generation [79].

12.3.2.2.3 Eco-Friendly Adsorption

Adsorption is an effective technique in advanced wastewater treatment; it can be used to reduce or even eliminate hazardous inorganic and organic pollutants present in the effluent [80]. This technique has been widely used and studied on a large scale due to its high efficiency, low cost, ease of use, and ability to not generate contaminated by-products [81]. The adsorption process is designed for the removal of many types of organic and inorganic substances (heavy metals, dyes, phenol, pharmaceutical waste, phosphates, nitrates, etc.).

This process is a surface phenomenon that consists of accumulating pollutants called adsorbate at the interface of a solid on which adsorption occurs, called adsorbent. Adsorption can occur by chemical or physical sorption. It consists of forming strong chemical associations, or even bonds, between the molecules or ions of the adsorbate and the adsorbent surface [80]. The excellent adsorption characteristics are mainly due to the solubility of solutes in water and the electrostatic force, the van der Waals force of attraction, or the chemical bonding force between the solute and the solution. Three main types of adsorption are generally distinguished: exchange adsorption, physical adsorption, and chemical adsorption [81].

Activated carbon is an effective adsorbent in water treatment, but its production method generates significant environmental emissions. Compared to commercial carbons, natural activated carbons are renewable and have low carbon emissions [82]. In general, natural bioadsorbents derived from plant, food, agricultural biomass, and industrial wastes are more cost-effective than commercial activated carbons, taking into account raw material, manufacturing, and regeneration costs [83]. Bioadsorption offers an environmentally sustainable, cost-effective, and promising method for removing pollutants from liquid waste.

12.4 INNOVATIVE APPROACHES TO SUSTAINABLE LIQUID WASTE TREATMENT

Research into ways of optimizing liquid waste management continues actively. While traditional methods have proved their worth, they are showing their limitations in terms of energy and environmental efficiency. New technologies offer a promising alternative, combining improved treatment performance with reduced environmental impact and operating costs.

12.4.1 Applying Nanotechnology

Innovative approaches to liquid effluent management and treatment include nanotechnology, which uses nanomaterials such as nanocomposites, nanoadsorbents, nanoparticles (ZnO, CuO, TiO_2), and nanomembranes [84]. Thanks to their characteristics, such as their small size, which confers a large specific surface area and a high degree of porosity, these nanomaterials enable high reactivity and excellent dispersibility [85]. Moreover, they feature high removal capacities, antimicrobial properties, and photocatalytic degradation. They offer several advantages: their high loading and efficient encapsulation capability, eco-friendly production, and non-toxicity, making them ideal for liquid waste treatment [3].

12.4.1.1 Nanocomposites

Nanocomposites are multiphase materials with nanometric sizes. They use nanoparticles or nanofibers dispersed in a matrix to improve mechanical strength, conductivity, and thermal stability. These materials offer higher performance compared to typical materials, thanks to their unique small-size properties [86]. Nanocomposites can effectively remove organic pollutants, heavy metals, dyes, and toxins. Nanocomposites can filter out microorganisms, solid particles, and nanomaterials from liquid waste. Their nanoscale small pores enable excellent filtration [87,88]. For example, to treat liquid waste from the textile industry, a study conducted by Shagufta Rasheed et al. [89] synthesized a bimetallic catalyst based on nickel–copper doped with zinc ferrite and embedded with graphene oxide using the hydrothermal method to demonstrate its efficiency and ability to remove methylene blue dye with 99% efficiency in 3h. The addition of graphene oxide reduces the band gap energy, thus enhancing light absorption. Another noteworthy and remarkable aspect is that this nanocomposite is stable and can be reused multiple times without any loss of efficiency. Also, to remove cooking oils from polluted water, a magnetically recoverable biological nanocomposite is manufactured from photosynthetic bacteria, hull biochar, and Fe_3O_4 nanoparticles (photosynthetic bacteria/Fe_3O_4/HB). The bacteria remove cooking oils while producing commercially valuable products such as microbial lipids, lipase, and carotenoids. The biological nanocomposite showed significant results in reducing oil by 70.31%. Furthermore, there was a rise of 65.0% in biomass, 84.9% in carotenoid synthesis, and 31.5% in lipase activity. Another research study was conducted on the development of magnetic nanocomposites based on cellulose and chitosan, called MCCh-N (chitosan and norvaline) and MCCh-8H (chitosan and 8-hydroxyquinoline), respectively, to capture zirconium and hafnium ions from two real effluent streams. An acidic effluent from the hydrometallurgical treatment, from which both nanocomposites showed remarkable ion recovery performance. MCCh-N demonstrated a 100% zirconium recovery rate, whereas MCCh-8H attained a 99% recovery rate of Zr (IV) ion recovery. For the second effluent, a nuclear materials authority laboratory effluent was 74% with MCCh-8H and 96% with MCCh-N. In addition, MCCh-N demonstrated 100% recovery for Hf (IV) ions, and MCCh-8H likewise recovered Hf (IV) ions with 100% efficiency. Demonstrating the feasibility of this nanocomposite for liquid waste treatment [90]. The combination of carboxymethylated chitosan with iron oxide (Fe_2O_3) forms a nanocomposite using the top-down ball milling technique in order to treat wastewater with reactive Blue 19 dye and also to possess antibacterial properties, acting against bacteria such as *E. coli* and *Pseudomonas aeruginosa*. The results of this work show that the adsorption capacity of the reactive Blue 19 dye is 140 mg g^{-1}, with an optimum pH of 5. In addition, the nanocomposite has high antibacterial activity, inhibiting the growth of bacteria, with inhibition zones reaching 28.4 mm against *E. coli* and 23.2 mm against *Pseudomonas aeruginosa*. The nanomaterial is durable, as it can be reused after five adsorption–desorption cycles with 90% efficiency [91]. A recent study in 2024 shows that a $SnO_2/CoFe_2O_4$ nanocomposite produced by the co-precipitation method is used as a photocatalyst for wastewater treatment under solar irradiation. Thanks to the properties of this photocatalyst, the degradation of rhodamine 6G reaches 92%, and the elimination of Cr (VI) reaches 87%. The recycling study over five cycles highlights the stability and easy regeneration of $SnO_2/CoFe_2O_4$, thereby reinforcing its relevance for practical use in the field of wastewater treatment [92].

All of these studies demonstrate the importance, effectiveness, and viability of nanocomposite technology in liquid waste treatment, with positive effects on the environment and the economy.

12.4.1.2 Nanofiltration

A nanofiltration filter is commonly used in liquid waste treatment and has pores of approximately 0.001 microns. Nanofiltration removes the majority of organic molecules, almost all viruses, a significant proportion of natural organic matter, and a wide range of salts that are difficult to extract using traditional methods [93,94]. Numerous nanomembranes have been developed for liquid waste treatment, for example, polyether sulfone/gelatin nanomembranes to treat effluent from the textile industry while eliminating Congo red dye. Results show that cationic dye removal reaches 98%. This nanofiltration method is also used to treat real polluted water from the textile industry, demonstrating its feasibility thanks to removal rates of 67.6% for COD, 60.7% for Total Dissolved Solids, 86.6% for Total Suspended Solids (TSS), 55% for oils and fats, 45.5% for ionic salts, and 85% for colorants, all under a pressure of 2 bar [95]. Another nanomembrane is the DADPM-TMC membrane (membrane based on 4,4′-diaminodiphenylmethane and 1,3,5-benzenetricarbonyl trichloride) manufactured by Chengyu Yang et al. [96] for the removal of dyes from wastewater; according to the study, this nanomembrane displays a significant rejection rate (99%) for the Congo red dye and over 95% for other negatively charged dyes. Filtration testing of a bovine serum albumin solution demonstrated a membrane flux recovery percentage of 89.85%, testifying to excellent stability and notable performance against clogging.

The NF90 and NF27 membranes demonstrated high selectivity for norfloxacin in effluents simulating wastewater from the pharmaceutical industry. The data showed that both membranes were highly attractive to this antibiotic, with removal rates ranging from 87% to 99.5%. These impressive results are particularly notable at a pH of 6.5. These findings highlight the ability of nanofiltration to effectively remove pharmaceutical pollutants from wastewater [97]. Based on these studies, we can conclude that nanofiltration is an effective and promising method for treating and managing liquid waste to remove various pollutants.

12.4.1.3 Nano-Adsorbent

A nano-adsorbent is a material at the nanoscale that has the capacity to take up pollutant materials from liquid effluents. The key properties of these nanomaterials, due to their small size, result in a very large surface area, multiple sorption sites, and high porosity, facilitating the adsorption mechanism [98,99]. Nanoparticles such as MgO [100], ZnO [101], nanofibers [102], nanotubes, and nanocomposites [103] have been utilized as adsorbents. Nanoadsorbents have high-affinity binding sites for contaminants, which effectively remove them from aqueous solutions [104]. To treat different pharmaceutical effluents from several pollutants, such as perfolaxicin [105], amoxicillin [106], metronidazole, and levofloxacin [107], the carbon nanotube is used thanks to its chemical, physical, and electronic properties, as well as its large surface area and the high degree of porosity. For instance, to treat wastewater from an electronics industry in New Cairo, Egypt, which discharges 6,251 $m^3\ day^{-1}$ into the public sewage system, Abdelmegeed et al. [108] have synthesized magnetic nano-adsorbents Fe_3O_4, Fe_3O_4/SiO_2, Fe_3O_4/NH_2, and Fe_3O_4/SiO_2-NH_2 core/shell to remove heavy metal ions (Al^{3+}, Cr^{3+}, Cu^{2+}, Fe^{3+}, and Zn^{2+}). The results show that 0.5 $g\ L^{-1}$ of Fe_3O_4/SiO_2-NH_2 showed the best performance, reaching a 95% removal rate for Al^{3+} in a short 15 min. Maximum capacity reached 229 $mg\ g^{-1}$. These nanomaterials retained an efficiency of over 90% even after five adsorption–desorption cycles, thanks to their magnetic properties facilitating separation and recycling.

To keep up with the trend toward green and blue chemistry, liquid waste treatment is moving toward the use of nano-biosorbents. At the same time, valorizing agricultural waste to extract nanocellulose [109], lignin [110], chitin, and chitosan [111] is the most studied for their ability to adsorb and eliminate waste due to their nanometric scales and also the presence of functional groups [112]. A study conducted by Meriem Kasbaji et al. [113] used coffee grounds to extract cellulose modified with sodium alginate double-network hydrogel beads to remove Methylene Blue and Congo Red dyes from wastewater. The results of this study indicate that the maximum adsorption capacities are 400.50 $mg\ g^{-1}$ for Methylene Blue and 411.45 $mg\ g^{-1}$ for Congo Red dyes, making these materials highly effective for water purification.

12.4.2 Combining Multiple Treatment Methods

Conventional methods of treating liquid waste, including chemical or biological degradation, often prove inadequate, underlining the growing need to create new, more effective methods for the remediation and recycling of such waste. Due to their complexity and heterogeneous composition, a combination of treatment methods may be required to ensure their safe reuse and disposal. A number of research projects are currently underway, focusing on the combination of various liquid waste treatment methods.

12.4.2.1 Physicochemical Combination Processes

The olive oil industry generates massive quantities of liquid waste. The marines are highly polluting; the COD of this waste is 200 times higher than that of wastewater. This poses environmental problems because of their composition, which is loaded with phenolic compounds, heavy metals, fats, and colorants; the variety of these compounds and their complexity make a single treatment insufficient to purify the water and give poor results. This is why researchers have combined treatment methods to maximize purification yield [114]. For example, one study proposes a combination of coagulation/flocculation and filtration; the first stage is divided into two phases using lime and alum to reduce COD by 67%. The resulting water is then treated by filtration using activated carbon and gravel. This phase reduces conductivity by 51%, solids by 37%, and COD by 26%. Final results show that the removal rate after combined treatment (coagulation/flocculation and filtration) for COD, turbidity, color, lead, total solids, and phenolic compounds is 92%, 99%, 95%, 93%, 89%, and 91%–97%, respectively; what's also remarkable is that the treated effluent is reusable for olive grove irrigation ($170\,m^3 ha^{-1}$) [115]. Further research has investigated seven different methods of treating bathroom gray water for non-potable reuse. The best method is the coagulation–flocculation (by ferric chloride) combination followed by sand filtration. Removal efficiency is BOD_5: 87% and COD: 93%. TOC: 79% TURB: 98% [116]; this shows that the coagulation–flocculation and filtration combination is effective, relevant, economical, and suitable for liquid waste treatment.

12.4.2.2 Chemical–Biological Combination

Poor treatment of black rural wastewater causes environmental and economic problems. To overcome this challenge, a solution is proposed: combining chemical and biological treatment. An integrated process combines two stages: the first stage is pretreatment by coagulation/flocculation, followed by inoculation with an indigenous microalgae-bacteria consortium, which reduces toxic compounds such as humic acid-like substances between 33.3% and 52.4% and hexadecenoic acid methyl ester: 96.5%, as well as nitrogen recovery from between 17.6% and 22.6%. In summary, coagulation/flocculation–microalgae-bacteria consortium treatment proved effective in improving the quality of black rural wastewater treatment while increasing nitrogen recovery [117]. Another study about the management of hospital wastewater, laden with recalcitrant compounds such as carbamazepine, requiring COD, TSS, ammonia, and pathogenic microorganisms, is achieved by means of a pilot-scale hybrid process (3000 L/day). Biological treatment includes two phases: strong removal of turbidity, COD (72%), TSS (93%), and ammonia (99%). In phase 2, performance dropped slightly (COD to 57%) due to the unavailability of modified bio-sorption sequential treatment, but intermittent aeration promoted better conversion of ammonia to nitrogen (denitrification). Although the biological units showed good removal of conventional parameters, the average efficiency for carbamazepine was low (~30%). Chemical treatment by photocatalysis, an advanced oxidation technology based on the activation of a semiconductor by light, resulted in high removal of carbamazepine (~85%) [118]. Treatment of liquid waste using a combination of chemical and biological approaches is promising, effective, and economical.

12.4.2.3 Combining Coagulation with an Electrochemical Method

To manage wastewater from the adhesives manufacturing industry while complying with environmental standards and ressducing pollution, two treatment methods were implemented consecutively to optimize the purification rate of these liquid residues. The coagulation–flocculation method was used as a preliminary step to remove the high concentration of suspended particles and reduce the organic load. Following a coagulation–flocculation process, the COD concentration of the wastewater, which was initially 9,884 mg L^{-1}, was reduced to 1,532 mg L^{-1}. Also, the removal efficiency for other pollutants was 95.2% for TSS, 90.6% for UV254 absorbance, and 96.7% for turbidity. Then, using an advanced electrochemical oxidation method (AOP), this value was reduced to 88.9 mg L^{-1} by electro-peroxidation, and removal efficiencies close to 100% were achieved for TSS and UV254 absorbance [119]. Changchao Dai et al. [120] have investigated a new combined electrochemical oxidation–coagulation method, using a boron-doped diamond anode and aluminum cathode, which has been suggested for the simultaneous removal of suspended particulates and antibiotics. The coupled system achieved removal efficiencies of 67.4% for turbidity and 95.48% for tetracycline in simulated wastewater containing tetracycline. Similarly, 71.25% turbidity removal and 99.78% sulfamethoxazole removal were achieved in simulated wastewater containing sulfamethoxazole. Compared to independent electrocoagulation and electrochemical oxidation, the integrated system offers benefits such as space optimization, minimal energy consumption, and a wide selection of applications. It shows promise for ensuring effective, simultaneous purification of suspended solids and antibiotics in wastewater, all at reduced cost.

12.4.3 Zero-Liquid Discharge Approach

Wastewater that cannot be reused, consumed, or discharged is routed to an integrated wastewater management system, in which Zero-Liquid Discharge (ZLD) treatment is applied in a closed loop, enabling recycling and efficient internal use of wastewater. The ZLD system integrates various sophisticated water treatment technologies to maximize the efficient use and recovery of water resources, thereby avoiding any discharge into the environment. This contributes to the conservation of aquatic resources and the preservation of the ecosystem [121,122]. In recent years, ZLD technology has drawn increased interest; it was first developed in the USA and then spread to China, India, and other countries. It is a preferred method since industrial effluent is recycled after being treated to fulfill regulations, preventing any liquid discharge to the environment [123].

ZLD technology has evolved through several phases. Initially, the approach focused on the use of evaporators and evaporation basins for wastewater treatment. Thanks to technological advances, membrane concentration methods were introduced to reduce the treatment load on evaporators. Today, ZLD technology has progressed beyond its initial function as a simple water purification technique to become

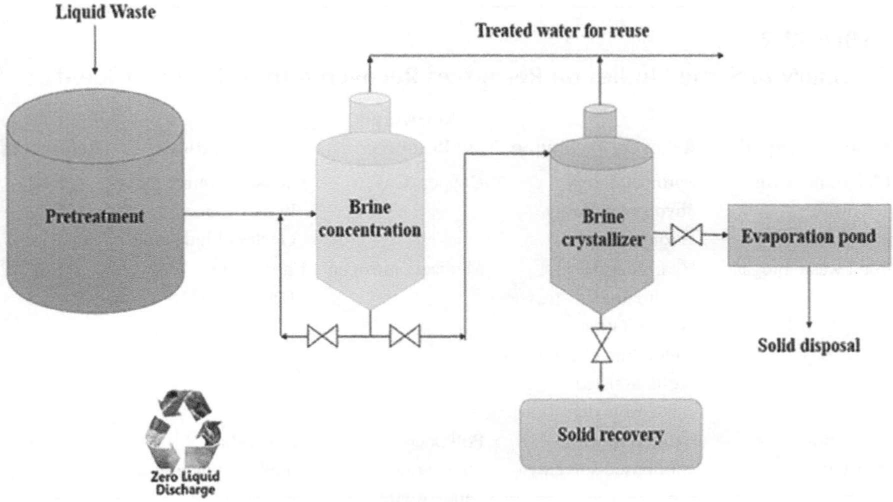

FIGURE 12.5 Path to ZLD: process steps.

a coherent set of wastewater treatment technologies. The contemporary approach to ZLD wastewater treatment is based on four major components: a pretreatment section, a concentration reduction section, an evaporative desalination section, and, finally, a solid waste management section [124,125] (Figure 12.5).

12.4.4 Resource Recovery from Liquid Waste

Liquid waste from the industrial, urban, and agricultural sectors represents an enormous source of materials recovery. This would bring both environmental and economic benefits. To this end, the application of recent technological advances is appropriate for the efficient extraction of metals and organic compounds from waste [126–128] and the recovery of nutrients such as nitrogen and phosphorus [129] (Table 12.2). These recovery methods are part of a global approach to sustainable development. They transform what was once considered waste into a veritable resource of secondary raw materials. For example, calcium carbonate-modified biogas residue biochar (CaBC) is used as an adsorbent to recover nitrogen and phosphorus from Moutai flavor liqueur production wastewater. The maximum removal efficiency of NH_4^+-N and TP was 67.60% ± 3.46% and 99.65% ± 3.60%, respectively. The addition of the adsorbent (CaBC) containing adsorbed elements TP and NH_4^+-N effectively promoted plant growth and soil fertility while optimizing existing soil microbial communities. This research suggested an innovative method for the sophisticated treatment and utilization of liquor wastewater and sludge residues. Following further research and development, this method could find wide application in plants treating high-concentration organic wastewater from the brewing, production of starch, and food processing industries [130]. Another research study exploits waste acid containing several elements such as Co^{2+}, H^+, and NO_3^- and alkaline solutions including Li^+, Na^+, OH^-, and CO_3^{2-} from lithium-ion battery production to extract metal cations from steel slag (which contains Ca^{2+}, Mg^{2+},

TABLE 12.2
Summary of Some Studies on Resources Recovered from Liquid Effluents

Source of Liquid	Recovery Technique	Recovered Resource	Applications	References
Electronic waste (E-waste)	Pyrometallurgy, hydrometallurgy, bioleaching	Copper (Cu)	Copper refining, reuse in electronics, electrical industries	[134]
Food waste biogas slurry	Hydrothermal pretreatment, struvite precipitation, microbial electrolysis cells-assisted anaerobic digestion	Methane, nitrogen	Energy, soil fertilization	[135]
Olive mill wastewater	Extraction with bio-based solvents	Polyphenols, flavonoids, antioxidants	Antioxidant applications in food, cosmetic, pharmaceutical industries	[136]
Chinese Moutai flavor liquor production wastewater	Adsorption by calcium carbonate-modified biogas residue biochar (CaBC)	Total phosphorus (TP) et azote ammoniacal (NH4+-N)	Soil fertilization	[130]
Industrial wastewater	Adsorption and electrodeposition coupling process	Trace gold	-	[137]
Waste liquid from the distillation of high-concentration nonylphenol polyethoxylates	Hydrothermal carbonization	Porous carbon	Filtration, energy storage, catalysis, adsorption, environmental applications	[138]

Al^{3+}, Fe^{3+}, and Mn^{2+}) to prepare high-entropy CoCaMgMnAlFe-LDHs catalysts for controlled hydrogen generation via $NaBH_4$ hydrolysis. The final product, free of any residual hazardous metal cations, can be safely reintegrated into the natural environment. The inclusion of Co^{2+} in the CoCaMgMnAlFe-LDHs catalyst gives it rapid photo-thermal conversion capability, enabling it to heat up rapidly on exposure to infrared radiation and efficiently catalyze $NaBH_4$ hydrolysis. The rate of hydrogen production rises to 1.12 mol h^{-1} g^{-1} W^{-1}. This research suggests an innovative method for the ecological treatment and high-level valorization of steel slag and liquid residues from lithium-ion batteries [131].

In addition to nutrients and precious metals, researchers have also emphasized the importance of energy resource recovery through the "waste-to-energy" approach [132]. A study conducted in 2025 examines the recovery of liquid waste from the anaerobic digestion of food products by recovering energy in the form of methane through hydrothermal treatment, converting 23% of the COD into biogas with a yield of 227 mL g^{-1} COD. What is also noteworthy is that the methane produced offsets the treatment energy, allowing for a 46% energy conservation. This will help reduce energy consumption [133]. Overall, the application of these strategies allows industries to move toward a more sustainable future by reducing their ecological footprint and implementing responsible management practices for liquid waste treatment.

12.4.5 Application of Sensors and Artificial Intelligence to Optimize Processing of Liquid Waste

Industry 4.0 technologies encompass IoT, cloud computing, big data analytics, adsorption index (AI), blockchain, robots, drones, VR/AR, and digital twins [139]. Industry 4.0 aims to automate and optimize industrial processes in real time by connecting equipment, systems, and products via intelligent networks (Figure 12.6). This will increase production flexibility, speed, and customer-orientedness. This not only improves efficiency and performance but also allows for faster, more informed decisions through live data analysis [140]. In recent years, much research has gone into developing less expensive and easier-to-use multi-parameter sensor networks for monitoring water ecology and the quality of liquid waste treatment, which requires in-depth processing due to its complex and varied compositions, making it particularly difficult to treat [141]. To overcome these challenges, live monitoring powered by real-time data becomes essential to rapidly detect anomalies and variations in this process, for example, pH and dissolved oxygen sensors [142], sensors for specific chemicals (e.g., ammonia, nitrates) [143], fluorescence sensors [144], heavy metals sensors [145], etc.

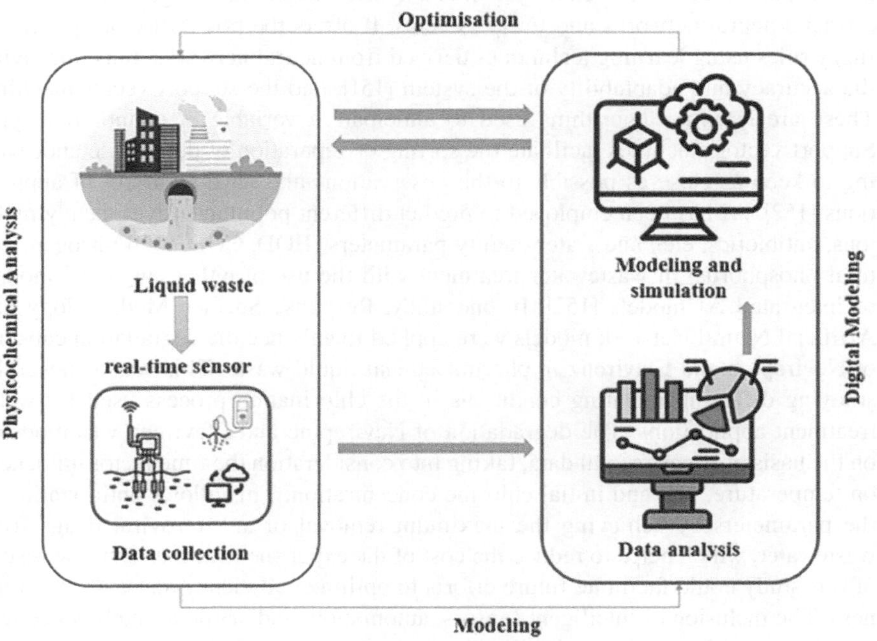

FIGURE 12.6 Optimization of liquid waste treatment through AI and real-time monitoring.

To overcome these challenges, activated sludge models are used in wastewater treatment and can also be tuned to model other forms of liquid waste. However, these traditional models are constrained by their theories and simplifications, such as specific temperature and pH ranges, which limit their performance under changing conditions. Moreover, these conventional models don't always adequately explain the complex processes involved in liquid waste [146]. This is where techniques based on artificial intelligence and machine learning take over. These approaches, which may detect non-linear relationships between variables, are especially useful for predicting and forecasting liquid waste behavior. AI algorithms enable systems to process massive amounts of data from real-time sensors, allowing for more dynamic management and accurate decision-making [147]. These models can overcome the constraints of conventional deterministic models and offer greater flexibility in managing different types of liquid waste. The use of artificial intelligence and machine learning facilitates decision-making, predicts, and reduces operational expenses in the wastewater treatment process by lowering costs and optimizing chemical consumption, improving operational efficiency, and eliminating human error [148,149]. Several artificial intelligence models have successfully anticipated the effects of pollutant removal in various liquid waste treatment procedures. Artificial intelligence technologies, such as artificial neural networks, are mathematical models that accurately model and predict the behavior of complex systems whose dynamics are both simple and complex [150]. They are useful for solving non-linear and complex problems, facilitating learning and recognition of different models, and enabling parameter adjustment. Their function is influenced by their global contributions. The adaptive network-based fuzzy inference system is a hybrid model that merges the benefits of artificial neural networks and fuzzy systems. It offers the possibility of optimizing fuzzy rules using learning techniques derived from neural networks, thus improving the accuracy and adaptability of the system [151], and the support vector machine. These are learning algorithms used to anticipate a variable of quantitative type. Support vector machines facilitate the sorting or separation of data while endeavoring to keep as close as possible to the observations and have a variety of applications [152]. AI has been employed to predict different pollutants (dyes, heavy metal ions, antibiotics, etc.) and water quality parameters (BOD, COD, total nitrogen, and total phosphorus) in wastewater treatment with the use of either single AI models or integrated AI models [152]. In one study, Response Surface Methodology and Artificial Neural Network models were applied to enhance the degradation capacity of Nevirapine and Efavirenz in pharmaceutical liquid waste. This was achieved by studying different operating conditions in the chlorination process used for water treatment applications. The degradation of Nevirapine and Efavirenz was modeled on the basis of experimental data, taking into consideration the time factor influences on temperature, pH, and initial chlorine concentration. This allows optimization of the parameters by achieving the maximum removal of antiretroviral drugs from wastewater, which helps to reduce the cost of the experiments. Moreover, the results of this study could facilitate future efforts to optimize efficiency and cost-effectiveness. The inclusion of intelligent sensors, automation, and artificial intelligence represents a significant advance in liquid waste management [153]. These innovations not only promote more efficient and sustainable management of water resources but also reduce environmental impact and operational costs. In the face of ever-increasing demand for drinking water, these technological advances are essential responses for a greener, more resilient future.

12.5 LCA AND ENVIRONMENTAL IMPACT ASSESSMENT

An efficient liquid waste management system is both a crucial challenge and a requirement for many industrial sectors. These systems must minimize water and energy consumption, prevent the generation of secondary waste, be cost-effective, and maximize pollutant removal under ideal conditions. It is essential to control and manage these processes to reduce their negative impact on the environment while integrating high-performance, sustainable, and environmentally friendly technologies into all industrial sectors [154].

Recently, there has been a growing interest in applying LCA to liquid waste management and treatment. This methodology serves as a valuable tool for evaluating treatment technologies by collecting and compiling data related to inputs and outputs, as it enables various environmental and economic elements to be studied and evaluated, such as water consumption, energy expended, greenhouse gas emissions, acidification, terrestrial ecotoxicity, etc. [155]. Also, it offers a structured approach to evaluation, including determining the objectives and scope of the study, conducting a life cycle inventory, analyzing life cycle impacts, and understanding the results. In addition, economic assessments can be carried out using cost-benefit analysis, life-cycle costing, and techno-economic analysis [156,157]. These economic assessments can frequently be integrated into LCA to provide a comprehensive analysis at the system level, helping to improve the efficiency of liquid waste treatment. Certain LCA guidelines have been defined by the International Organization for Standardization; however, these standards offer a wide latitude and interpretation to LCA experts and professionals. As a result, comparing results from different studies is sometimes complex and requires considerable care for a number of reasons, including methodological divergences (such as functional unit, allocation technique, and impact assessment procedure) and the inventory data involved.

Several studies have been conducted to analyze the life cycle of treatment technologies for various liquid wastes and often polluted waters from different sources. For instance, a study conducted by Omid Sedaghat et al. [158] examined the environmental impact of treating wastewater from the textile industry through electrocoagulation using the LCA methodology. To properly address this work, they chose the ReCiPe midpoint methods, which focus solely on environmental effects such as climate change, acidification, eutrophication, and fossil resource scarcity, as well as the endpoint (H) method, which is subdivided into three categories: the effects on human health, biodiversity, and resource scarcity [159]. After selecting the methodology, the functional unit set for this study was 1 kg of COD, allowing for the analysis of different environmental impacts associated with the studied scenarios. The results of this research showed that the significant impacts of the electrocoagulation process were associated with climate change (40%), terrestrial ecotoxicity (33%), fossil resource scarcity (14%), and non-carcinogenic human toxicity (6%). The data indicate that COD removal has the greatest influence on global warming. The endpoint analysis (H) demonstrated that the most significant effect of the electrocoagulation process was related to human health (93.2%), followed by ecosystems (3.6%) and natural resources (3.14%). The effects on human health are mainly attributed to air pollutant emissions and the increase in greenhouse gases during electricity production. This highlights the importance of selecting an appropriate energy source to minimize environmental and health impacts. In addition, the analysis also highlights significant economic results, with three key cost parameters included in the process. Construction costs (31.9%) represent the highest expense due to the necessary materials and infrastructure, followed by equipment costs (7.3%), including reactors and electrodes, and, finally, energy costs (4.7%). These results provide a general overview of the efficiency of the processes and their adaptability to environmental standards. In addition, waste liquids also include oils and fats, such as used cooking oil, which can cause considerable environmental damage when dumped or discharged into sewer systems [160]. With this in mind, their conversion aligns with the circular economy strategy and zero-waste principles. Extensive research has been conducted on converting these oils into biodiesel. To assess the viability of this innovation, specialists use the LCA method to compare biodiesel production with traditional diesel [161]. For example, research carried out in China to analyze economic competitiveness and environmental impacts using the ReCiPe 2016 method, which integrates midpoint and endpoint methodologies in a coherent way. Based on the cradle-to-gate methodology, this method examines the environmental impact of a product from the moment it is first extracted until it leaves the factory, encompassing all stages of production and finishing [162]. The data shows that, generally, a biodiesel derived from household consumption unit (HCU) has a superior environmental performance compared with conventional diesel. However, due to the significant use of coal and the use of methanol in the transesterification process, biodiesel production from HCU tends to have a greater environmental impact in certain categories, such as climate change, generation of solid particulates, and human toxicity. Furthermore, from an economic point of view, this study reveals that the cost of biodiesel from HCU is 31% higher than that of diesel from fossil fuels. This is due to the collection of HCU, which accounts for the bulk of the cost (83%), mainly because of the high price of HCU, influenced by competition with the illegal use of oils. This is the first research to measure the ecological and economic effects of biodiesel produced from HCU in China. Future research directions include the development of a policy framework for the supply of HCU, the development of efficient energy conversion technologies, and the creation of a beneficial business environment [163].

In conclusion, LCA is an important tool for comparing and evaluating the environmental and economic effects of liquid waste treatment systems. Although the methodology is best suited to systematic and integrated procedures, its implementation remains a challenge due to methodological differences and data inventory decisions. The research reviewed, presented in Table 12.3, highlights the use of renewable energy sources and the development of technologies to limit negative effects on the environment and human health. In addition, the use of economics, such as cost estimation and cost-effectiveness analysis, illustrates the importance of decisions regarding the treatment of liquid waste. Consequently, improving the treatment process, combined with optimization measures, promotes the circular economy.

TABLE 12.3
LCA Analysis of Liquid Waste Treatments: Methodologies and Key Impacts

Research Area	Functional Unit	Description of the System Investigated	Purpose of Liquid Waste Treatment	Impact Assessment Method	Impact Categories	References
Spain	1 m^2 of urban garden	Cradle-to-gate	Recover nutrients for soil fertilization	ReCiPe 2016: midpoint and endpoint (H)	Ecosystem quality, human health, and resource scarcity	[164]
India	1 m^3 of treated wastewater	Cradle-to-gate	Recovering irrigation water	Impact 2002+, midpoint, and endpoint	Global warming potential, energy consumption, chemical dosing, eutrophication, and freshwater ecotoxicity.	[165]
South Brazil	1 m^3 of domestic effluent	Gate-to-gate	Extraction of sludge for the production of bio-based compounds	SimaPro 8.4, ReCiPe 2016 midpoint, dashboard of sustainability	Economic, social, and environmental impacts, including freshwater eutrophication, terrestrial acidification, global warming potential, ozone formation, stratospheric ozone depletion, terrestrial ecosystems, terrestrial ecotoxicity, freshwater ecotoxicity, and human non-carcinogenic toxicity.	[166]
Hérault, France	1 ha of vineyard	Gate-to-gate	Irrigation water recovery	SimaPro	The eight midpoint impact categories include climate change, ozone depletion, human toxicity (which includes both human carcinogenic toxicity and human non-carcinogenic toxicity), ionizing radiation, acidification, freshwater eutrophication, marine eutrophication, and freshwater ecotoxicity.	[167]

(Continued)

TABLE 12.3 (*Continued*)
LCA Analysis of Liquid Waste Treatments: Methodologies and Key Impacts

Research Area	Functional Unit	Description of the System Investigated	Purpose of Liquid Waste Treatment	Impact Assessment Method	Impact Categories	References
Mexico City	1 kWh of electricity produced	Cradle-to-gate	using waste cooking oil for the production of biodiesel	CML 2001	Climate change, acidification, formation of photo-oxidants, depletion of abiotic resources, and potential for water scarcity.	[168]
Mexico	1 kg of crops (cereal crops)	Cradle-to gate	Recovery of nutrients from biosolids for soil amendment	TRACI 2.1 (U.S. E.P.A.)	Eutrophication potential, global warming potential, ecotoxicity potential, and acidification potential.	[169]
Iran	1 kg COD removal of textile industry wastewater	Gate-to-gate	Electrocoagulation for water reuse	ReCiPe midpoint and endpoint (H)	Global warming, terrestrial ecotoxicity, fossil resource scarcity.	[158]

12.6 STRENGTHS, WEAKNESSES, OPPORTUNITIES, AND THREATS (SWOT) ANALYSIS

The SWOT analysis is used to identify strengths, weaknesses, opportunities, and risks in order to guide strategic decisions and improve performance and flexibility in the treatment and management of liquid waste [170], as illustrated in the graph (Figure 12.7). The current research takes into account several elements influencing the progress of this management, starting with identified strengths such as the development of eco-responsible approaches or technologies that preserve the environment and reduce the ecological footprint. In addition, the incorporation of the circular economy through the reuse of resources such as nutrients, precious metals, and energies (such as biodiesel and electricity…), as well as automation through the use of real-time sensors and data analysis via artificial intelligence [171]. However, these strengths face challenges and obstacles, not least the variety and complexity of new-generation pollutants. Despite the high cost of energy for infrastructure treatment and the use of chemicals, there are many opportunities and prospects for development, such as in the field of nanotechnology. Despite promising prospects, the integration of renewable energies and the support of politicians and the private sector are essential. Threats remain, including regulatory constraints and public reluctance to reuse wastewater. This requires a balance between technology, economic viability, and social acceptance.

FIGURE 12.7 SWOT analysis of liquid waste management and treatment.

12.7 CONCLUSION

Population growth implies an increase in human consumption of natural resources, which generates enormous liquid waste (municipal, industrial, and agricultural), posing major problems and challenges in waste management, as well as environmental imbalances such as climate change, acidification, warming, human health, and marine life. What is alarming is the necessity to demand regulations and raise awareness among industries to implement environmental policies in order to preserve the environment and integrate the green economy approach. This chapter highlights and describes various aspects of liquid waste management, including conventional methods and their limitations, sustainable approaches, innovative technologies, and the economic and environmental life cycle study of this management. Conventional methods have their limitations on several levels: high costs in terms of energy consumption, chemicals, maintenance, etc., as well as the generation of by-products and the complexity of processing. These methods are becoming insufficient, thus necessitating more effective alternatives that bring together different ecologically sustainable treatments, which can be subdivided into several biological, physical, and chemical categories (phytoremediation, adsorption on activated carbon, and advanced oxidation) that make it possible to reduce economic and environmental impacts by cutting costs and reducing by-product generation.

In addition to sustainable treatments, scientific research and technical expertise aim to develop new innovative approaches to improve liquid waste management. This includes the incorporation of various environmental alternatives. The rise of nanotechnology paves the way for advanced solutions capable of improving the performance of purification systems while reducing energy consumption, the integrity of various treatments, and their combination, which increases the efficiency of treatments. The integration of the ZLD approach enables the recovery of valuable resources (nutrients, energy...) and their applications in different fields. The improvement of procedures involves automation and the implementation of cutting-edge technologies, such as the use of real-time sensors and their simulation and modeling by artificial intelligence, in order to optimize the management of liquid waste, while facilitating tasks, reducing time, and minimizing human errors.

The use of assessment tools such as LCA, environmental impact assessment, and economic impact assessment helps to improve these solutions by measuring their benefits and reducing their harmful effects. These assessments are essential to the transition to more sustainable liquid waste management, as they identify best practices and encourage innovation in the sector. The SWOT analysis shows the strengths, weaknesses, opportunities, and threats of this innovation, which allows us to conclude that the sustainability of liquid waste treatment goes beyond mere technological improvements, requiring a comprehensive re-evaluation of current approaches, incorporating eco-concept, energy optimization, and circular economy, and through research, innovation, and collaboration.

REFERENCES

[1] Hajian M, Jangchi Kashani S. 1- Evolution of the concept of sustainability. From Brundtland report to sustainable development goals. In: Hussain CM, Velasco-Muñoz JF, editors. *Sustain. Resour. Manag.*, Elsevier; 2021, pp. 1–24. https://doi.org/10.1016/B978-0-12-824342-8.00018-3.

[2] Odeibat AS, Mohammad R, Abu-Zreig M. Integrated environmental management and GPS-X modelling for current and future sustainable wastewater treatment: A case study from the Middle East. *Heliyon* 2024;10:e34164. https://doi.org/10.1016/j.heliyon.2024.e34164.

[3] Dulta K, Iwuchukwu FU, Kumari S, Narayanasamy S, Selvasembian R. Chapter 6- Traditional techniques of water purification in rural areas. In: Madhav S, Srivastav AL, Chibueze Izah S, Hullebusch E van, editors. *Water Resour. Manag. Rural Dev.*, Elsevier; 2024, pp. 65–76. https://doi.org/10.1016/B978-0-443-18778-0.00007-6.

[4] Unesco World Water Development Report 2020 – Water and Climate Change n.d. https://unesdoc.unesco.org/ark:/48223/pf0000372985 (accessed January 4, 2025).

[5] Connor R, Ortigara A, Koncagül E, Uhlenbrook S, Cherfane C, Karnib A, Qadir M. *The United Nations World Water Development Report 2017.* Wastewater: The Untapped Resource. 2017.

[6] Amandi Thilakarathna PT, Fareed F, Athukorala SNP, Jinadasa R, Premachandra T, Noordeen F, Gamage CD, Makehelwala M, Weragoda SK, Fernando BR, Zhang Y, Wei Y, Yang M, Karunaratne SHPP. Spatio-temporal variation of microbial indicators of river water and treatment efficiencies of drinking water treatment plants along the upper Mahaweli River segment of Sri Lanka. *Environ Pollut* 2025:125628. https://doi.org/10.1016/j.envpol.2025.125628.

[7] Gonçalves AL, Pires JCM, Simões M. A review on the use of microalgal consortia for wastewater treatment. *Algal Res* 2017;24:403–415. https://doi.org/10.1016/j.algal.2016.11.008.

[8] Zohbi GA. *Review on Sustainable Technologies for Liquid Waste Treatment and Management.* Waste Recovery Manag., CRC Press; 2023.

[9] Balbinoti JR, dos Santos Junior RE, de Sousa LBF, de Jesus Bassetti F, Balbinoti TCV, de Matos Jorge LM, Jorge RMM. Plant-based coagulants for food industry wastewater treatment. *J Water Process Eng* 2023;52:103525. https://doi.org/10.1016/j.jwpe.2023.103525.

[10] Singh A, Shah SS, Sharma C, Gupta V, Sundramoorthy AK, Kumar P, Arya S. An approach towards different techniques for detection of heavy metal ions and their removal from waste water. *J Environ Chem Eng* 2024;12:113032. https://doi.org/10.1016/j.jece.2024.113032.

[11] Muhammad-aree S, Teepoo S. On-site detection of heavy metals in wastewater using a single paper strip integrated with a smartphone. *Anal Bioanal Chem* 2020;412:1395–1405. https://doi.org/10.1007/s00216-019-02369-x.

[12] Kasbaji M, Mennani M, Grimi N, Oubenali M, Mbarki M, EL Zakhem H, Moubarik A. Adsorption of cationic and anionic dyes onto coffee grounds cellulose/sodium alginate double-network hydrogel beads: Isotherm analysis and recyclability performance. *Int J Biol Macromol* 2023;239:124288. https://doi.org/10.1016/j.ijbiomac.2023.124288.

[13] Zhou Y, Meng J, Zhang M, Chen S, He B, Zhao H, Li Q, Zhang S, Wang T. Which type of pollutants need to be controlled with priority in wastewater treatment plants: Traditional or emerging pollutants? *Environ Int* 2019;131:104982. https://doi.org/10.1016/j.envint.2019.104982.

[14] Arita S, Kristianti D, Komariah LN. Effectiveness of biomass-based fly ash in pulp and paper liquid waste treatment. *South Afr J Chem Eng* 2022;41:79–84. https://doi.org/10.1016/j.sajce.2022.05.004.

[15] Sharma R, Verma N, Lugani Y, Kumar S, Asadnia M. Chapter 1- Conventional and advanced techniques of wastewater monitoring and treatment. In: Inamuddin, Boddula R, Asiri AM, editors. *Green Sustain. Process Chem. Environ. Eng. Sci.*, Elsevier; 2021, pp. 1–48. https://doi.org/10.1016/B978-0-12-821883-9.00009-6.
[16] Muñoz-Martinez MX, Macías-Quiroga IF, Sanabria-González NR. Adsorption of Cr(VI) using organoclay/alginate hydrogel beads and their application to Tannery effluent. *Gels* 2024;10:779. https://doi.org/10.3390/gels10120779.
[17] Gao J, Han H, Zhu J, Zhou J. Navigating challenges in electroplating wastewater management: A study on pollutant removal characteristics and economic impacts by physicochemical treatment. *Colloids Surf C Environ Asp* 2025;3:100060. https://doi.org/10.1016/j.colsuc.2025.100060.
[18] Ingle J, Patel UD. Electrochemical treatment of metal-finishing wastewater using Ti/Co_3O_4 electrode: Efficient removal of nitrate and heavy metals, and recovery of struvite. *J Environ Manage* 2025;377:124659. https://doi.org/10.1016/j.jenvman.2025.124659.
[19] Ali I, Asim Mohd, Khan TA. Low cost adsorbents for the removal of organic pollutants from wastewater. *J Environ Manage* 2012;113:170–183. https://doi.org/10.1016/j.jenvman.2012.08.028.
[20] Wang Y, Wang J, Long Z, Sun Z, Lv L, Liang J, Zhang G, Wang P, Gao W. MnCe-based catalysts for removal of organic pollutants in urban wastewater by advanced oxidation processes - A critical review. *J Environ Manage* 2024;370:122773. https://doi.org/10.1016/j.jenvman.2024.122773.
[21] Sun S, Jiang T, Lin Y, Song J, Zheng Y, An D. Characteristics of organic pollutants in source water and purification evaluations in drinking water treatment plants. *Sci Total Environ* 2020;733:139277. https://doi.org/10.1016/j.scitotenv.2020.139277.
[22] Vo TP, Danaee S, Chaiwong C, Pham BT, Poddar N, Kim M, Kuzhiumparambil U, Songsomboon C, Pernice M, Ngo HH, Ralph PJ, Vo PHN. Microalgae-bacteria consortia for organic pollutants remediation from wastewater: A critical review. *J Environ Chem Eng* 2024;12:114213. https://doi.org/10.1016/j.jece.2024.114213.
[23] Öztürk E, Ateş AE, Selçuk H, Ateş S. Treatment of real pharmaceutical industry wastewater by photo-fenton oxidation using the response surface methodology, evaluation of diclofenac degradation and toxicity. *Eng Sci Technol Int J* 2025;64:102004. https://doi.org/10.1016/j.jestch.2025.102004.
[24] Kanafin YN, Abduvalov A, Kaikanov M, Poulopoulos SG, Atabaev TSh. A review on WO_3 photocatalysis used for wastewater treatment and pesticide degradation. *Heliyon* 2025;11:e40788. https://doi.org/10.1016/j.heliyon.2024.e40788.
[25] Marizzi del Olmo A, López-Doval JC, Hidalgo M, Serra T, Colomer J, Salvadó V, Escolà Casas M, Medina JS, Matamoros V. Holistic assessment of chemical and biological pollutants in a Mediterranean wastewater effluent-dominated stream: Interactions and ecological impacts. *Environ Pollut* 2025;370:125833. https://doi.org/10.1016/j.envpol.2025.125833.
[26] Bôto ML, Dias SM, Crespo RD, Mucha AP, Almeida CMR. Removing chemical and biological pollutants from swine wastewater through constructed wetlands aiming reclaimed water reuse. *J Environ Manage* 2023;326:116642. https://doi.org/10.1016/j.jenvman.2022.116642.
[27] Adeola AO, Paramo L, Fuoco G, Naccache R. Emerging hazardous chemicals and biological pollutants in Canadian aquatic systems and remediation approaches: A comprehensive status report. *Sci Total Environ* 2024;954:176267. https://doi.org/10.1016/j.scitotenv.2024.176267.
[28] Ma H, Shen M, Tong Y, Wang X. Radioactive wastewater treatment technologies: A review. *Molecules* 2023;28:1935. https://doi.org/10.3390/molecules28041935.

[29] Qian Z, Tao Y, Peng Y, Li Y, Ke L, Shi M. Capacitive deionization: An emerging electrosorption technology for efficient radioactive wastewater treatment. *J Environ Chem Eng* 2025;13:115104. https://doi.org/10.1016/j.jece.2024.115104.
[30] Zhang Q, Wang Y, Wen Y, Wang R, Zhang Y, Zeng Q. Co/N-codoped carbon nanoplate array coated carbon fiber cathode for solar driven uranium extraction from complex radioactive wastewater. *Desalination* 2024;591:118014. https://doi.org/10.1016/j.desal.2024.118014.
[31] Sangamnere R, Misra T, Bherwani H, Kapley A, Kumar R. A critical review of conventional and emerging wastewater treatment technologies. *Sustain Water Resour Manag* 2023;9:58. https://doi.org/10.1007/s40899-023-00829-y.
[32] Nadal-Bach J, Bruno JC, Farnós J, Rovira M. Solar stills and evaporators for the treatment of agro-industrial liquid wastes: A review. *Renew Sustain Energy Rev* 2021;142:110825. https://doi.org/10.1016/j.rser.2021.110825.
[33] Crini G, Lichtfouse E. Advantages and disadvantages of techniques used for wastewater treatment. *Environ Chem Lett* 2019;17:145–155. https://doi.org/10.1007/s10311-018-0785-9.
[34] Gupta S, Mittal Y, Panja R, Prajapati KB, Yadav AK. Chapter 3- Conventional wastewater treatment technologies. In: Kumar S, Kumar R, Pandey A, editors. *Curr. Dev. Biotechnol.* Bioeng., Elsevier; 2021, pp. 47–75. https://doi.org/10.1016/B978-0-12-821009-3.00012-9.
[35] Zhang D-Q, Jinadasa KBSN, Gersberg RM, Liu Y, Tan SK, Ng WJ. Application of constructed wetlands for wastewater treatment in tropical and subtropical regions (2000–2013). *J Environ Sci* 2015;30:30–46. https://doi.org/10.1016/j.jes.2014.10.013.
[36] Vymazal J, Zhao Y, Mander Ü. Recent research challenges in constructed wetlands for wastewater treatment: A review. *Ecol Eng* 2021;169:106318. https://doi.org/10.1016/j.ecoleng.2021.106318.
[37] Gupta B, Gupta AK, Ghosal PS, Lal S, Saidulu D, Srivastava A, Upadhyay M. Recent advances in application of moving bed biofilm reactor for wastewater treatment: Insights into critical operational parameters, modifications, field-scale performance, and sustainable aspects. *J Environ Chem Eng* 2022;10:107742. https://doi.org/10.1016/j.jece.2022.107742.
[38] Saidulu D, Majumder A, Gupta AK. A systematic review of moving bed biofilm reactor, membrane bioreactor, and moving bed membrane bioreactor for wastewater treatment: Comparison of research trends, removal mechanisms, and performance. *J Environ Chem Eng* 2021;9:106112. https://doi.org/10.1016/j.jece.2021.106112.
[39] Eshamuddin M, Zuccaro G, Nourrit G, Albasi C. The influence of process operating conditions on the microbial community structure in the moving bed biofilm reactor at phylum and class level: A review. *J Environ Chem Eng* 2024;12:113266. https://doi.org/10.1016/j.jece.2024.113266.
[40] Zhang X, Liu Y. Resource recovery from municipal wastewater: A critical paradigm shift in the post era of activated sludge. *Bioresour Technol* 2022;363:127932. https://doi.org/10.1016/j.biortech.2022.127932.
[41] Hreiz R, Latifi MA, Roche N. Optimal design and operation of activated sludge processes: State-of-the-art. *Chem Eng J* 2015;281:900–920. https://doi.org/10.1016/j.cej.2015.06.125.
[42] Mainardis M, Buttazzoni M, Goi D. Up-Flow Anaerobic Sludge Blanket (UASB) technology for energy recovery: A review on state-of-the-art and recent technological advances. *Bioengineering* 2020;7:43. https://doi.org/10.3390/bioengineering7020043.
[43] Boiocchi R, Zhang Q, Gao M, Liu Y. Modeling and optimization of an upflow anaerobic sludge blanket (UASB) system treating blackwaters. *J Environ Chem Eng* 2022;10:107614. https://doi.org/10.1016/j.jece.2022.107614.

[44] Li M, Zhang H, Lemckert C, Roiko A, Stratton H. On the hydrodynamics and treatment efficiency of waste stabilisation ponds: From a literature review to a strategic evaluation framework. *J Clean Prod* 2018;183:495–514. https://doi.org/10.1016/j.jclepro.2018.01.199.
[45] Ho LT, Van Echelpoel W, Goethals PLM. Design of waste stabilization pond systems: A review. *Water Res* 2017;123:236–248. https://doi.org/10.1016/j.watres.2017.06.071.
[46] Gruchlik Y, Linge K, Joll C. Removal of organic micropollutants in waste stabilisation ponds: A review. *J Environ Manage* 2018;206:202–214. https://doi.org/10.1016/j.jenvman.2017.10.020.
[47] Zhou H, Smith DW. Advanced technologies in water and wastewater treatment. *Can J Civ Eng* 2001;28:49–66. https://doi.org/10.1139/l00-091.
[48] Marques DG, de Melo Franco Domingos J, Nolasco MA, Campos V. Textile effluent treatment using coagulation-flocculation and a hydrodynamic cavitation reactor associated with ozonation. *Chem Eng Sci* 2025;304:121094. https://doi.org/10.1016/j.ces.2024.121094.
[49] Sk MM, Yue CY, Ghosh K, Jena RK. Review on advances in porous nanostructured nickel oxides and their composite electrodes for high-performance supercapacitors. *J Power Sources* 2016;308:121–140. https://doi.org/10.1016/j.jpowsour.2016.01.056.
[50] Li J, Campos LC, Zhang L, Xie W. Sand and sand-GAC filtration technologies in removing PPCPs: A review. *Sci Total Environ* 2022;848:157680. https://doi.org/10.1016/j.scitotenv.2022.157680.
[51] Corbera-Rubio F, Goedhart R, Laureni M, van Loosdrecht MC, van Halem D. A biotechnological perspective on sand filtration for drinking water production. *Curr Opin Biotechnol* 2024;90:103221. https://doi.org/10.1016/j.copbio.2024.103221.
[52] Azam K, Shezad N, Shafiq I, Akhter P, Akhtar F, Jamil F, Shafique S, Park Y-K, Hussain M. A review on activated carbon modifications for the treatment of wastewater containing anionic dyes. *Chemosphere* 2022;306:135566. https://doi.org/10.1016/j.chemosphere.2022.135566.
[53] Sathya K, Nagarajan K, Carlin Geor Malar G, Rajalakshmi S, Raja Lakshmi P. A comprehensive review on comparison among effluent treatment methods and modern methods of treatment of industrial wastewater effluent from different sources. *Appl Water Sci* 2022;12:70. https://doi.org/10.1007/s13201-022-01594-7.
[54] Ding GKC. Wastewater treatment and reuse—The future source of water supply. In: Abraham MA, editor. *Encycl. Sustain.* Technol., Oxford: Elsevier; 2017, pp. 43–52. https://doi.org/10.1016/B978-0-12-409548-9.10170-8.
[55] Chen K-H, Wang H-C, Han J-L, Liu W-Z, Cheng H-Y, Liang B, Wang A-J. The application of footprints for assessing the sustainability of wastewater treatment plants: A review. *J Clean Prod* 2020;277:124053. https://doi.org/10.1016/j.jclepro.2020.124053.
[56] Priya AK, Muruganandam M, Suresh S. Bio-derived carbon-based materials for sustainable environmental remediation and wastewater treatment. *Chemosphere* 2024;362:142731. https://doi.org/10.1016/j.chemosphere.2024.142731.
[57] López J, Gibert O, Cortina JL. Integration of membrane technologies to enhance the sustainability in the treatment of metal-containing acidic liquid wastes. An overview. *Sep Purif Technol* 2021;265:118485. https://doi.org/10.1016/j.seppur.2021.118485.
[58] Sambursky G, Shkaredo V, Epstein A, Maximova A, Epifanova M. Unification of requirements for the use of reagents for the treatment of sewage sludge from urban sewage treatment plants. *E3S Web Conf* 2024;486:04008. https://doi.org/10.1051/e3sconf/202448604008.
[59] Pandian AMK, Rajamehala M, Singh MVP, Sarojini G, Rajamohan N. Potential risks and approaches to reduce the toxicity of disinfection by-product – A review. *Sci Total Environ* 2022;822:153323. https://doi.org/10.1016/j.scitotenv.2022.153323.

[60] Maddalon A, Pierzchalski A, Krause JL, Bauer M, Finckh S, Brack W, Zenclussen AC, Marinovich M, Corsini E, Krauss M, Herberth G. Impact of chemical mixtures from wastewater treatment plant effluents on human immune cell activation: An effect-based analysis. *Sci Total Environ* 2024;906:167495. https://doi.org/10.1016/j.scitotenv.2023.167495.

[61] Longo S, Hospido A, Mauricio-Iglesias M. Energy efficiency in wastewater treatment plants: A framework for benchmarking method selection and application. *J Environ Manage* 2023;344:118624. https://doi.org/10.1016/j.jenvman.2023.118624.

[62] Ateunkeng JG, Boum AT, Bitjoka L. Enhancement of energy and cost efficiency in wastewater treatment plants using hybrid bio-inspired machine learning control techniques. *J Environ Chem Eng* 2024;12:112496. https://doi.org/10.1016/j.jece.2024.112496.

[63] Copeland C, & Carter NT. *Energy-water nexus: The water sector's energy use* (CRS Report No. R43200). Congressional Research Service.

[64] Cardoso BJ, Rodrigues E, Gaspar AR, Gomes Á. Energy performance factors in wastewater treatment plants: A review. *J Clean Prod* 2021;322:129107. https://doi.org/10.1016/j.jclepro.2021.129107.

[65] Manahan SE. *Hazardous Waste Chemistry, Toxicology, and Treatment.* CRC Press; 1990.

[66] Demirbas A, Edris G, Alalayah WM. Sludge production from municipal wastewater treatment in sewage treatment plant. *Energy Sources Part Recovery Util Environ Eff* 2017;39:999–1006. https://doi.org/10.1080/15567036.2017.1283551.

[67] Geng H, Xu Y, Dai X, Yang D. Abiotic and biotic roles of metals in the anaerobic digestion of sewage sludge: A review. *Sci Total Environ* 2024;912:169313. https://doi.org/10.1016/j.scitotenv.2023.169313.

[68] Lou T, Yin Y, Wang J. Improving fermentative hydrogen production from sewage sludge by ionizing radiation treatment: A mini-review. *Int J Hydrog Energy* 2023;48:37124–37133. https://doi.org/10.1016/j.ijhydene.2023.06.096.

[69] Liu R, Han D, Zhang H, Ma Y, Hao X. Reviewing and diagnosing upgradation strategies of wastewater treatment plants in China. *J Water Process Eng* 2025;71:107267. https://doi.org/10.1016/j.jwpe.2025.107267.

[70] Sustainability Assessment for Wastewater Treatment Systems in Developing Countries n.d. https://www.mdpi.com/2073-4441/14/2/241 (accessed March 11, 2025).

[71] Du R, Li C, Liu Q, Fan J, Peng Y. A review of enhanced municipal wastewater treatment through energy savings and carbon recovery to reduce discharge and CO_2 footprint. *Bioresour Technol* 2022;364:128135. https://doi.org/10.1016/j.biortech.2022.128135.

[72] Ghimire U, Sarpong G, Gude VG. Transitioning wastewater treatment plants toward circular economy and energy sustainability. *ACS Omega* 2021;6:11794–11803. https://doi.org/10.1021/acsomega.0c05827.

[73] Sarfaraz A, Sumbal S, Qin Y, Faqir Y, Zveushe OK, Zhou L, Zhang W, Li J, Lv Z, Han Y, Dong F. Bioaugmentation-assisted phytoremediation of petroleum hydrocarbon-contaminated soils. *J Environ Chem Eng* 2025;13:115895. https://doi.org/10.1016/j.jece.2025.115895.

[74] Lwanga IC, Bhattacharya S, Dey S, Mandal S, Gupta K, Biswas JK, Sengupta S, Watts M. Phytoremediation of arsenic: A state-of-the-art review with special emphasis on modern biotechnological approaches. *Total Environ Eng* 2025;2:100014. https://doi.org/10.1016/j.teengi.2025.100014.

[75] Pandis PK, Kalogirou C, Kanellou E, Vaitsis C, Savvidou MG, Sourkouni G, Zorpas AA, Argirusis C. Key points of Advanced Oxidation Processes (AOPs) for wastewater, organic pollutants and pharmaceutical waste treatment: A mini review. *ChemEngineering* 2022;6:8. https://doi.org/10.3390/chemengineering6010008.

[76] Gągol M, Przyjazny A, Boczkaj G. Wastewater treatment by means of advanced oxidation processes based on cavitation – A review. *Chem Eng J* 2018;338:599–627. https://doi.org/10.1016/j.cej.2018.01.049.
[77] Li Y, Duan Y, Wang S, Zhang F, Li J, Dai Z, Li Z, Zhang Y, Wang Y. Supercritical water oxidation for the treatment and utilization of organic wastes: Factor effects, reaction enhancement, and novel process. *Environ Res* 2024;251:118571. https://doi.org/10.1016/j.envres.2024.118571.
[78] Gong Y, Guo Y, Sheehan JD, Chen Z, Wang S. Oxidative degradation of landfill leachate by catalysis of $CeMnO_x/TiO_2$ in supercritical water: Mechanism and kinetic study. *Chem Eng J* 2018;331:578–586. https://doi.org/10.1016/j.cej.2017.08.122.
[79] Li J, Wang S, Li Y, Wang L, Xu T, Zhang Y, Jiang Z. Supercritical water oxidation of semi-coke wastewater: Effects of operating parameters, reaction mechanism and process enhancement. *Sci Total Environ* 2020;710:134396. https://doi.org/10.1016/j.scitotenv.2019.134396.
[80] Vieira Y, Netto MS, Lima ÉC, Anastopoulos I, Oliveira MLS, Dotto GL. An overview of geological originated materials as a trend for adsorption in wastewater treatment. *Geosci Front* 2022;13:101150. https://doi.org/10.1016/j.gsf.2021.101150.
[81] Kasbaji M, Mennani M, Oubenali M, Ait Benhamou A, Boussetta A, Ablouh E-H, Mbarki M, Grimi N, El Achaby M, Moubarik A. Bio-based functionalized adsorptive polymers for sustainable water decontamination: A systematic review of challenges and real-world implementation. *Environ Pollut* 2023;335:122349. https://doi.org/10.1016/j.envpol.2023.122349.
[82] Vilén A, Laurell P, Vahala R. Comparative life cycle assessment of activated carbon production from various raw materials. *J Environ Manage* 2022;324:116356. https://doi.org/10.1016/j.jenvman.2022.116356.
[83] Badamasi H, Naeem Z, Antoniolli G, Praveen Kumar A, Olaleye AA, Sadiq IS, Durimin-iya NI, Khan MS. A review of recent advances in green and sustainable technologies for removing 4-nitrophenol from the water and wastewater. *Sustain Chem Pharm* 2025;43:101867. https://doi.org/10.1016/j.scp.2024.101867.
[84] Poornima S, Manikandan S, Karthik V, Balachandar R, Subbaiya R, Saravanan M, Lan Chi NT, Pugazhendhi A. Emerging nanotechnology based advanced techniques for wastewater treatment. *Chemosphere* 2022;303:135050. https://doi.org/10.1016/j.chemosphere.2022.135050.
[85] Wu Y, Pang H, Liu Y, Wang X, Yu S, Fu D, Chen J, Wang X. Environmental remediation of heavy metal ions by novel-nanomaterials: A review. *Environ Pollut* 2019;246:608–620. https://doi.org/10.1016/j.envpol.2018.12.076.
[86] Panigrahy SK, Nandha A, Chaturvedi M, Mishra PK. Novel nanocomposites with advanced materials and their role in waste water treatment. *Sustain* 2024;4:100042. https://doi.org/10.1016/j.nxsust.2024.100042.
[87] Naseem T, Waseem M. A comprehensive review on the role of some important nanocomposites for antimicrobial and wastewater applications. *Int J Environ Sci Technol* 2022;19:2221–2246. https://doi.org/10.1007/s13762-021-03256-8.
[88] Obi CC, Abonyi MN, Ohale PE, Onu CE, Nwabanne JT, Igwegbe CA, Kamuche TT, Ozofor IH. Adsorption of antibiotics from aqueous media using nanocomposites: Insight into the current status and future perspectives. *Chem Eng J* 2024;497:154767. https://doi.org/10.1016/j.cej.2024.154767.
[89] Rasheed S, Nadeem S, Nawaz R, Razzaque S, Mahmood A, Zidan A, Al-Masry W, Javed M, Ditta NA, Bahadur A, Iqbal S, Mahmood S, Akhter T. Revolutionizing textile wastewater treatment: Enhanced degradation of dyes using bimetallic Zinc Ferrite-GO nanocomposites. *J Mol Struct* 2025;1320:139480. https://doi.org/10.1016/j.molstruc.2024.139480.

[90] Mubark AE. Sequestration of zirconium and hafnium content using magnetic cellulose/chitosan nanocomposites from waste effluents. *J Water Process Eng* 2024;58:104759. https://doi.org/10.1016/j.jwpe.2023.104759.

[91] Abdulaziz F, Alanazi A. Tailoring of magnetite nanocomposite of carboxymethyl chitosan impregnated with iron (III) oxide for enhanced degradation of reactive blue 19 dye and inactivation of harmful microbes in wastewater. *Int J Biol Macromol* 2024;282:137004. https://doi.org/10.1016/j.ijbiomac.2024.137004.

[92] Venkatesh D, Kanchana V, Vasanthan S, Kistan A, Rajeswaran P, Arunkumar K, Siva Karthik P. Design and fabrication of magnetically recoverable $SnO_2/CoFe_2O_4$ nanocomposite for enhanced visible light-driven wastewater treatment. *Inorg Chem Commun* 2024;166:112657. https://doi.org/10.1016/j.inoche.2024.112657.

[93] Abdel-Fatah MA. Nanofiltration systems and applications in wastewater treatment: Review article. *Ain Shams Eng J* 2018;9:3077–3092. https://doi.org/10.1016/j.asej.2018.08.001.

[94] Joseph TM, Al-Hazmi HE, Śniatała B, Esmaeili A, Habibzadeh S. Nanoparticles and nanofiltration for wastewater treatment: From polluted to fresh water. *Environ Res* 2023;238:117114. https://doi.org/10.1016/j.envres.2023.117114.

[95] Nasr RA, Ali EA. Polyethersulfone/gelatin nano-membranes for the Rhodamine B dye removal and textile industry effluents treatment under cost effective condition. *J Environ Chem Eng* 2022;10:107250. https://doi.org/10.1016/j.jece.2022.107250.

[96] Yang C, Xu W, Nan Y, Wang Y, Chen X. Novel negatively charged nanofiltration membrane based on 4,4′-diaminodiphenylmethane for dye removal. *Sep Purif Technol* 2020;248:117089. https://doi.org/10.1016/j.seppur.2020.117089.

[97] De Souza DI, Dottein EM, Giacobbo A, Siqueira Rodrigues MA, De Pinho MN, Bernardes AM. Nanofiltration for the removal of norfloxacin from pharmaceutical effluent. *J Environ Chem Eng* 2018;6:6147–6153. https://doi.org/10.1016/j.jece.2018.09.034.

[98] Umar W, Zia Ur Rehman M, Umair M, Ayub MA, Naeem A, Rizwan M, Zia H, Karri RR. Use of nanotechnology for wastewater treatment: potential applications, advantages, and limitations. *Sustain. Nanotechnol. Environ. Remediat.*, Elsevier; 2022, pp. 223–272. https://doi.org/10.1016/B978-0-12-824547-7.00002-3.

[99] Thamarai P, Kamalesh R, Saravanan A, Swaminaathan P, Deivayanai VC. Emerging trends and promising prospects in nanotechnology for improved remediation of wastewater contaminants: Present and future outlooks. *Environ Nanotechnol Monit Manag* 2024;21:100913. https://doi.org/10.1016/j.enmm.2024.100913.

[100] Perera HCS, Gurunanthanan V, Singh A, Mantilaka MMMGPG, Das G, Arya S. Magnesium oxide (MgO) nanoadsorbents in wastewater treatment: A comprehensive review. *J Magnes Alloys* 2024;12:1709–1773. https://doi.org/10.1016/j.jma.2024.05.003.

[101] Carmel Aruja T, Sharma R, Dhiman V, Chand S, Prashant, Kondal N. ZnO based nanoadsorbents for purification of lead contaminated water. *Mater Today Proc* 2022:S221478532206953X. https://doi.org/10.1016/j.matpr.2022.11.105.

[102] Almutairi ST. Fabrication and catalytic activity of TiO_2/Fe_3O_4 and Fe_3O_4/β-cyclodextrin nanocatalysts for safe treatment of industrial wastewater. *Heliyon* 2024;10:e35400. https://doi.org/10.1016/j.heliyon.2024.e35400.

[103] Hosseini SG, Pasikhani JV. Kinetic and thermodynamic investigation on the adsorption of hexavalent chromium pollution by $Fe_3O_4/AC/TiO_2$ nanotubes as a novel ternary magnetic nanocomposite. *Desalination Water Treat* 2019;152:351–365. https://doi.org/10.5004/dwt.2019.24008.

[104] Kasbaji M, Mennani M, Barhoumi S, Esshouba Y, Oubenali M, Ablouh E-H, Kassab Z, Moubarik A, El Achaby M. Synergy of magnetic nanoparticles and sodium alginate-coated lignin for effective pollutant remediation, simple recovery, and cost-effective regeneration. *Langmuir* 2024;40:20657–20678. https://doi.org/10.1021/acs.langmuir.4c02734.

[105] Zhou Y, He Y, Xiang Y, Meng S, Liu X, Yu J, Yang J, Zhang J, Qin P, Luo L. Single and simultaneous adsorption of pefloxacin and Cu(II) ions from aqueous solutions by oxidized multiwalled carbon nanotube. *Sci Total Environ* 2019;646:29–36. https://doi.org/10.1016/j.scitotenv.2018.07.267.

[106] Balarak D, Mostafapour F, Bazrafshan E, Saleh TA. Studies on the adsorption of amoxicillin on multi-wall carbon nanotubes. *Water Sci Technol* 2017;75:1599–1606. https://doi.org/10.2166/wst.2017.025.

[107] Kariim I, Abdulkareem AS, Abubakre OK. Development and characterization of MWCNTs from activated carbon as adsorbent for metronidazole and levofloxacin sorption from pharmaceutical wastewater: Kinetics, isotherms and thermodynamic studies. *Sci Afr* 2020;7:e00242. https://doi.org/10.1016/j.sciaf.2019.e00242.

[108] Abdelmegeed AF, Sayed M, Abbas M, Abdel Moniem SM, Farag RS, Sayed AZ, Naga SM. Synthesis of functionalized superparamagnetic nanoparticles as highly efficient nanoadsorbents: Removal of heavy metals from industrial wastewater sample as a case study. *Inorg Chem Commun* 2023;155:111094. https://doi.org/10.1016/j.inoche.2023.111094.

[109] Pandey A, Kalamdhad A, Chandra Sharma Y. Recent advances of nanocellulose as biobased adsorbent for heavy metal ions removal: A sustainable approach integrating with waste management. *Environ Nanotechnol Monit Manag* 2023;20:100791. https://doi.org/10.1016/j.enmm.2023.100791.

[110] Essekri A, Laabd M, Albourine A. Efficient adsorption of crystal violet dye using functionalized Argan shell: Experiments and statistical optimization modeling. *Colloids Surf Physicochem Eng Asp* 2024;687:133401. https://doi.org/10.1016/j.colsurfa.2024.133401.

[111] Lin X, Chan K, Kingkhambang K, Hayashi H, Zinchenko A. Hydrothermal preparation of pharmaceuticals adsorbents from chitin and chitosan: Optimization and mechanism. *Bioresour Technol* 2024;414:131583. https://doi.org/10.1016/j.biortech.2024.131583.

[112] Chong ZT, Soh LS, Yong WF. Valorization of agriculture wastes as biosorbents for adsorption of emerging pollutants: Modification, remediation and industry application. *Results Eng* 2023;17:100960. https://doi.org/10.1016/j.rineng.2023.100960.

[113] Kasbaji M, Mennani M, Grimi N, Oubenali M, Mbarki M, El Zakhem H, Moubarik A. Adsorption of cationic and anionic dyes onto coffee grounds cellulose/sodium alginate double-network hydrogel beads: Isotherm analysis and recyclability performance. *Int J Biol Macromol* 2023;239:124288. https://doi.org/10.1016/j.ijbiomac.2023.124288.

[114] Tolun A, Altintas Z. Optimizing antioxidant and thermal stability in olive mill wastewater through hydroxypropyl-β-cyclodextrin complexation: A comprehensive study. *Carbohydr Polym Technol Appl* 2025;9:100631. https://doi.org/10.1016/j.carpta.2024.100631.

[115] Fleyfel LM, Matta J, Sayegh NF, El Najjar NH. Olive mill wastewater treatment using coagulation/flocculation and filtration processes. *Heliyon* 2024;10:e40348. https://doi.org/10.1016/j.heliyon.2024.e40348.

[116] Szabolcsik-Izbéki A, Bodnár I, Fábián I. The removal of pollutants from synthetic bathroom greywater by coagulation-flocculation and filtration as a fit-for-purpose method. *J Environ Chem Eng* 2024;12:114250. https://doi.org/10.1016/j.jece.2024.114250.

[117] Li M, Chen Z, Zhou D, Xu S, Qiu S, Ge S. Coagulation pretreatment coupled with indigenous microalgal-bacterial consortium system for on-site treatment of rural black wastewater. *Sci Total Environ* 2024;913:169728. https://doi.org/10.1016/j.scitotenv.2023.169728.

[118] Majumder A, Otter P, Röher D, Bhatnagar A, Khalil N, Gupta AK, Bresciani R, Arias CA. Combination of advanced biological systems and photocatalysis for the treatment of real hospital wastewater spiked with carbamazepine: A pilot-scale study. *J Environ Manage* 2024;351:119672. https://doi.org/10.1016/j.jenvman.2023.119672.

[119] Can-Güven E, Bayat ME, Yazici Guvenc S, Varank G. Refractory pollutants removal from industrial wastewater by a coupled process of coagulation-flocculation and alternative electrochemical advanced oxidation processes. *Colloids Surf Physicochem Eng Asp* 2025;707:135916. https://doi.org/10.1016/j.colsurfa.2024.135916.

[120] Dai C, Chen D, Wu J, Liu J, Shi S, Zhang J, Feng Y. Construction of a novel integrated electrochemical oxidation-coagulation system for simultaneous removal of suspended solids and antibiotics. *Chem Eng J* 2022;447:137505. https://doi.org/10.1016/j.cej.2022.137505.

[121] Muhammad Yaqub, Lee W. Zero-liquid discharge (ZLD) technology for resource recovery from wastewater: A review. *Sci Total Environ* 2019;681:551–563. https://doi.org/10.1016/j.scitotenv.2019.05.062.

[122] Abdelfattah I, El-Shamy AM. Review on the escalating imperative of zero liquid discharge (ZLD) technology for sustainable water management and environmental resilience. *J Environ Manage* 2024;351:119614. https://doi.org/10.1016/j.jenvman.2023.119614.

[123] Tong T, Elimelech M. The global rise of zero liquid discharge for wastewater management: Drivers, technologies, and future directions. *Environ Sci Technol* 2016;50:6846–6855. https://doi.org/10.1021/acs.est.6b01000.

[124] Zhang J, Peng L, Zhao L, Li R, Liu Q, Yang L, Wu C, Liu H. Microbial osmotic pressure desalination system for desulfurization wastewater from coal-fired power plants: A pilot study. *J Water Process Eng* 2024;64:105620. https://doi.org/10.1016/j.jwpe.2024.105620.

[125] Moharram NA, Konsowa AH, Shehata AI, El-Maghlany WM. Sustainable seascapes: An in-depth analysis of multigeneration plants utilizing supercritical zero liquid discharge desalination and a combined cycle power plant. *Alex Eng J* 2025;118:523–542. https://doi.org/10.1016/j.aej.2025.01.091.

[126] Zhang Y, Dong H, Du W, Dong C, Xiong M, Yang N, Zhao S, He H, Nie Z. An ionic liquid-modified PVC nanofiber facilitates gold recovery from wastewater by a light-enhancing effect. *Chem Eng J* 2025;505:159419. https://doi.org/10.1016/j.cej.2025.159419.

[127] Wang Y, Zhu W, Cheng H, Zhong J, Li X, Qi J, Wang S, Zhu Z, Wang Y, Cui P. Enhanced reaction extraction distillation via multi-objective optimization and 4E analysis for sustainable recovery of organic compounds from wastewater. *Process Saf Environ Prot* 2024;186:884–894. https://doi.org/10.1016/j.psep.2024.04.064.

[128] Jeong S, Jeong H, Park C, Gu B, Jeong S. Effective lithium recovery from battery wastewater via nanofiltration and membrane distillation crystallization with carbon nanotube spacer. *Chem Eng J* 2025;503:158315. https://doi.org/10.1016/j.cej.2024.158315.

[129] Sniatala B, Kurniawan TA, Sobotka D, Makinia J, Othman MHD. Macro-nutrients recovery from liquid waste as a sustainable resource for production of recovered mineral fertilizer: Uncovering alternative options to sustain global food security cost-effectively. *Sci Total Environ* 2023;856:159283. https://doi.org/10.1016/j.scitotenv.2022.159283.

[130] Liu Y, Dai J, Lv X, Mai W, Shi Y, Liang J, Wan D. Novel calcium-modified biochar prepared from liquor industrial biogas residue for wastewater resource recovery and soil improvement. *J Environ Chem Eng* 2025;13:115403. https://doi.org/10.1016/j.jece.2025.115403.

[131] Li Q, Wang F, Liu X, Xu W, Zhang K, Zhang S, Teng P, Zhang L. An infrared radiation-responsive high-entropy CoCaMgMnAlFe-LDHs for controllable hydrogen generation via NaBH4 hydrolysis: Utilization of steel slag and lithium-ion battery liquid waste. *Chem Eng J* 2024;500:157380. https://doi.org/10.1016/j.cej.2024.157380.

[132] Li K, Duan H, Wang S, Wu Z, Wardrop P, Lloyd J, Christy N, De Jong P, Ye L. Balancing energy recovery and direct greenhouse gas emissions in wastewater treatment. *Water Res X* 2025;28:100328. https://doi.org/10.1016/j.wroa.2025.100328.

[133] Chen Y, Zhang Y, Xu S. Carbon and nitrogen recovery from biogas slurry via hydrothermal treatment and microbial electrolysis cells assisted anaerobic digestion system. *Bioresour Technol* 2025;425:132316. https://doi.org/10.1016/j.biortech.2025.132316.
[134] Yang F, Wu Y, Zhang Q. Towards resource regeneration: A focus on copper recovery from electronic waste. *J Clean Prod* 2024;484:144286. https://doi.org/10.1016/j.jclepro.2024.144286.
[135] Chen Y, Zhang Y, Xu S. Carbon and nitrogen recovery from biogas slurry via hydrothermal treatment and microbial electrolysis cells assisted anaerobic digestion system. *Bioresour Technol* 2024;425. https://doi.org/10.1016/j.biortech.2025.132316.
[136] Pareja-Sánchez E, García-Moreno AJ, Martínez-García M, Pérez-Colodrero L, García-Zapata L, García-Ruiz R. Bio-based solvents for polyphenol recovery: Transforming olive mill wastewater into high-value resources. *J Water Process Eng* 2025;71:107238. https://doi.org/10.1016/j.jwpe.2025.107238.
[137] Wang D, Zeng Y, Liang Y, Chen Y, Liu C, Zhan C, Chen P, Sun K, Song S, Jia F. Pilot-scale, cost-effective waste resource recovery: Efficient reduction of trace Au(I) from industrial wastewater via an adsorption and electrodeposition coupling process. *Sep Purif Technol* 2025;361:131641. https://doi.org/10.1016/j.seppur.2025.131641.
[138] Hao J, Zhang W, Xue G, Rao P, Wang R. Treatment of distillation residue waste liquid from NPEOs by hydrothermal carbonization process for resource recovery. *Desalination Water Treat* 2018;125:26–31. https://doi.org/10.5004/dwt.2018.22924.
[139] Kumar R, Maurya A, Raj A. Emerging technological solutions for the management of paper mill wastewater: Treatment, nutrient recovery and fourth industrial revolution (IR 4.0). *J Water Process Eng* 2023;53:103715. https://doi.org/10.1016/j.jwpe.2023.103715.
[140] Bai C, Dallasega P, Orzes G, Sarkis J. Industry 4.0 technologies assessment: A sustainability perspective. *Int J Prod Econ* 2020;229:107776. https://doi.org/10.1016/j.ijpe.2020.107776.
[141] Uwamungu JY, Kumar P, Alkhayyat A, Younas T, Capangpangan RY, Alguno AC, Ofori I. Future of water/wastewater treatment and management by industry 4.0 integrated nanocomposite manufacturing. *J Nanomater* 2022;2022:5316228. https://doi.org/10.1155/2022/5316228.
[142] De Vleeschauwer F, Caluwé M, Dobbeleers T, Stes H, Dockx L, Kiekens F, Copot C, Dries J. A dynamic control system for aerobic granular sludge reactors treating high COD/P wastewater, using pH and DO sensors. *J Water Process Eng* 2020;33:101065. https://doi.org/10.1016/j.jwpe.2019.101065.
[143] Nadi F, Hossain S, Rahmat RF, Kasan NA, Ikhwanuddin M, Rastegari H. Detection of ammonia in aquaculture wastewater using mango leaf extract-immobilized paper sensors and smartphone colorimetric analysis. *Microchem J* 2024;207:112257. https://doi.org/10.1016/j.microc.2024.112257.
[144] Marino L, Todesco R, Gagliano E, Santoro D, Roccaro P. Real-time wastewater quality monitoring by fluorescence sensors: Validation for COD and CEC monitoring and implication for carbon footprint reduction. *Sci Total Environ* 2025;963:178464. https://doi.org/10.1016/j.scitotenv.2025.178464.
[145] Yang J, Yuan Y, Zhang R, Liu C, Zhu D, Ji D. Rapid detection of Cu^{2+} and Cr^{3+} ions in industrial wastewater using gelatin-poly-ε-caprolactone nanofiber fluorescent sensors. *J Mol Liq* 2025;418:126718. https://doi.org/10.1016/j.molliq.2024.126718.
[146] Wang D, Thunéll S, Lindberg U, Jiang L, Trygg J, Tysklind M, Souihi N. A machine learning framework to improve effluent quality control in wastewater treatment plants. *Sci Total Environ* 2021;784:147138. https://doi.org/10.1016/j.scitotenv.2021.147138.
[147] Sheik AG, Kumar A, Srungavarapu CS, Azari M, Ambati SR, Bux F, Patan AK. Insights into the application of explainable artificial intelligence for biological wastewater treatment plants: Updates and perspectives. *Eng Appl Artif Intell* 2025;144:110132. https://doi.org/10.1016/j.engappai.2025.110132.

[148] Xu W-L, Wang Y-J, Wang Y-T, Li J-G, Zeng Y-N, Guo H-W, Liu H, Dong K-L, Zhang L-Y. Application and innovation of artificial intelligence models in wastewater treatment. *J Contam Hydrol* 2024;267:104426. https://doi.org/10.1016/j.jconhyd.2024.104426.
[149] Oruganti RK, Biji AP, Lanuyanger T, Show PL, Sriariyanun M, Upadhyayula VKK, Gadhamshetty V, Bhattacharyya D. Artificial intelligence and machine learning tools for high-performance microalgal wastewater treatment and algal biorefinery: A critical review. *Sci Total Environ* 2023;876:162797. https://doi.org/10.1016/j.scitotenv.2023.162797.
[150] Hafeez S, Ishaq A, Intisar A, Mahmood T, Din MI, Ahmed E, Tariq MR, Abid MA. Predictive modeling for the adsorptive and photocatalytic removal of phenolic contaminants from water using artificial neural networks. *Heliyon* 2024;10:e37951. https://doi.org/10.1016/j.heliyon.2024.e37951.
[151] Mullai P, Vishali S, Sobiya E. Experiments and adaptive-network-based fuzzy inference system modelling in a hybrid up-flow anaerobic sludge blanket reactor to assess industrial azadirachtin effluent quality. *Bioresour Technol* 2022;358:127395. https://doi.org/10.1016/j.biortech.2022.127395.
[152] Di Loreto S, Serpilli F, Lori V. Application of the SVM algorithm for the development of a model classification of the visual and sound landscape. *Inter-Noise Noise-CON Congr Conf Proc* 2023;265:189–196. https://doi.org/10.3397/IN_2022_0032.
[153] Kang Z, Duan L, and Zahmatkesh S Optimizing removal of antiretroviral drugs from tertiary wastewater using chlorination and AI-based prediction with response surface methodology. *Science of the Total Environment* 2024;934. https://doi.org/10.1016/j.scitotenv.2024.172931.
[154] Schmitt A, Mendret J, Roustan M, Brosillon S. Ozonation using hollow fiber contactor technology and its perspectives for micropollutants removal in water: A review. *Sci Total Environ* 2020;729:138664. https://doi.org/10.1016/j.scitotenv.2020.138664.
[155] Nematollahi H, Ghasemzadeh R, Tuysserkani M, Aziminezhad M, Pazoki M. Comparative life cycle assessment of hospital waste management scenarios in Isfahan, Iran: Evaluating environmental impacts and strategies for improved healthcare sustainability. *Results Eng* 2024;24:102912. https://doi.org/10.1016/j.rineng.2024.102912.
[156] Aziz NIHA, Hanafiah MM. Application of life cycle assessment for desalination: Progress, challenges and future directions. *Environ Pollut* 2021;268:115948. https://doi.org/10.1016/j.envpol.2020.115948.
[157] Escobar N, Laibach N. Sustainability check for bio-based technologies: A review of process-based and life cycle approaches. *Renew Sustain Energy Rev* 2021;135:110213. https://doi.org/10.1016/j.rser.2020.110213.
[158] Sedaghat O, Bahramifar N, Nowrouzi M, Younesi H. Life cycle assessment of industrial wastewater treatment: Evaluating the environmental impact of electrocoagulation technologies. *J Water Process Eng* 2025;71:107257. https://doi.org/10.1016/j.jwpe.2025.107257.
[159] Huijbregts MAJ, Steinmann ZJN, Elshout PMF, Stam G, Verones F, Vieira M, Zijp M, Hollander A, Van Zelm R. ReCiPe2016: a harmonised life cycle impact assessment method at midpoint and endpoint level. *Int J Life Cycle Assess* 2017;22:138–147. https://doi.org/10.1007/s11367-016-1246-y.
[160] Hosseinzadeh-Bandbafha H, Nizami A-S, Kalogirou SA, Gupta VK, Park Y-K, Fallahi A, Sulaiman A, Ranjbari M, Rahnama H, Aghbashlo M, Peng W, Tabatabaei M. Environmental life cycle assessment of biodiesel production from waste cooking oil: A systematic review. *Renew Sustain Energy Rev* 2022;161:112411. https://doi.org/10.1016/j.rser.2022.112411.
[161] Caldeira C, Queirós J, Freire F. Biodiesel from waste cooking oils in Portugal: Alternative collection systems. *Waste Biomass Valorization* 2015;6:771–779. https://doi.org/10.1007/s12649-015-9386-z.

[162] Byles D, Mohagheghi S. Sustainable power grid expansion: Life cycle assessment, modeling approaches, challenges, and opportunities. *Sustainability* 2023;15:8788. https://doi.org/10.3390/su15118788.

[163] Zhao Y, Wang C, Zhang L, Chang Y, Hao Y. Converting waste cooking oil to biodiesel in China: Environmental impacts and economic feasibility. *Renew Sustain Energy Rev* 2021;140:110661. https://doi.org/10.1016/j.rser.2020.110661.

[164] Estévez S, González-García S, Feijoo G, Moreira MT. How decentralized treatment can contribute to the symbiosis between environmental protection and resource recovery. *Sci Total Environ* 2022;812:151485. https://doi.org/10.1016/j.scitotenv.2021.151485.

[165] Dewalkar SV, Shastri SS. Environmental and economic assessment of proposed on-site wastewater management system in multi-storey residential building. *Water Sci Technol* 2020;82:3003–3016. https://doi.org/10.2166/wst.2020.548.

[166] Amaral KGCD, Aisse MM, Possetti GRC. Sustainability assessment of sludge and biogas management in wastewater treatment plants using the LCA technique. *Ambiente E Agua - Interdiscip J Appl Sci* 2019;14:1. https://doi.org/10.4136/ambi-agua.2371.

[167] Kalboussi N, Biard Y, Pradeleix L, Rapaport A, Sinfort C, Ait-mouheb N. Life cycle assessment as decision support tool for water reuse in agriculture irrigation. *Sci Total Environ* 2022;836:155486. https://doi.org/10.1016/j.scitotenv.2022.155486.

[168] Viornery-Portillo EA, Bravo-Díaz B, Mena-Cervantes VY. Life cycle assessment and emission analysis of waste cooking oil biodiesel blend and fossil diesel used in a power generator. *Fuel* 2020;281:118739. https://doi.org/10.1016/j.fuel.2020.118739.

[169] Lam KL, Solon K, Jia M, Volcke EIP, Van Der Hoek JP. Life cycle environmental impacts of wastewater-derived phosphorus products: An agricultural end-user perspective. *Environ Sci Technol* 2022;56:10289–10298. https://doi.org/10.1021/acs.est.2c00353.

[170] Canaj K, Mehmeti A. Unveiling drivers and barriers in advancing agricultural wastewater reuse in Southern Italy: A SWOT analysis informed by stakeholder insights. *Clean Water* 2024;1:100008. https://doi.org/10.1016/j.clwat.2024.100008.

[171] Al-Baldawi IA, Mohammed AA, Mutar ZH, Abdullah SRS, Jasim SS, Almansoory AF, Ismail N 'Izzati. Application of phytotechnology in alleviating pharmaceuticals and personal care products (PPCPs) in wastewater: Source, impacts, treatment, mechanisms, fate, and SWOT analysis. *J Clean Prod* 2021;319:128584. https://doi.org/10.1016/j.jclepro.2021.128584.

Index

For Product Safety Concerns and Information please contact our EU representative GPSR@taylorandfrancis.com Taylor & Francis Verlag GmbH, Kaufingerstraße 24, 80331 München, Germany